This book is an invaluable contribution to thinking r  Yellowstone and well beyond. Susan Clark draws on de a new path, one that weaves physical, biological, and et..... ___ whole. This is a book that should be read by all who are interested in the future of healthy ecosystems and species.

Mary Evelyn Tucker, Yale University
Senior Lecturer and Research Scholar
Yale School of the Environment
Cofounder and Codirector, Yale Forum on Religion and Ecology
Coauthor, *Journey of the Universe*

Susan Clark has made an invaluable contribution to the literature on human–nature coexistence. "Yellowstone's Survival: A Call to Action for a New Conservation Story" is a masterful work on the Greater Yellowstone Ecosystem (GYE), but it is also an exploration of the human condition and how people think and feel in their interactions with one another and with nature. In drawing on more than half a century of experience in the GYE as well as international experience on six continents, Dr. Clark uncovers insights about human–nature coexistence in contexts large and small. Her analysis is an extraordinarily valuable support to all of us interested in addressing the myriad human problems facing large ecosystems and the species they support.

In my 30-year career in species and ecosystem conservation—in government, higher education, and the NGO community—I have worked in many contexts, including the GYE and large marine and coastal ecosystems in all regions of North America. The lessons Dr. Clark impart from her half-century of professional practice have proven invaluable to me in all the places and contexts in which I work. This is due to her inimitable insight into human character and behavior and her skill at helping her readers attain a better understanding of our fellow humans. The wisdom Dr. Clark has shared will improve our chances of solving the wicked problems facing human–nature coexistence.

Richard L. Wallace, PhD, Editor-in-Chief, Frontiers in Ecology and the Environment
Ecological Society of America

Yellowstone's Survival

The Anthem Environment and Sustainability Initiative (AESI) seeks to push the frontiers of scholarship while simultaneously offering prescriptive and programmatic advice to policymakers and practitioners around the world. The program publishes research monographs, professional and major reference works, upper-level textbooks, and general interest titles. Professor Lawrence Susskind (MIT) acts as the general editor of AESI and oversees our book series, each featuring scholars, practitioners, and business experts keen to link theory and practice. Our series editors include Brooke Hemming (US EPA), Shafiqul Islam (Tufts University), Saleem Ali (University of Delaware), and Richardson Dilworth (Center for Public Policy, Drexel University).

Strategies for Sustainable Development Series
Series Editor: Professor Lawrence Susskind (MIT)

Climate Change Science, Policy and Implementation
Series Editor: Dr. Brooke Hemming (US EPA)

Science Diplomacy: Managing Food, Energy and Water Sustainably
Series Editor: Professor Shafiqul Islam (Tufts University)

International Environmental Policy Series
Series Editor: Professor Saleem Ali (University of Delaware)

Big Data and Sustainable Cities Series
Climate Change and the Future of the North American City
Series Editor: Richardson Dilworth
(Center for Public Policy, Drexel University, United States)

Included within the AESI is the Anthem EnviroExperts Review. Through this online micro-review site, Anthem Press seeks to build a community of practice involving scientists, policy analysts, and activists committed to creating a clearer and deeper understanding of how ecological systems—at every level—operate, and how they have been damaged by unsustainable development. This site publishes short reviews of important books or reports in the environmental field, broadly defined. Visit the website: www.anthemenviroexperts.com.

Yellowstone's Survival

A Call to Action for a New Conservation Story

Susan G. Clark

ANTHEM PRESS

Anthem Press
An imprint of Wimbledon Publishing Company
www.anthempress.com

This edition first published in UK and USA 2021
by ANTHEM PRESS
75–76 Blackfriars Road, London SE1 8HA, UK
or PO Box 9779, London SW19 7ZG, UK
and
244 Madison Ave #116, New York, NY 10016, USA

British Library Cataloguing-in-Publication Data
A catalogue record for this book is available from the British Library.

Library of Congress Control Number: 2021935006

ISBN-13: 978-1-78527-731-3 (Hbk)
ISBN-10: 1-78527-731-6 (Hbk)
ISBN-13: 978-1-78527-999-7 (Pbk)
ISBN-10: 1-78527-999-8 (Pbk)

This title is also available as an e-book.

Cover image: Painting by Denise Casey

Dedicated to

Ana Lambert

Avana Andrade

Hanna Jaicks

Jesse Oppenheimer

Rosalie Chapple

Katie Christiansen

Lindsey Larson

Marian Vernon

Mariana Sarmiento

Carlie Kierstead

Anna Reside

Patty Ewing

Sandy Shruptine

Lloyd Dorsey

Dorian Baldes

The future is not fixed.

It is not a set point over the horizon

that we are all running toward,

helpless to do anything about it.

The future is built everyday by the actions of people.

—Brian David Johnson in "How to Invent the Future"

CONTENTS

LIST OF ILLUSTRATIONS

Figures

Tables

FOREWORD

The Greater Yellowstone Ecosystem (GYE) is one of the finest examples of natural resources our nation has to choose to protect. It is home to millions of acres of preeminent public lands including Yellowstone and Grand Teton National Parks, National Wildlife Refuges, National Forest, and Bureau of Land Management lands. The GYE encompasses places whose names resonate as unique, beautiful, wild, remote, exciting and protected, while being accessible for a wide array of recreational and, yes, development opportunities. Fortunately, the GYE is an ecosystem that has the amenities and diversity to support a wide range of lifestyles for residents and visitors alike.

The GYE is currently experiencing increased development to accommodate the migration of people wanting to live in the area and is experiencing an equally dramatic increase in visitation from around the world. More people means more pressure on the ecosystem, which then calls for new, innovative, and coordinated approaches to managing these resources.

At over 20 million acres, the GYE is the largest nearly intact ecosystem in the continental United States. Fifteen million of these acres are public lands managed by four federal agencies. The remainder are managed by three state governments, numerous county and local governments, and private landowners. While these entities have organizations in place to accomplish some mutual goals, they are often not ecosystem goals. The rapid social and environmental changes are causing unintended harmful impacts and to truly solve these issues, they must be considered on an ecosystem basis. The conflicting missions of preservation and use and the differing priorities of the entities often put the long-term preservation of our ecosystems and the opportunities they provide at risk.

In this book, Dr. Susan Clark lays out a road map of productive ways to discuss, learn about, and act upon these challenges. Her approach to seeking solutions is unique in that she reaches far beyond the traditional boundaries of her field of ecology to integrate common sense, geography, psychology, sociology, anthropology, philosophy, policy sciences, and hope. She lays out our need for a rapid, new, colearning network to move us forward quickly in a responsible way to conserve greater Yellowstone. Her position throughout the book recognizes the need for understanding and a reinvigorated civil dialogue, cooperative problem solving, environmental education and strategic leadership, and new specific management and policy actions. She uses concrete examples to illustrate new ways of thinking that can help federal, state, and local officials, professionals, and the public face the challenges head-on.

To do this in the GYE requires that we make use of the reservoir of scientific knowledge, practical experience, and wisdom that we have now, but it also calls for creating new knowledge taking new and different approaches, all working together to articulate the vision that will protect the GYE for future generations. This task is urgent.

Typically, when challenged by the range of problems we now face and those just over the horizon, we tend to fall back on the old stories of humans versus nature and preservation versus use, where our wild places are nothing more than resources for us to exploit for profit and pleasure. We are learning the hard way that this old narrative will not serve us or nature well. I have both participated in and observed the function and dysfunction of decision-making by federal and state managers in the GYE and the effect of those decisions have had on the long-term health and preservation of the ecosystem. Dr. Clark illustrates such difficulties with excellent research and firsthand observations, documenting both its agency management and policy shortfalls and opportunities in our current system. She then reaches beyond the system to suggest a better way to address growing challenges.

She gives us a much deeper and broader understanding of the challenges faced by managers through her wide range of experience and knowledge into a framework that can be applied across specific cases, scientific and management issues, and public and leadership problems. Though the challenges discussed are focused on the GYE, the ways of thinking, planning, and acting could be of use in any part of the world facing such challenges. It is clear that the case made by Dr. Clark for a new paradigm of thought and management is crucial at this time for the welfare of the GYE.

From her background of 50 years of study and observation, Dr. Clark illustrates the meaning of Yellowstone National Park, public lands, and the GYE to the many individuals who live in, visit, enjoy, or simply appreciate Yellowstone from a distance. Dr. Clark's book brings together many subjects, experience, and wisdom from many fields seldom put together in an integrated, overarching way for practical application. It is as strategic and farsighted as it is encompassing and practical. I hope it is read widely and discussed by officials, professionals, and the public, and used to guide us toward a healthier future.

Dan Wenk

Yellowstone National Park, Superintendent, 2011–18, Retired

PREFACE

Experience is a hard teacher
because she gives the test first,
the lessons afterwards.
                    —Unknown

Yellowstone is a national park, a concept of a greater ecosystem, and, even more so, an "idea" about the wild, wildlife, and people. This "idea" is in our stories of hope for a future where humans live sustainably with nature—something we're presently not doing. Yellowstone as a reference point for nature is widely recognized no matter where one lives on our home planet.

What is this book about and why did I write it? These two questions have complex and incomplete answers, as explained in this preface and the text of all chapters. First, about this book: it is focused on Yellowstone National Park as a place and an idea with a long history, a present to be visited, and a contested future that we have long argued over. Yellowstone as a theme and symbol grounds this book in a place, time, and set of issues. I give scientific data, use systems thinking, and talk about grounded case examples concerning Yellowstone. Beyond that, most importantly, this book concentrates attention on people in general and on those who care about Yellowstone, including ourselves.

In so doing, I bring in knowledge from many fields seldom used currently by people worried about Yellowstone's future. The present focus of attention by most people is on biophysical things "out there" (e.g., wolves and bears, elk migrations, diseases, and fires). Consequently, it misses a huge component in Yellowstone's future—people, us. I organize the information I offer around people and work to give us context—historic, social, and political—for all things Yellowstone. Because Yellowstone's future has yet to be written, we can use this information in new integrated ways to create a healthy, thriving future for greater Yellowstone and ourselves.

This book's contents go well beyond Yellowstone as a geologic and biologic phenomenon to concentrate upon human beings. It is about the individuals who support conservation and sustainability in their thinking, feelings, and actions. It is also a reflection on our society and an exploration of a new way forward for better outcomes in all areas. The proposal here is to advance sustainability in the present in Yellowstone and outward far beyond that special place to other places worldwide. Many of us are alarmed by the ways in which environments, other living forms, and peoples are being treated. This book offers new thinking, solutions, and actions as responses to calls for us to rethink our behavior and learn how to work together, which are goals being vigorously sought or at least ostensible expressed. This book is meant for anyone interested in Yellowstone, wildlife, and larger conservation and sustainability matters. Readers may want to skip around through the chapters depending on their interests. Perhaps, a reader new to the subjects in this book should take a quick perusal of the chapters and headings and first dip into parts of interest to them.

Though this book references Yellowstone and its problems, it goes much further to advance answers to the following questions: What are the principal, essential attributes and actions of individuals who are the people who can bring about the conservation and transformative change needed? What actions do we need to undertake? Answers come from clues provided by what is known about human beings in the humanities, social sciences, biological sciences, traditional knowledge, and in everyday conventional life. I extract a sample from these sources relevant at this crucial moment for Yellowstone and as well to address our bigger existential challenges. Literary, philosophic, and psychological sources of knowledge and insight are offered. I do promote certain insights and actions. In short, the real reason for our growing environmental and social dilemma is our mental perceptions, thinking, or metacognitive constraints.

It is obvious to me that conservation and sustainability, when seen through the personalities and actions of the people, groups, and organizations, and our institutions, is much more than only a contemporary physical or biological challenge. Rather, the task before us is affected by what has plagued, puzzled, and pleased us over the centuries, and will continue to do so—ourselves and our presence in the world. Thus, this book transcends what many people might imagine conservation and sustainability should be in Yellowstone or elsewhere. I answer another question: What do we need to do and what can current individuals, groups, and organizations, especially educationally and through media-wise communications, do to promote conservation and sustainable behavior? I hope my inquiry and suggestions can influence, increase, and magnify what are urgent commonsense actions.

Yellowstone's Levels

Yellowstone exists in our world in many different levels. First, it is a physical place full of interesting things such as geothermal features. This is our first level of entity. Yellowstone is a landscape of mountains, a high plateau, and rivers with a unique geological past. The physical realities of Yellowstone set constraints on how we interact with it. Yellowstone can best be dealt with at that level of engagement initially, before we move on to other levels to consider, including all the wildlife there and importantly ourselves in the story of Yellowstone and nature.

On the second level, Yellowstone is about all the life there—communities and ecosystems full of wolves, elk, bears, grasslands, and forests. It is also about the interactions among these living beings and about the lives they live. When we visit Yellowstone, we can marvel at all the wildlife and be awed by the big vistas across meadows, forests, and alpine communities. At this biological level (though really a level about physics and chemistry), Yellowstone is truly a lifetime experience. These two levels and their constrains—the physical and the biological—are what most people consider Yellowstone to be. This too sets constraints on how we interact with it.

But there is a third level, our mental world—our thinking and imaging. This level is often called our metacognitive world. With each visit to Yellowstone, we individually experience and engage with it in different ways. In so doing we make sense and meaning of our visit. We do this as a product of our personalities, previous experiences, and our culture, its institutions. While in Yellowstone we may begin to explore questions about what that physical place and those biological entities mean. Conscious or not, this is the level that we use to engage with nature and wildlife in Yellowstone. This is both personal and cultural. Our personal answers to these concerns or questions become part of a broader intergenerational cultural story about the park and the wild that we share. What does Yellowstone represent to us? Why? The narratives of Yellowstone as nature writ-large, uninterrupted and pristine like a living museum, emerge from the stories that we have told to ourselves and to each other about what our experiences mean. This is our mental or metacognitive level in action. It too sets constraints on how we interact with it.

As we engage all things Yellowstone, our mental world brings out not only everyday issues but also the basic and fundamental philosophic questions about the mental world level itself. These three levels of Yellowstone and derivative questions are about our story of humans in relation to nature and what we will do in the future. Today, we are of many competing minds offering different views on Yellowstone and our future, and these either contest or complement one another. We are now in fierce competition over control of Yellowstone, our culture, and our perceived scarce resources, such as space, quiet, and simply personal enjoyment of wildlife and the wild. This is a clash between mental representations or stories of the world. This is perhaps the root cause of the complex and unsustainable situation we find ourselves in presently. For me, in short, this is the real reason for our complicated and complex unsustainable living that we find ourselves in presently. My pragmatic concern is, what should we do?

In this book, I address all three levels—the physical, the biological, and our mental or metacognitive world of experiences and stories of meaning. For me, recognition of all three levels at the same time is the best way to deal with all things Yellowstone. For this book, I emphasize the mental level, survey various knowledge domains, and seek to integrate it with the world "out there," the other two levels. This is the key integrative endeavor at the heart of this book. The text moves between everyday experiences and stories and across academic sections, and perhaps less obvious subjects. I suggest that you the reader move through the text and focus on sections that interest you first then go on to read other sections and whole chapters.

Our mental level especially imposes constraints to how we integrate at all levels through our stories of meaning. Many of these constrains flow out of our personalities and our institutions. These interrelated subjects often remain inscrutable to many people operating well within them—like fish and water. Our personalities and culture generally operate slowly, subtly, and outside conscious awareness. People rarely understand how or why they and their institutions work or see that they "do" anything. In daily life, people's explicit views or theories about their own behavior and society's operations are generally post hoc and often wrong. This book brings these constraints into view drawing on the humanities and the social sciences.

This book aspires to help us all integrate and find deeper meaning in Yellowstone, nature and wildlife, and in our living. I bring together many ideas and experiences from diverse areas of human endeavor that might help us best address the unsustainable future we all face. It calls for new thinking. I recognize that I, like everyone else, have a mental or metacognitive world I live in and that communicating our personal views across other people's mental worlds for common cause is a struggle for all of us. With these consideration, I used this book in an attempt to contribute to resolution of our foreseeable challenges—environmental, social, and personal. At the least, I hope this book helps people to think about and engage at all three levels in whatever task is before them.

My effort here is not so much to reject dominant, existing views and actions but to lay some deeper sociological and psychological footings in the shifting sands that most determine our current thinking and behavior when it comes to wildlife and nature. Our foundational understandings of our (post)modern world are built upon a deep foundation laid down over recent centuries and our long evolutionary history. One shift underway currently is a spurt toward analytic thinking and a rethinking of our relationship to nonhuman life and nature. This shift is more and more entering into our collective cultural "brain" or consciousness. A popular way to talk about this shift presently is through the language of "sustainability" and innovation. I want to accelerate innovation and transformation.

In sum, our mental world has a very long evolution. Today, it makes up our language, concepts, belief systems, stories, and more—culture and institutions. For example, over history, we have not always had a "land ethic" or "animal rights ethic." These are relatively new to our thinking and actions as we come to learn about ecosystems and animals and human impacts on them. From this mental world, we as a society made and set aside Yellowstone as a national park long ago. More recently, we have come to understand the region as a large, open, and complex evolving ecosystem. This development is a new understanding, a new metacognitive world that we continue to evolve, experience, and make new meaning about it that was not possible earlier. At present, we are struggling to enlarge our mental world—our collective consciousness—to meet demands of our precarious situation (e.g., climate change, biodiversity loss, human health challenges). A complete story of Yellowstone, people, and our evolving mental world is beyond any single author's telling, including this book. Yet, I try to integrate all three levels—physical, biological, mental—in a pragmatic way to learn from experience and many disciplines about our own stories of meaning. My goal is to move toward a new story of meaning for Yellowstone and nature more broadly and to secure a sustainable human–nature relationship. My goal, above all else, is healthy people in healthy environments.

This Book as the Author

Now why did I write this book? The simplest explanation is that I care about Yellowstone! As I tell you about this region, you can come to understand the importance of Yellowstone, as symbol, place, and an idea, and come to care about it too. For many people, visiting Yellowstone is transformative. Most importantly though, you the reader can come to see why we should care about Yellowstone in the first place.

The more nuanced explanation about why I wrote this book is that my understanding of people and nature has changed since I first came to Wyoming in the mid-1960s studying wildlife biology as a biophysical scientist. I have come to realize that the complicated social phenomena I was (and still am) studying had been playing out for centuries. I needed to get through decades of the leaf litter of convention, dig through the topsoil, and reach the bedrock of the issues about Yellowstone and our mental world. Doing so has taken me to the deepest roots of our meaning making—to our psychology, philosophy, and policy. I strive for an integrative way of seeing the world. This book is about me trying to make sense of what I have experienced, sought to understand, and am currently struggling to communicate adequately. It is a window into my mental world.

This book is about the greater Yellowstone as a microcosm of the challenges we face about the human condition and our relationship to nature. These subjects are difficult to integrate simply and pragmatically and to communicate given our current everyday conventional understanding and language. This is because the subjects of my interest involve personal experience, scientific details, philosophic traditions, psychological subjects, and more, all at the same time. Learning to think integratively across all scales and with multiple lenses is essential for effective problem solving to get us out of our present mess.

We have made quantum leaps forward in our understanding and capacities for thought before—imagine living at the time when microscopes were first invented and seeing microscopic life for the first time or seeing stars and celestial bodies for the first time with the invention of the telescope. Both the microscope and telescope opened up a whole new context for our human lives and allowed us to ask new, bigger, and more fundamental questions about ourselves and the world. I argue that Yellowstone is a place and idea that can function similarly for us today. My hope is that Yellowstone, as a place deeply embedded with sentiment and story, might inspire us to work toward a new way of thinking and knowing about our human relationship to nature. We might be on a cusp, a paradigm shift to see, learn, and transform our present unsustainable situation. It just might move us on to a new worldview or story—a new mental construct that is much more humane and sustainable. This book is full of pragmatic hope for that journey.

In this book, I use an integrative mental framework, combined with long experience and the disciplinary knowledge available to us from the work of preceding generations. I include selected examples but tried not to overburden the text with detailed description and analysis. It seems to me, overall, we can most easily deal with concrete cases more so than ideas whether small or large. However, a rigorous command of the knowledge from philosophy, anthropology, psychology, sociology, policy, ecology, and other disciplines is essential for solving conservation challenges in a successful and enduring way. Therefore, in this book I attempt to lay out some of the theoretical and intellectual basis of these critical fields of study and then provide examples for what application of those ideas looks like on the ground in Yellowstone.

Still, the reality is that we are not all prepared to engage constructively with complexity or the mental, analytic, and pragmatic tools at our disposal. We do each have our own experiences, views on Yellowstone, and personal stories about ourselves, communities, and the world. Melding many conventional and functional worldviews into a workable formula that addresses the common cause of humanity's sustainable living is the challenge of our time. To do so requires evolving our mental world. Now is our opportunity to help one another. I hope that this book may serve as a guide for understanding one another more deeply and for solving complex conservation problems in ways that uphold principles of human dignity and respect.

My Standpoint

As an author, analyst, and citizen, I took on the task of looking at all things greater Yellowstone. I have worked and lived in the region for about a third of the time Yellowstone as a park has existed. Soon Yellowstone will celebrate its 150th year. At first, I was not fully clear about what I was trying to do. I took a step back from daily life to observe and listen systematically, to contemplate, and to synthesize what I saw. As the book came together, I learned a lot about myself, human issues—our metacognitive world—and the patterns at play at all three levels in greater Yellowstone. I learned about a great diversity of subjects, both technical and social. In turn, I tried to integrate it all into an insightful picture of greater Yellowstone—people, society, and nature. I drew on complex ideas, metacognitive tools, and analytic frameworks about society and nature, history and people. I surveyed many disciplines for their knowledge and insight. I combined them into the way I see greater Yellowstone, its future, and ourselves in that picture. Although I present this as a new paradigm, I realize that it is not really new.

In this book, I combine three major threads in my life to look at the greater Yellowstone situation. First, I have lived and worked in the region for over five decades. Greater Yellowstone is my home and I have come to know the people and the landscape very closely. Second, I have researched and taught at multiple institutions, including the Yellowstone Institute, Tetons Sciences Schools, Idaho State University, University of Wyoming, University of Michigan, and Yale University, and practiced applied conservation including conducting grizzly bear surveys, leading endangered species recovery efforts, and developing widely used techniques for management and policy. Third, I always use a meta-framework to help me grasp, organize, and interpret experiences and information. The meta-framework I ascribe to was distilled from human experience by some of the greatest social and integrative scientists of our time. I learned about this approach decades ago and still use it to understand all that has happened, is happening now, and may happen in the foreseeable future. I have taught and used this framework to students and with colleagues from over forty different countries. This combined experience, work, perspective, and skill all come together in my formation of this book.

Before going on, a word about our world, as many of us see it. First, things are getting more complex. Complexity is not the same thing as complicated. The approach in this book can help us best deal with complexity. What we are striving for is an enduring relationship to one another within the "people-animals-nature" nexus. What I see missing in too many educational efforts, activists movements, and management policy initiatives is this expansive, yet well-grounded understanding, dialectic, and integrated application to actual problems. This book seeks to help address this omission to these concerns.

Second, the best way to address complexity is to be grounded in foundational knowledge and experience. This includes diverse personal and professional experience, as well as diverse academic disciplines—philosophy, anthropology, psychology, sociology, and many others. This book brings these knowledge traditions to bear in the Yellowstone case helpful to other situations well beyond GYE. It does so by using an integrative approach throughout, wherein some chapters emphasize the benefits of various disciplines and pragmatic approaches.

Third, importantly, the best way to address complexity is to possess and use comprehensive concepts and analytic tools, whatever the case you are interested in. Such concepts and tools have been known and used for over 70 years and longer. This set of concepts and tools goes by many names in the literature—the configurative approach, the New Haven School of Jurisprudence, and the Policy Sciences—to name a few. It is amazing to me to see that many of my colleagues and some of the literature converge on and in fact recreate some of these concepts and tools. It seems that the policy sciences are constantly being reinvented based on experience, yet they are not often taught in our colleges and universities. There are many examples of experience leading to the operations of the policy sciences—problem orientation (rationality), contextual understanding (politics, social, and decision processes), and practical applications (pragmatics, morality). There is a huge convergence on these concepts and tools by relatively successful people working across diverse fields. This book explicates these concepts and tools and encourages readers to become skilled in applying them to whatever situations they choose.

Over decades of my work on six continents, including the more than half a century in Great Yellowstone, I have witnessed a convergence in case after case on these basic policy science concepts and tools. This convergence is not an outcome of chance. Experience in the field and, better yet, a mix of experience and academics quickly bring practitioners to this functional set of concepts and tools. Why? Because it works, relatively speaking compared to traditional approaches.

I am interested not only in plausibly explaining the past and present but also in looking at what is coming in the future. I seek to think beyond convention and accepted conversation about these matters. To explore future concerns, I argue that we must start by examining ourselves and our society and environment now and dig to the roots of our culture—our collective representations. We are presently experiencing a confusing situation with mixed signals. What does the future hold for Yellowstone? What is likely to happen? What do we want to happen? What must we understand and do now to bring about the future we want? In the end, I hope a comprehensive public dialogue about these matters will lead to a healthy greater Yellowstone for ourselves and our children.

The Book's Organization and Readership

Taken altogether, this book is my progress report about my own grappling with the interconnected realms of Yellowstone (its physical and biological levels), our mental world, and all three different levels. It offers my own experiences and inquires. Perhaps this book will advance a rigorous dialogue and cooperative effort for on-the-ground work to advance our understanding, responsibility, and living in places like greater Yellowstone for the benefit of all.

In many ways, this book falls short. I cannot offer the full clarity, grounding, and answers that I would like. I am also limited by my own constructs and ways of understanding. Fortunately, the challenge of conservation in Yellowstone and other beloved places is not one for individual people and individual solutions. The current challenge necessitates a cultural reckoning and rigorous conversation to draw on the vast experiences and beliefs of many people for creative and insightful ways forward. I have faith that we can do this and do it well. Consider this book my own personal contribution to this journey—it is the path that I see forward.

This book does not a have a single home—academically, professionally, or in any single governmental, business, or activist approach. It is way too expansive for something so clean, simple, and clear. Instead, this book is part of a widespread counteroffensive to the long trends that have accentuated only or largely technical, disciplinary, and ideological education and management policy down to our present. Over last decades, we have undermined the humanities, especially in the cultivation of an expansive education that allows us to attend to the many growing complex challenges we individually and collectively face today. In contrast to these trends, this book examines and offers a practical guide to real-world concerns at all three levels—physical, biological, and mental—focused on our living and resource uses. I feel that a "new" integrated approach is necessary to the operation of our lives, society, and relationship to the environment, given our high-tech culture and the rapid changes we are experiencing. The comprehensive policy analytic approach that I use in this book and recommend understands these trends, complexity, and it offers a way to come to grips with it in our social and decision-making and actions.

We all want to live healthy rewarding lives in dynamic communities and environments. The aim of this book is to try to help all people best visualize what is required for addressing problem solving in complex situations as we now find ourselves. It also seeks to help readers best grasp the concepts and tools as I lay them out in this book. In turn these can be applied in their own situation. Finally, these concepts and tools can help practitioners rapidly harvest the lessons of their experience, communicate them widely, and further accelerate their successes.

This aim, grasp, and overall focus of this book fits well with the worldwide call for sustainability, however understood. I use the metaphor of stories throughout this book and try to build the case for a new story of ecosystem conservation. The stories that I use capture in diverse ways the complexity we face, as well as illustrating the concepts and tools that I recommend.

The book has two thematic sets: (1) the Greater Yellowstone situation as a case showing common challenges paralleled in many other cases and (2) a set of concepts and tools for problem solving. My own personal, value, and intellectual standpoint comes out in this book. I tried to use a style that mixes, blends the three levels of concern—physical, biological, and our mental worlds—into an overall readable and helpful account. You are the judge to how well I did, given you own experiences, standpoint, and goals.

This is not a workbook or receipt of sequential steps for us to carry out. It is about ideas, knowledge, and actions we might use as we work toward a sustainable future.

Part 1 focuses on the physical and biological levels. Parts 2 and 3 explore the mental level, its management and policy implications, and pragmatic ways forward. As the subjects of this book are diverse, no doubt readership will also be diverse. My hope is that this book can speak to many different readers and their friends. My only request is that my readers take this as a good-faith effort to improve our current standing, and that they find within these pages my pragmatic hope that we have it within ourselves to secure a better future for people and the environment. It is my case for a new paradigm and a new relationship with nature. *In the end, this book is an invitation to you to engage with all that is Yellowstone and simultaneously think about how you experience and make sense of what you see visiting there—your mental world.*

January 15, 2021

ACKNOWLEDGMENTS

I owe a great debt of gratitude to many people who have guided me to where I am today in my personal and professional pursuits. For decades, I have drawn on the experience, ideas, energy, and inspiration of countless folks through unending conversation and interaction. I acknowledge many of them below but recognize that there are far more than I can explicitly enumerate here.

First, I am grateful to the many researchers, scholars, and theorists, who by their thoughts and work embody a broad and varied sweep of knowledge, ideas, and accomplishment. Their individual and combined contributions to our understanding of the world, nature, and people are astounding. They have greatly enlightened me over the years and have encouraged me to write this book. These scholars include Lasswell, Tarnas, Kegan, Rorty, Sellars, Dewey, McDougal, Yalom, and many others. Their works are now part of who I am and what I think. I am not sure where one ends and the other begins any longer.

Second, my collaborations with friends and colleagues including Richard Wallace, Tim Terway, and Gao Yufang in recent years have been rewarding. The continuing friendship of Christina M. Cromley, David Cherney, Quint Newcomer, Richard P. Reading, Murray Rutherford, Seth Wilson, Doug Clark, Toddi Steelman, and Mike Gibeau is greatly appreciated. My friendships and discussions with Ronald Brunner, Garry Brewer, Bill Ascher, and Andrew Willard have been critical to my personal development. These people are my colleagues in the modern policy sciences analytic movement that seeks a flourishing world for humans and nature—a yet to be achieved placeholder for the idea of environmental sustainability and justice and an idea and practice.

I have greatly benefited from many excellent students and colleagues at the Yale School of Forestry & Environmental Studies and the Institution for Social and Policy Studies. I would like to recognize students Josh Fein and Marian Vernon, among them, and faculty Herb Bormann, Bill Burch, Steve Kellert, Leonard Doob, and Charles Remington, with whom I have worked closely. I owe a deep thanks to the many guest speakers in my courses, seminars, and field trips, from whom I have learned a great deal.

Third, I have benefited from personally knowing many hardworking and dedicated individuals, including Franz Camenzind, Todd Wilkinson, and many others. The help of dedicated public servants Dan Wenk, Doug Smith, Ann Rodman, P. J. White, Tim Reid, Charissa Reid, and Ryan Atwell in Yellowstone National Park is deeply appreciated. There are many hard workers in other agencies too who have been helpful.

In GYE, many citizens and friends have been a source of inspiration and support, including Peyton Curlee Griffin, Bill Barmore, Patty Ewing, Sandy Shurptrine, Jason

Wilmot, Jim Halfpenny, Dick Baldes, Jason Baldes, Lance Craighead. Mike Whitfield, Steve Primm, Katie Christensen, Avana Andrade, Deb and Susan Patla, Susan Marsh, Sue Lurie, Gary Kofinas, Molly Loomis Tyson, and Hannah Jaicks. The Northern Rockies Conservation Cooperative team has been especially encouraging, so I extend a warm thank you to Cathy Patrick, Maggie Schilling, and Ben Williamson.

No one could ask for more intelligent, graceful, or skilled editors and readers than I have had—Marie Gore, Liz Naro, Anna Reside, and Molly Loomis Tyson. I thank my friends and colleagues, beyond those listed here who generously agreed, and even asked in some cases, to read earlier parts of this book. Most important of all is Denise Casey who aided the writing of this book by her support over years. She is an accomplished artist and life partner, whose encouragement and help allowed me to do this book. Her art work is on the cover.

In the end, I have stood on the shoulders of so many great minds and close friends that I am no longer sure what, if anything, in this book is really mine and what is born of others. I have tried to reference the many authors and sources influential and important in this writing, and if I failed to acknowledge or property cite any person or idea, I sincerely apologize. Lastly, I want to express my gratitude for the land, wildlife, and spirit of the Greater Yellowstone Ecosystem that has sustained, nurtured, challenged, and inspired me throughout all of the years that I have lived in the region, and that has served as an enduring touchstone for me.

Chapter 1

STORIES OF PEOPLE, NATURE, YELLOWSTONE

Yellowstone is loved and overrun because it offers a glimpse of what life used to be like, or what we miss and want from it.[1]

The Yellowstone we know and love is endangered. Many of us have a sense of the vast challenges facing the region given the environmental and social changes underway. We see the growing litany of threats in newspapers and on social media. Suburban development, road and trail expansion, front and back country recreation, record-breaking visitation, traffic and wildlife jams, disruption of ecological processes, fragmentation of wildlife habitat, changes in migration patterns, threatened species, invasive species, oil and gas development, loss of wildness, and wildlife disease are all salient challenges. With multiple geographic and political divisions in the region, integrated management across artificial boundaries to address problems is itself problematic, as is the lack of farsighted, coordinated, overarching leadership. Though these interconnected problems are diverse in nature and impact, they all derive from the same underlying source: us.

For people who live in the Greater Yellowstone Ecosystem (GYE) today, the rapid rate of change is cause for considerable concern. Many people are worried that we are losing the habitats and species that have defined our stories about pristine wildness in Yellowstone. Superintendent Dan Wenk said, "This is an extraordinary region [that] isn't like [everywhere else] nor should we accept that its slow, steady disintegration is [in] evitable."[2]

When asked about the most ominous threats facing Yellowstone today, Dennis Glick of *Future West* in Bozeman, Montana, said, "Number one would be the effects of climate changes: droughts, big wildfires, and diseases that were never here before [...] [and today] humankind may very well be loving this place to death."[3] While for many people, the threats of climate change seem to loom far in the future, it is already easy to see the miles of dead and dying trees from the main roads, and the effects are only going to become more dramatic. For example, the next few decades of change will very likely cause more fires, less forest, larger grasslands, smaller and hotter waterways, increased invasive plants, and fewer big animals.[4] Many native species may not be able to adapt to this rapid change.

The GYE is experiencing dramatic change to be sure, and consequently its historic and cultural meaning is changing, too. Dave Hallac, former chief scientist for Yellowstone National Park (YNP), said, "I think we're losing this place [...] slowly. Incrementally. In a cumulative fashion [...] I call it a sort of creeping crises."[5] Off the record, many citizens, scientists, managers, and public officials living in the area express serious concern about

Yellowstone's future. Mushrooming changes are taking the GYE to a place it has not been before, and it's taking us with it. We are the ones affecting change, pulling Yellowstone along in our wake.

In this first chapter, I explore our meaning of Yellowstone through individual and cultural stories from experiences in YNP and the encompassing greater ecosystem. The GYE is a special place that plays many roles and has diverse meanings—immersion in nature, reflection, renewal, adventure, and appreciation. Stories describing these meanings are expressed through literature, poetry, art, religion, sports, and science.[6] The GYE is a crucible for story making at many levels because it is nature in the largest sense— or so we tell ourselves. But our individual stories of meaning often conflict as some stories are utopian and others dystopian. How do we reconcile vastly different stories derived from different individual experiences and social influences? Can we get a story most of us can share, one grounded in reality? I end this chapter with an introduction to my methods I used in this story of Yellowstone.

Yellowstone and Our Stories

The GYE is a story about ourselves as people, our understanding of nature, and how we relate the two to make meaning for ourselves as individuals and societies (Figure 1.1). According to Dr. Timothy Terway, a policy scientist, "meaning is an activity of mind, integral to thinking itself. It is no wonder that it is so hard to get our heads around: we make meaning in our very attempts to understand, dismiss, or ignore meaning itself."[7] Stories are the way we make meaning for ourselves. Meaningfulness is the fount of motivation, commitment, and satisfaction. So, what does Yellowstone mean to us? How and why do we value Yellowstone? Is Yellowstone just a symbol in our mind, a stand in, a surrogate, or reference for real nature? If so, what is nature and what does it mean to us?

Individual stories of Yellowstone

Whatever the future holds for Yellowstone, it will be determined largely by our stories and actions today. YNP is home to the Hayden and Lamar valleys, the iconic backdrops to many visitors' experiences in the park. Visitors flock to Artist's Point, a vista offering a striking view of the Lower Falls in the Grand Canyon of Yellowstone. The sweeping Lamar Valley teems with Yellowstone's wildlife: wolves, bison, and bears. What is a person to think and feel when taking in such sights? Sixteen-year-old Helen Mettler, who visited what is now Grand Teton National Park (GTNP) in 1926, declared in her diary, "God bless Wyoming and keep her wild."[8] Her words express a majestic and powerful experience. The stories we create from experiencing such places have hard consequences in the real world—a cause and effect we will explore in the following chapters.[9]

Two opposing experiences

Take these two true stories. "It's full of life," Jane exclaimed, almost in tears. She was a newcomer to Yellowstone. "Before when I lived in California," she said, "I was

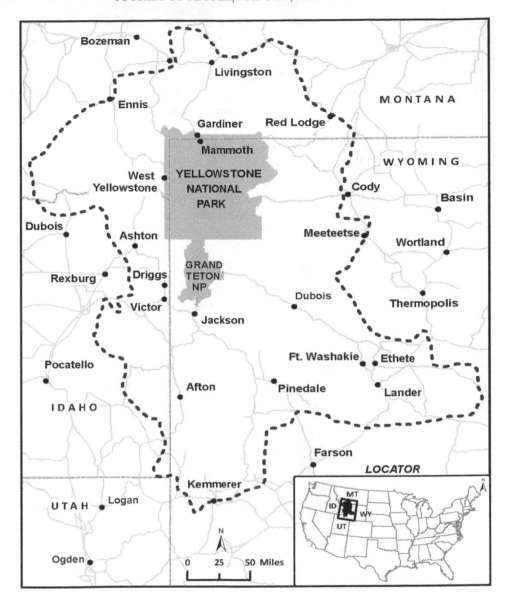

Figure 1.1 Map of the Greater Yellowstone Ecosystem in relation to Yellowstone National Park, the three states, and within the United States.

uncomfortable, out of sorts, felt something was wrong, something was missing."[10] She felt her life was out of control. A community facilitator and psychologist, Jane said of her first visit to Yellowstone, "I saw all that life, animals by the hundreds and thousands. I realized that California, like most of America, and the world—is empty—it has lost most of its life." Soon thereafter, she moved to Yellowstone permanently. It proved to be a quantum leap in personal development for her. Jane regrounded herself, emotionally and physically, in Paradise Valley, Montana. She found a profound sense of meaningfulness, a connectedness unknown to her before. In her own words, she was "revitalized" with a new sense of belonging with nature. Moving to the Yellowstone region brought Jane to utopia and transcendence.

In contrast, longtime Yellowstone resident and biologist Deb said, "I have been feeling awfully glum lately about human population growth swallowing up most of what I love about this planet. Even Yellowstone has changed a lot in my [25 year] career, and the frog population I studied for my science degree has been extirpated!"[11] She went on, "Being an old-timer seems to bring a lot of sadness. Every generation makes a new threshold, thinking it should compromise what remains for yet more [economic] growth." The current generation of "adventure recreationists" and supporting businesses is just one contemporary example. Deb's experience, like so many longtime residents, gives a sense of dissolution, fragmentation, and loss. For Deb, and other "old-timers," the Yellowstone experience was turning into *dystopia* and pain.

Is GYE's future *utopian*, full of life thriving in a landscape much like today? Or is it one of creeping, slow decay invisible to the average visitor, official, or recreationist, a *dystopian* future of degraded landscapes and lost wildlife and migrations? What can we do to affect the outcome for the better?

Excerpts from personal accounts

In her book, *A Hunger for High County: One Woman's Journey to the Wild in Yellowstone Country*, Susan Marsh eloquently captures over 30 years of experience in Yellowstone:

[I] was always drawn to the wild from an early age. Animals were [my] primary conduit to the beauty and mystery of forests and mountains. [...] As I dropped into the dark gorge of the canyon, the mountain breeze followed me down the trail, stirring at the back of my neck as if to encourage this blissful mood. Once again, high country came to my aid and I felt whole again. Joyfully, I broke into a run [...] The beauty I find in forests and mountains never fails to move me, and I have never strayed far from places that inspire me.

[I was] saddened to realize many tourists did not share the same level of love and appreciation for nature. "Isn't that pretty," people would remark at national park overlooks, just before collecting their kids and moving on [...] [Tourists threw] distracted glances or the kind of impatient shuffling that one saw at Old Faithful as the crowd waited, only to disperse before the geyser finished its eruption. [...] The rattling of a garbage truck lifting a bear-resistant dumpster [...] errant car alarms bayed urgently [...] vehicles idled with windows opened and radios cranked [...] irony floated like an overheard joke: this was a national park, the nation's flagship park, two million acres of wilderness in the far corner of Wyoming. Visitors came for wildlife, scenery, and memorable outdoor experiences, but most of what lay in front of me could have been replicated in front of a suburban Wal-Mart.[12]

Mary Beth Baptiste's *Altitude Adjustment: A Quest for Love, Home, and Meaning in the Tetons* tells another highly personal story from years of exploring, being outdoors, and living life in the GYE.

Promising a new freedom previously unimaginable to me, the serrated peaks challenge me to take the risks that will bring purpose and depth to my humdrum suburban life. I belong here. But it's bigger than me, orders of magnitude bigger. Through my head flows with words of Aldo Leopold: "A thing is right when it tends to preserve the integrity, stability and beauty of the biotic community." I strive to live this tenet. As a mountain woodswoman, I could. [...] Windfall or not, I'm here [...] I'm here. The mountains called, and I came. I found my way home. I created my own piece, albeit a scarred and dimpled one, in the wilderness puzzle. In this world of mountain, snow, and sky, I finally feel the power of my life, and it matters.[13]

Most people only know about Yellowstone from afar, and yet it is meaningful to them, too, at least in a distant sort of way. They may never get to Yellowstone country, and never experience it personally, yet it is important to people worldwide, simply to know that it exists.

Cultural stories of Yellowstone

Our stories reflect how we, as different people, are each unique at one level, but at the same time similar at the most basic level.[14] We all live inside of our society and its culture. As such, Yellowstone is a cultural story, too. It is both a personal and institutional story at the same time. It is an historic and contemporary story that we all share as part of our society. It is a story that tells others in the world who Americans are, what we stand for and value.

Yellowstone in historic culture

One of our first GYE stories occurred over a century ago as we were finishing off the last of the great bison herds, decimating the vast flocks of passenger pigeons, and destroying the free-roaming Plains Indians. In 1872, YNP came into existence, just a few years before the frontier was officially declared closed. Stories of Yellowstone from these early years depict a pristine wildness or wilderness. Yellowstone was designated to protect the region's geothermal features.

Richard Slotkin, a professor at Wesleyan College, details the story of the American Frontier myth from 1600 to the present in a trilogy on the frontier.[15] His first book, *Regeneration through Violence: The Mythology of the American Frontier, 1600–1860*, focuses on the cultural and institutional story of the American West. In the second installment, *The Fatal Environment: The Myth of the Frontier in the Age of Industrialization, 1800–1890*, he demonstrates how the myth of frontier expansion and subjugation of the American Indians helped to justify the course of America's rise to wealth and power. And, in the last installment of the trilogy, *Gunfighter Nation: The Myth of the Frontier in Twentieth-Century America*, he draws on a wide range of sources to examine the pervasive influence of Wild West myths on American culture and politics of today.

Yellowstone in contemporary culture

Yellowstone is a placeholder for the epic story of American cultural history, from the times of the frontier and settlement up through the world of today. Many people today hold that old story to some degree, perhaps with a mix of contemporary views. The old story—our cultural myth—tells us who we see ourselves to be. It tells us what we want to see as the essential American character. Our character came from our struggles on the frontier and is a narrative about individual freedom.

Each of us lives inside of our culture and its institutions. It is not possible to escape its conventional mindset. It holds us together, but many lack full awareness of its influence on our thinking. These psychological, sociological, and philosophical facts are part of our cultural story, a story that has been shared over generations and reaffirmed time and time again in our shared consciousness.

Our cultural stories are told not only in words but also in art, poetry, music, science, markets, policies, and actions. Diverse cultural representations or stories in circulation today span from author Terry Tempest William's evocative and emotive narratives, such as her book *The Open Space of Democracy*, to Christian Beckwith's utilitarian stories, formalized through his annual *SHIFT* conference promoting a techno-recreation-centered view of nature. Songwriter and performer LeeLee Robert writes in her song "WYOMING (Crown Jewel of the West)":

One starry night by a campfire light
in the heart of Yellowstone.
One man stood and said,
"Now this great place I'd like to own!"
But another man had a better plan:
a site for ALL to see.
And Yellowstone National Park became our legacy
Wyoming, Wyoming, Crown Jewel of the West.[16]

Personal experiences add to our cultural story—a sunrise in the Grand Canyon of Yellowstone, backlighting on a herd of frost-covered bison, the unfolding drama of a wolf pack at work. Katie Christensen's *The Artists Field Guide to Greater Yellowstone* is a nature guide to Yellowstone's wildlife told through the words and artwork of fifty of the region's most distinguished storytellers. Denise Casey's artistic renderings of the Grand Tetons and other scenes from Jackson Hole, Wyoming, are some of many expressions of natural beauty and wonder.[17] R. J. Turner and Tom Mangelsen's independent works on wildlife and nature photography capture a singular quality of Yellowstone, conveying images to the public that would otherwise be seldom appreciated.[18] The Museum of Wildlife Art in Jackson Hole hosts a world-class collection of nature and wildlife presentations covering over two hundred years of art. This historical art is especially important, as it offers a window into how nature and animals were viewed in the past. The museum also houses contemporary views of wildlife and nature. Comparing historic and modern subjects is enlightening as our cultural views of these subjects have changed over the centuries.

Finally, people experience the GYE through their work and living. The on-the-ground work of Albert Sommers, a rancher from Pinedale, Wyoming, and Matt Barnes of the Northern Rockies Conservation Cooperative seeks to minimize large carnivore conflict with livestock in the upper Green River Basin, Wyoming. Penny Maldonado of the Cougar Fund is working to harmonize the relationship between mountain lions and humans. Mark Elbrock, a cougar researcher in Jackson Hole, does the same work now with Panthera, a conservation organization. Lisa Roberston seeks better conservation practices through her organization Wyoming Untrapped. These are just a few of the many unsung heroes, each working in their own way and inspired by their own stories to improve conservation and sustainability in the GYE.

Our story of the GYE today comes from more than four centuries of evolved cultural narrative. It is a long history made up of a great many individual stories, all woven into our overall cultural story of meaning. Our personal stories and the overarching cultural story have hard consequences when played out on the ground, as recent history and current events show.

Nature and People

When viewing our individual and cultural stories, their meaningfulness, and their consequences for management policy in the GYE, we should acknowledge that today we know much that was unknown to previous generations. We humans are a species with vastly more mind power and inventiveness than any other species. We can use these attributes, should we decide to do so, to gain a contextual overview to guide our actions toward a healthy future for the GYE.

Yellowstone as nature

Engaging with Yellowstone—nature writ large, the wild, the animals there—can bring out great questions about life, meaning, and the world. What is nature? Is YNP "nature" and downtown Jackson, Wyoming, not "nature?" Importantly, what is the relationship that we conceive between nature and humans? What are the consequences of that conception?

People and nature

Many people have a love for nature. Some are worried about the destruction of nature—as they should be, given our heavy hand on the wild and wildlife. Yet, there are many who think that nature needs to be "tamed." They want to control nature—human and nonhuman. To understand the GYE, we need to ask basic and foundational questions about our individual and collective views of nature.

To start, consider Rick Bass, a regional author from Montana, who asked, "What drew me here?" This is what he wrote:

I can identify, fairly easily, those predispositions toward which I was already canted. It's a big question for me, why the Yaak [a mountain range in Montana]? There are similarities between the South [of the U.S.] and the Yaak, but the differences, as well—particularly between the loss of individuality, loss of voice, in the suburban, homogenous, petrochemical mall-land of Houston where I grew up, and the gnarly fecundity of Yaak. I believe strongly that the antithesis of a thing can shape, define, sculpt, the boundaries of the other thing, the thing-to-come; and maybe it was, and is, that simple. Perhaps growing up in Houston is what created in me the place and space for loving—needing—so deeply this rank wild mountain valley so unlike where I was raised.[19]

He goes on to ask, "Do our personal destinies lie below us, with powerful predispositions, or do they exist more strongly in this world-above, with its scour of wind and ice, its tongues of flame, and all its other magnificent testings and fashionings of our spirit?" His musing concluded,

I think the answer is an unsatisfying both: a unique and authentic, time-crafted land will certainly sculpt your fit to it, but there can also be a summons—an unseen tendril of logic and grace—of fittedness—that has not yet been achieved but which yearns to be achieved, with every bit as much of the power and unseen subtle elegance of mere electrons and neutrons and protons (if such things even go by those olden names anymore). That as there is a chemistry of soil and stone, so too might there be a chemistry of spirit, a kind of awareness or trembling pre-awareness, even between objects separated by great distances. [...] All of which is to say, the summons, in the beginning, does not have to be large, nor the first step of a journey dramatic. It can be but a whisper, a thought, a question, even a dream.

Sense of nature

Presently, you have a sense of "nature" that is sculpted by your experiences and what you know of the experiences of others. We talk of Yellowstone as nature or natural, but what do we mean? Diversity in use of the word "nature" is significant. Are genetically modified fish that grow rapidly natural? Is ecological restoration natural? What do we mean when we say that bulldozing a marsh full of rare plants is a "crime against nature?" Is nature the same as the wild or wilderness? What is human nature?

Nature (as in the GYE) is a concept. It is expressed linguistically in words and in other ways (e.g., art). Using the lens of semiology (e.g., the study of language), we see that "Yellowstone" is a concept and the word is a signifier and referent for nature that is applied to a very wide array of ordinary things (e.g., bears in Yellowstone or wildness in GYE). It is never clear what people mean when they invoke the word "Yellowstone" because the word signifies a plurality of things depending on the speaker.

Nature seems to be one of a combination of several things to people (Figure 1.2).[20] Is nature: (1) the entire physical world, (2) the nonhuman world, (3) the essence of something, or (4) an inherent force? Figure 1.2 illustrates the concept of nature and its meaning and references. These four possibilities (and combinations thereof) are all signifiers of something. At this point, we will leave open the possibilities of what "nature" *is*. However, as you read this book, think about how you conceive of these matters.

As humans, we need the concept of nature—something that is "out there"—so that we can talk about something in us or "in here." We need to conceive of who we are in relation to nature. We cannot live without such a concept as nature, as it is central to our meaning making and a vehicle to address our deepest existential challenges. In short, the

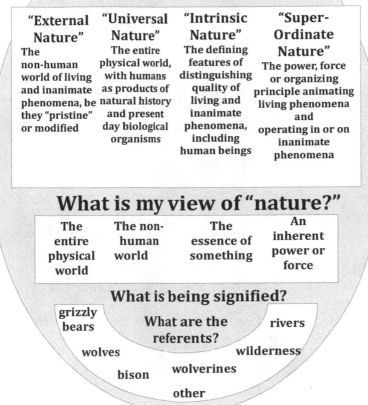

Yellowstone as NATURE?

What are collateral terms?
ecosystems, environment, animals, genes, other

Four views of "nature"

"External Nature"	"Universal Nature"	"Intrinsic Nature"	"Super-Ordinate Nature"
The non-human world of living and inanimate phenomena, be they "pristine" or modified	The entire physical world, with humans as products of natural history and present day biological organisms	The defining features of distinguishing quality of living and inanimate phenomena, including human beings	The power, force or organizing principle animating living phenomena and operating in or on inanimate phenomena

What is my view of "nature?"

The entire physical world	The non-human world	The essence of something	An inherent power or force

What is being signified?

grizzly bears

What are the referents?

rivers

wolves

wilderness

bison

wolverines

other

Figure 1.2 Four major meanings of the concept/word nature in today's Western world. All these distinctions and signifier words were invented by humans through culture over the last 100,000 years of evolutionary time.

concept of "nature," and all its collateral notions, is vitally important and essential to our lives, both individually and collectively, for the work of meaning making.

Nature and wild

How we act on our conceptions of nature has life-changing consequences for the land, for nonhuman animals, and for humans. Take Jack Turner's big questions in *Travels in the Greater Yellowstone* and *The Abstract Wild*. Both books speak to nature, naturalness, and the wild, and our relations and responsibilities to them.[21] Turner is a world-class trekker, mountain climber, and keen observer. He talks of nature in the context of the postmodern world, where nature and wildness are rarely valued. In *The Abstract Wild*, Turner sees that we are losing touch with places such as Yellowstone and the GYE, places that represent the wild for him. Today, he says, because most people do not personally experience the wild in the way he uses the word, they are not prepared to engage nature as such. Instead, we are "abstracting wild," ritualizing it through superficial adventure sports, techno-recreation, and "selfies." In short, we are reducing nature to hollow symbols while living our self-interested lives, much to the detriment of places like the GYE and its wildlife.

We can discern several vital contextual trends in our culture and in our individual stories.

To summarize Turner, these trends include

1. our diminished personal experience of nature, the concomitant devaluation of that experience, and the attendant rise of mediated experience;
2. our preference for an artifice, copy, simulation, and surrogate, for the engineered and the managed instead of the natural;
3. our increasing dependence on experts to control and manipulate a natural world we no longer know;
4. our addiction to economics, recreation, and amusement at the expense of other values;
5. the homogeneity that flattens not only biodiversity but also cultural and linguistic diversity as Western thought, perceptions, production, and social structure spread across the globe; and
6. our increasing ignorance of what we have lost in sacrificing our several-million-year-old intimacy with the natural world.

He concludes, "[what we need] is a radical transformation that revalues the wild earth—its mystery, order, and essential harmony"—a new story.[22]

Yellowstone—a nexus for colliding stories

The GYE's diverse stories of meaning are colliding in today's postmodern, deconstructed world. At the heart of the matter is conflict among the differing stories we tell ourselves about who we are, what nature is, and what our relationship to nature is and ought to be.[23] As such, this conflict is over foundational questions about what is reality, truth, knowing, rightness, and more.

A collision of stories

The dominant story line today about how we should relate to nature is an overpowering narrative of recreation and exploitation. The glossy magazines out of Jackson Hole show the world from that viewpoint—unlimited outdoor recreational opportunities, as long as one is outfitted with the latest skis, bikes, boats, snowmobiles, bicycles, and stylish attire, all available for purchase from local businesses. The images tell us to believe that life is best enjoyed at one of the trendy dining spots or beer halls, all the while staying constantly wired up electronically. In our dominant techno-rationalist culture, virtually everything is commodified, monetized, or marketized. Images and experiences are sold for profit to tourists and recreationists who are encouraged to just "have a good time."[24] Meanwhile in the GYE, Jackson Hole and other gateway communities to YNP are all selling experiences (at dude ranches, on adventure trips, backcountry biking) and trinkets (rubber tomahawks, stuffed bears, T-shirts, technical outdoor gear) to those who can pay. The dominant, neoliberal view is pushing for year-round recreation regardless of the consequences. This mode of operating sends a clear message: short-term personal enjoyment is more important than harmful consequences to wildlife, other people, or the environment.

In contrast to this dominant progressive story is the minority story that is often buried and typically ignored. The minority story is about "attentive participation"— environmental concern and responsible living. This minority myth is problematic for the dominant view because it calls our attention to our responsibilities, which the dominant view tends to overlook. Perhaps Mardy and Olaus Murie, Jackson Hole's late and famous conservation couple, best embody this minority paradigm.[25] Today there are only faint traces of that worldview left in Jackson Hole because the dominant current view emphasizes individuality and commerce and powerfully overrides it.

The two cultural stories—dominant and minority—are colliding in the GYE. They represent two important ways to find meaningfulness. The struggle over whose stories of meaningfulness will dominate in our culture and markets in the GYE is visible in newspapers, literature, art, science, politics, and policy. Take just one day in the *Jackson Hole Daily* (March 31, 2016) to get a flavor of the conflict.[26] One headline told about a federal official's violation of the Endangered Species Act when the number of grizzly bears that could be killed by hunters in GTNP was increased. Another story told the case of developers who wanted to build 200 units in South Park, Jackson Hole, while battling county planning staff. Developers alleged that growth control regulations are capricious. Another article discussed animal road deaths, which are a significant problem, though decision makers continue to defer significant action.

Surrounding these local articles are others on court challenges to wetland protection and the biggest challenge of all: globalization. The world is becoming more integrated in many ways, especially economically. A quote in one article about globalization stated, "Nothing can be fixed or stopped until we come to terms with globalization as a profound psychological issue, not just a matter of economics or immigration patterns."[27] The GYE is now a global destination. It is populated and visited by world and business leaders.

Recycling endless conflict

The two contradicting cultural stories of Yellowstone offer conflicting views on how we are going to live in the GYE, and who should make the decisions about how we will behave. This conflict is not likely to be resolved anytime soon, as special interest groups vigorously contest for the advantage. Many of the high-profile surrogate issues—bison, bears, wolves, recreation, development—are ongoing after decades of debate. We are simply recycling conflict as a shortsighted diversion from actually addressing the deep basic and foundational questions about our relationship and responsibility to nature.

The bison case is just one among hundreds of cases in the last decade that highlights the conflicting views of these two cultural stories. Consider Scott Turner's paper that explored the heart of the issue: *Cauldron of Democracy: American Pluralism and the Fight over Yellowstone Bison*. Turner wrote,

> Given [bison's] iconic status, it is surprising to discover that within the Greater Yellowstone Area bison are the subject of heated political controversy. At issue is the threat posed by free-roaming bison to the cattle ranching business. On one side are ranchers and their interest groups, locally elected representatives, and the Montana Department of Livestock, while on the other are an array of environmental groups, Native American tribes, and the Montana Department of Fish, Wildlife and Parks. In between is the National Park Service, with its often-challenging mandate to conserve nature and wildlife and to "leave them unimpaired for the enjoyment of future generations."[28]

The array of people Turner mentioned makes the management of YNP bison an instructive case study in American pluralism—democracy (or lack thereof) in action. Two decades earlier, Christina Comely, at the time a PhD student at Yale, dug into the bison case and offered a way to clarify, secure, and sustain the common interest.[29] This subject is discussed below. Chief biologist in YNP, Dr. P. J. White, recently offered his perspective on bison management stating, "I realize this is a difficult task in modern society, but it should be feasible [...] [to] reduce the [bison] reliance on Yellowstone [...] to preserve the species in the wild."[30] The bison case is just one of scores of cases in replay, all based on some view of nature and our relationship to it.

Values and politicization

It is clear to me that all the values people strive for are either already politicized or are becoming more so in the GYE and in the management policy dialectic over its future. People wanting respect feel disrespected, those wanting power or influence feel powerless, and people who want principled standards or rectitude feel slighted. Those wanting enlightenment feel unable to fulfill their quest for knowledge and apply it to help matters. People who have practical skills feel their talents are not appreciated. Some people gain wealth, while others work for low wages, are underemployed, or lose money. Well-being is politicized as people suffer mental and physical problems. At the same time, others are indulged, benefiting from the conflict. In such situations, trust vanishes, social capital evaporates, and it becomes difficult or impossible to work cooperatively to find and secure the common interest.

The personal stories above, the newspaper reports, and our shared collective cultural story all show us how people relate to the GYE differently—to nature and to each other. For some people, YNP is a kind of utopia, as it is for Jane. Experiencing utopia creates feelings of wonder, awe, and fulfillment. But for others, Yellowstone is at best just a photo stop, something to post on social media. For a few "critical theorists," it is about racism, genocide, and colonialism. For many, the GYE may be simply a place that is superficially and immediately experienced as a self-centered recreation, a fun event. I once heard a tourist say about Yellowstone, "It's nothing special," followed immediately with an older woman grabbing the arm of her teenage daughter and dragging her back into a large motorhome. "Wow!" I thought as they sped away. These diverse views make clarifying, securing, and sustaining our common interest elusive at best and impossible at worst. If these individual and cultural trends and conditions continue, then politicization and conflict will compound in the future. Under such conditions, the common interest goes unqualified and unsecured.

Yellowstone's Future

It is increasingly clear to a growing number of people that we need a new story about the GYE's future, one truly committed to a healthy, sustainable environment, rich in other life forms, and one that affords dignity for all—human and nonhuman. The new story must be about living such a commitment to the future of the GYE in real time. In this section, I look at these concerns. I ask large, profound questions about our living in the GYE and our relationship to wildlife and nature. In doing so, I ask basic and foundational questions. I also introduce concepts, theories, and my methods that I use throughout this book. This maybe new to some readers, yet these ideas and approaches are well grounded in the humanities and social sciences.

Our place in nature

We live inside a box of our own creation. It is called the everyday, ordinary, taken-for-granted, or "convention" of our living—the GIVEN (to us by our culture and its institutions). This worldview often consists of misguided assumptions, and we have to live with the consequences of our acting on such assumptions.[31] It is time that we become aware and conscious of these assumptions—the trap of convention—so that we can manage the implications in a more sustainable manner.

Grounding issues

Our everyday, conventional understanding of the world (and of ourselves within that world) is evident in our self-generated stories of meaningfulness—personal and cultural. We are typically not fully aware or conscious of the powerful grip that these stories have on us or the worldviews we hold.

    We intuitively grasp everyday convention. In fact, we anchor our lives in it all the time. Our stories tell us that the GYE is nature, that grizzlies and wolves are embodiments of what it means to be wild. Typically, we take conventional stories literally and blindly accept these narratives as truth. Conventional stories function to tell us what is real, what we know, and what is right. This is the problem of "bounded rationality" or the "conventional trap."[32]

People seek **Meaning** through **Society**
using/affecting the **Environment**

MOST OBVIOUS   ROUTINE

EVERYDAY QUESTIONS
(What is convention, the ordinary, the GIVEN?)
1. How do I get a job and things I need?
2. How do I take care of spouse, kids, relatives?
3. How do I get a social life, have fun, and all?

BASIC QUESTIONS
(What do I think about the world and people?
1. What is "nature?"
2. Who are people?
3. What is or should be the relationship?

FOUNDATIONAL QUESTIONS
(How do I perceive, experience, believe, be?)
1. What is? – reality, truth
2. How do we know? – knowing, with confidence?
3. What is right? – rightness, correct
4. What is sense of things? – coordination, location of self
5. What is practical? – pragmatic, doing

LEAST OBVIOUS   EXCEPTIONAL

Figure 1.3 Ordinary (everyday, conventional), basic, and foundational lens, levels, or considerations in the Greater Yellowstone Ecosystem that we can use to ask and answer key questions.

Basic and fundamental questions

We need to come to view the world through multiple lenses and see more deeply. Some lenses are basic and foundational, well outside of convention. Appreciating these multiple lenses and what they offer us in terms of "seeing" is critical. Perhaps more difficult is integrating these multiple views into a composite whole. This is the only route to deeper meaning, awareness, and insight needed for sustainability. Encouraging inquiry into the deeper meaning of the GYE, or our lives, is a difficult argument to carry out successfully, given the powerful hold of convention.

In the dominant view, it is widely believed that dualism of nature versus humans is simply the way that it is. I contend that this view makes it difficult to find deeper meaning of the GYE. We need serious attention to ontological (what is?), epistemological (how we know?), and axiology (what is right?) concerns—lenses to get to that deeper meaning (Figure 1.3). These questions are explored below.

The assumptions we take for granted in convention are formidable barriers to getting to that deeper meaning. By using these multiple and integrative lenses or layers, we can come to a new understanding of nature and ourselves in it. We can move closer to the kind of world we really want to live in—a world that is healthy, thriving, and sustainable—by using these multiple lenses. Dr. Shawn Achor, a psychologist and author, said, "It's not reality that shapes us, but the lens through which your brain views the world that shapes your reality. And, if we study what is merely average, we will remain merely average."[33]

Some of the multiple lenses through which we must observe the world if we hope to understand essential realities are the ecological, geographic, sociological, psychological, anthropological, philosophical, and policy perspectives. And perhaps the most useful lens for orienting our thinking is that of the policy sciences, a very old meta-discipline that is invaluable in our overall task of understanding and sustainability.

I argue that we need a more fundamental understanding of the GYE, and ourselves as meaning-making creatures, than the one popular view that conventional culture allows. To get a more fundamental understanding requires that we dig much deeper into our present situation, including our history and thoughts about nature. We need to put our spade into the soil and dig down to bedrock, so to speak, in our investigation to find the needed new GYE story for sustainability. We need to go to the very basis and foundation of our thinking, culture, and behavior, and reappraise matters.

The basic questions that I strive to answer are in the following chapters:

- What is my conception of nature? (Part 1: Yellowstone)
- What is my conception of people? (Part 2: People)
- What do I conceive the relationship between my people and my nature to be? (Part 3: Work Ahead)

These are not stand-alone questions, but closely interconnected ones, and as such, threads of each will be explored both directly and indirectly throughout every part of this book.

Below those three basic questions are deeper foundational questions, truly important ones (Figure 1.3). Answers to these questions determine how we live and how society functions:

- What is reality and truth? (Ontology)
- How do we know and what is the mind's way of knowing? (Epistemology)
- What is right and what is correctness? (Axiology)
- What is our sense of things—coherence, agreement, and concurrence? (Ordination)
- What is practical? (Pragmatism)

Neither basic nor foundational questions seem to be asked in our daily conversations. Nevertheless, we address them indirectly and subconsciously when, for example, we conflict over grizzly bears or wolves management in the GYE. We address the three basic and five fundamental questions unknowingly all the time, in one form or another (Figure 1.3). Our implicit answers to these questions are built into our society and its culture in such a way that we almost never think about them. Consequently, these sets of questions are largely invisible as important concerns in most people's daily lives. This is convention at work. The fact is, our culture conditions us with certain modes of thinking that give us answers that we tend to live with unquestionably and unconsciously.

At this point, you, the reader, might want to see how you would answer the questions in Figure 1.3. Think about the consequences of living out your answers in the GYE. Consider the effects of your answers on other life forms, the environment, and other people. Do you think that your friends, and diverse others, would give the same responses as you? Why or why not?

Stories that materialize

Among the basic and foundational philosophic issues undergirding our stories is our tendency to materialize and objectify everything around us, especially other life forms and even nature itself. Most of our language presently works to create a division between people and nature. This dualistic view has been highly beneficial to us given our material and energy needs in the last few hundred years. However, we are now seeing that the dualistic story and language supporting it has a dark down side—pollution, degradation, resource depletion, and biodiversity loss.

The material world in which we live is not only a place full of things and objects to be instrumentally manipulated for our immediate gains. There are alternatives to conventional understanding and conventional harmful practices. Finding these alternatives gives us pragmatic hope for a better future. Yet, a question remains: is there something in our core culture or constitution that causes us to hold on to convention and the dualistic story of nature versus people? Perhaps the core problem is how we think and use knowledge. We think in words and those words make up our stories of meaning. Each word in this book, everything you think and say, our entire culture, is only possible because of the outpouring of our thoughts. Maybe there is something about how we think that plays out in our individual and cultural stories of meaning that is fundamentally problematic. If so, then we have a much deeper kind of problem with the GYE's future than convention shows.

Seeking meaningfulness

I wrote this book for people who want to actively control their future rather than just accept it passively. I assume readers are interested in finding meaningfulness in their lives and working for a healthier GYE, or some other place that they love. There are many other places experiencing similar changes and showing similar concerns. To sufficiently attend to any of these places, we need a realistic and practical heuristic—a problem-solving approach—to examine matters. That heuristic has to be larger than what we are examining or using it for now. Fortunately for us, such a heuristic exists.

The problems of Yellowstone

We have three types of interconnected problems that, taken together, pose a dilemma for us. These problems are roaring along like a runaway truck heading down Teton Pass, out of control, straight into downtown Wilson, Wyoming. But not everyone is aware of the looming danger.

What are these three kinds of problems? First is *technical (ordinary) problems*, such as the cub of bear #399 killed in a road collision in GTNP in June 2016.[34] These problems are easily understandable. Second is *systemic (governance) problems*, which are due to issues inherent in the overall management policy system, especially in decision-making or governance, rather than to a specific or isolated factor, incident, or case. These are also evident in grizzly bear management. Third is *cultural (constitutive) problems*, which are about the persistent patterns of conduct, the cultural rules/norms underlying those systemic patterns. These are harder to see, yet they lie at the foundation of the two kinds of problems above. Examples of constitutive problems are well illustrated in grizzly bear science, management, and policy over the last five decades. Systemic and constitutive problems remain largely invisible to most people; yet they are the core problems that need our immediate attention.

Because our present condition arises from a mix of these three kinds of problems— and our gross inability to identify them—we are growing out of sync with nature in the GYE. Glenn Albrecht labels this problem as "solastalgia,"[35] which he calls a "form of psychic or existential distress caused by environmental change." He was researching the consequences of long-term drought and large-scale mining on local communities in New South Wales, Australia, and concluded that no word existed to describe his findings—a growing unhappiness of people whose landscapes were being transformed all around them by forces and factors outside of their control. Albrecht concluded that "a worldwide increase in ecosystem distress syndromes [is] matched by a corresponding increase in human distress syndromes."[36] This captures what many people are experiencing in the GYE today. We need to be able to see the forest for the trees in the GYE, so to speak, and do something pragmatic about these three kinds of interrelated problems if we are ever to put ourselves on a path to sustainability.

Further, as Gabriela Lichtenstein said, "conservation has not failed. It is we who have failed in promoting the unity between the social and ecological systems."[37] We have failed in addressing complex issues with our limited and conventional views.[38] Our failure to address problem(s) is due to our myopic, self-interested focus on the here and now. We seem to be lurching from incident to crisis, then moving on to the next one. In short, the problem is with us, not with the mule deer being killed on the highway. It is we who will keep failing, unless we dare to address the ultimate causes of biodiversity decline and environmental decay—our unsustainable thinking and unbalanced patterns of resource appropriation, production, and consumption, unplanned economic growth, and social inequity.

A heuristic

A heuristic is an approach to problem solving, learning, or discovery that employs a practical method and offers foundational insights. The heuristic that I use in this book recognizes that all individual and cultural stories share four major themes, offered here in general terms as below:

People seek **Meaning** through **Society** using/affecting the **Environment**
others   values   institutions   resources
self   dignity   organizations   culture

This heuristic (or relationship) considers both content (biophysical substance) and process (people, relations, procedures, and decision-making patterns) of conservation and sustainability in our communities and in our daily living. It also simultaneously gives us a practical focus and a way to understand context. I give examples of this throughout this book. This heuristic in this book focuses on four major variables and their functional relationships throughout.

Conventionally, many of these four variables are left out of analysis and discussion in the GYE. Many people, especially technical experts, consider only the "environment" part of this heuristic and seek facts on energy, forests, and biodiversity.[39] Convention often blindly assumes that once knowledge about raw natural resources has been rendered legible, good decisions will follow. This is erroneous. Over time, we have institutionalized the assumptions behind the conventional approach (see questions in Figure 1.3). By doing so, we misunderstand reality and the limits of our rationality. But when used in full, the heuristic can free us from the limitations of the conventional, everyday thinking that dominates our problem-solving approach today. By using this heuristic, we can come to see the competing stories and conflict over the GYE in a much wider and deeper "landscape" of knowledge production and meaning-making processes.[40]

My reasons for using this functional heuristic model are threefold. First, we cannot possibly form a view on the what, why, or how of the GYE unless we understand the broader context in which it presently exists. Second, we must pay close attention to the key challenges in all four variables included in this heuristic. Challenges are present in government, universities, NGOs, and private businesses, and elsewhere in society and culture. Third, I wish to be highly normative—to explore what *should* be and understand how that differs from what *is*. I am going to examine the present dominant paradigm about what ought to happen in the near future in the GYE, given the current situation. We cannot presume that current conventional actions constitute a sufficient response to how we should respond to the mushrooming, rapid social changes underway in the GYE.[41]

Using the heuristic

James Gustave Speth, former head of the UN Environmental Program and dean of the Yale School of Forestry and Environmental Studies, said, "I used to think the top environmental problems were biodiversity loss, ecosystem collapse, and climate change. I thought that with 30 years of good science we could address those problems. But I was wrong! The top environmental problems are selfishness, greed, and apathy. [...] We need a spiritual and cultural transformation, and we scientists don't know how to do that."[42] Thus, it is as Pogo said, "we have met the enemy and he is us."[43] I argue that we need to revisit the questions in Figure 1.3, using the heuristic, and come to new, pragmatic answers to the basic and foundational questions listed there. In so doing, we might come to a new story of sustainability for the GYE. It appears that we are facing a mix of conventional, systemic, and cultural problems all rolled together. The causes for them are apparent, if we know what to look for and have a heuristic that helps us see them, understand the root causes, and find effective solutions.

Transitioning

In thinking about the GYE's future, we should not forget, as Paul Schullery, the historian, said, "Yellowstone is invented in our head with each visit. We say that Yellowstone Park was established on March 2, 1872, but in fact we have never stopped establishing Yellowstone."[44] While some people still view Yellowstone as a place where certain things are right to do and others are not, a growing number of people recognize that the park is the site of something much more dynamic in human culture, a kind of perpetual experiment that will never end. That experiment is about figuring out a sustainable relationship with nature and wildlife. What we make of the experiment today, and what we do in the present, determines the GYE's future.

Direction is all

In thinking about a new cultural story for the GYE, one that permits humans and wildlife to coexist sustainably, we need to understand that the function of our stories is either solidarity or truth.[45] Solidarity stories are fundamentally about a desire for connection and unity with one another and the world, while truth stories seek to achieve knowledge or understanding about the world. Truth stories, unlike solidarity stories, are focused not on us but rather about the things in nature. It is interesting to me to see how we use both kinds of stories individually and sometimes together in making meaning of the GYE.

Solidarity stories illustrate the desire for connectedness, coherence, unity, and wholeness—a way of showing attachment and loyalty to one another and things in the world. Solidarity stories place our lives in a larger context. These stories are derived from our everyday, ordinary sensory experiences with the GYE. They are about the telling of those experiences to each other. In doing so, they contribute to self-awareness, understanding, and, most importantly, connection with a living community. We all try to place our individual contributions to our community in favorable light. Regardless of the place and time, all stories have heroes and heroines taken from history, our imagination, or both.

Truth stories exemplify our desire for objectivity, knowledge, and reality, or some kind of fundamental understanding and grounding in the world. As philosopher Richard Rorty notes, this kind of story "is immediate in the sense that it does not derive from a relation between such a realty and the tribe or their imagined band of comrades."[46] This kind of story is focused not on us but rather about the things in nature. It is about the things "out there" in the wild, not the comforting stories we generate to make us feel good, in charge, and at the center of the world. Such truth stories purport to tell us something fundamental about the world, about the GYE, and the larger biosphere itself. Both kinds of stories meet our needs for meaning.

Creating a new story

Robert Righter, the historian of GTNP said in closing one of his books:

> Grand Teton is home to wildlife, but also a place for human inspiration. The park's mountains, rivers, wildlife, and high country make it one-of-a-kind place. The fascinating story of human habitation adds to its allure. It is a story of human conflict, born of love of this land. Consequently, both the NPS [National Park Service] and the public are watchful and protective. This speaks well for the future.[47]

These are heartfelt and hopeful words, embodying in part the kind of new story we need. But what else is needed to ensure a healthy future for the GYE?

Over the last century, many people have given us the outlines of the needed new story to reverse the current destructive trend. This new story melds our growing understanding of life, our cognitive and social selves, and ecological knowledge with justice into one of hopefulness and pragmatism. The new story, and this book, focuses on seeing the world, and ourselves, as part of a whole, as a single network, a unitary system of "implicate order."[48] To arrive at a new story, we need to facilitate productive public engagement in all ways and at all levels as quickly as possible.

In creating a new story, we need to facilitate meaningful engagement with one another that leads to a new synthesis, new understanding, and new behaviors. The engagement should embody critical appraisals and lead to action. In facilitating engagement, I rely on the field of future studies as introduced by Wendell Bell, a sociologist at Yale University and leader in future studies scholarship. He writes, "With the future arriving ever faster with an ever increasing need for adequate and timely responses to a rapidly changing world, and with the future increasingly the result of human actions, foresight is more important today than ever before."[49] The future studies approach is about understanding the past and present, as a basis for projecting future trends and events. Unfortunately, this valuable approach has gone little used in the GYE to date.

The future is not just an extension of the past. Future studies concerns the "big picture" and a deep understanding of who we think or imagine we are, the kinds of stories we tell, and, simultaneously, the context and content of the complex regional and global system in which we live. It requires an insightful look at existing worldviews, stories, and myths that underlie how we currently live. Attending to these subjects can help us transition to a new story of the GYE and, more importantly, new practices, actions, and management policies.

Conclusion

In this book, I offer a new worldview, a story emerging out of the last 60+ years in the global dialogue about our times, its problems, and what to do about them. We need to find a way forward together, as a society. Wendell Berry reminds us, "The real sources of hope are personal and spiritual, not public and political."[50] Yet for the GYE to have a healthy future, that source of hope must be made public and, eventually, political.

Concerning our future, I end with Wendell Bell, the famous author, philosopher, and pragmatist:

> I live in a decade during which some of the most important choices in the history of human civilization will be made. I happily join others in facing the heroic challenges of this decade— to move from our present catastrophic path to a new path that will dramatically improve our prospects for a flourishing future.

Part 1

YELLOWSTONE AS A STORY

The few surveyed stories in the first chapter about the meaningfulness of the Greater Yellowstone Ecosystem (GYE) are only a small sampling. At the heart of each of the stories, perhaps even unknown to authors of those stories, is concern about three basic questions as noted below. The three chapters in Part 1 look at what the physical and biological sciences tell us about Yellowstone. It is important to start with these two levels before moving on to look at the human—our mental level—in Parts 2 and 3. This part grounds the rest of this book.

Please keep the following questions in sharp focus. These are all broad questions. First, what is Yellowstone as nature or as the wild? Do we go to Yellowstone to find and experience nature? More simply, why do we go to Yellowstone?

Second, what do you bring inside you (e.g., expectations), as one who experiences Greater Yellowstone? For sure we are each a human with an evolutionary, personal, and cultural history as we each make sense of nature while we experience Yellowstone. This sensemaking for many people may be completely unconscious and subjective. For other people it may be conscious and active.

And third, what is our relationship to nature, the wild, and the plants and animals with which we share the world? What should our relationship be? In the GYE, like almost nowhere else, we can witness, even participate in nature with hundreds of big animals (i.e., bison, elk, wolves, bears) all around us. We also get to experience vast landscapes relatively untouched by humans. This is a rare experience for visitors given our population size and land-use conversion on this planet. What are we to make of the panorama, all the animals, and ecology?

The GYE today is a marvel, but it is riddled with challenges—small and large, short- and long-term, and minor and deeply profound ones—all at the same time. The challenges in the region that we face are time urgent.

Chapter 2

YELLOWSTONE AND SIGNIFICANCE

To conserve the scenery and the natural and historic objects and the wildlife
therein and to provide for the enjoyment of the same in such a manner
and by such means as will leave them unimpaired for future generations.[1]

Just after daybreak on a November morning in 1975, headed north out of Jackson, Wyoming, I sat stunned watching hundreds of elk rushing across the shallow, icy Gros Ventre River. The migrating herd was 10–20 animals wide and a quarter of a mile long; the magnificent creatures were backlit by the rising sun, steam jetting from their nostrils, with Sleeping Indian Mountain behind them on the eastern horizon. I watched, riveted by the splendor and urgency of their passage. The long-distance migrations of elk and other ungulates are one of the great natural wonders of the Greater Yellowstone region. But nearly 50 years later, some migrations are now lost, and still more are threatened. Today, no one can see that migration I came across in 1975. Now there is a bike path bridge blocking the view from the highway.

This chapter looks first at the Greater Yellowstone Ecosystem's (GYE's) biotic significance, emphasizing migrations and large carnivores as ecological processes, and places this knowledge in the context of evolution, ecology, and ecologists. Second, the chapter examines the biotic communities and ecosystem that are the GYE today by modern science standards. This is largely a task of gathering, organizing, and summarizing facts and thinking in terms of complex evolving ecological systems and the contemporary challenges for humans living in such a place. Third, the chapter briefly recounts the region's human history or ecology from the nineteenth century to the present, including the many people, over two hundred years, who have committed themselves to protecting this land and life for us to enjoy. Finally, I survey the formal, authoritative goals for the GYE—past, present, and, of course, future. It is our task, in our time, to refine and live up to the legacy we have inherited and the responsibility we bear.

This chapter continues using the heuristic outlined in Chapter 1 about *people* seeking *meaning (values)* through *society (institutions)*, using and affecting *resources* (the *environment*). Looking at the GYE through an ecological lens gives us an understanding of GYE's wildlife, landscapes, and history. With this knowledge, we can best ensure a healthy future for the GYE, if we so choose. This chapter focuses on questions about nature and the wild, our understanding of them, and the challenges they reveal. Consequently, it focuses on the most dominant species in the GYE, with the broadest ecological niche—us as people and our behavior.

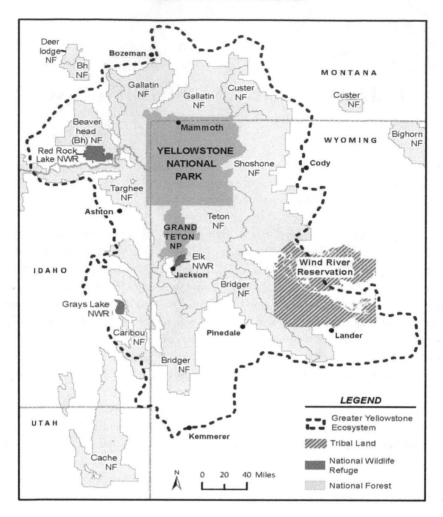

Figure 2.1 The Greater Yellowstone Ecosystem, showing major federal and state entities and the Wind River Indian Reservation.

Plants and Animals

The GYE as an ecosystem is a 34,375 sq. mi. geographic area in and around Yellowstone National Park (YNP) and Grand Teton National Park (GTNP) extending out to include all the US geological watershed and federal lands units that encompass the mountains and basin of this unique area (Figure 2.1). The GYE is a geographic (spatial) area and it has existed for a long time (temporarily). It includes many thousands of different species, each represented in turn from many millions to only a few individuals. The GYE can be understood by looking at the biotic communities and species, as partially reviewed below. In turn, all these communities and species are in constant interaction with one another and the physical environment. As an open-ended process of species and environmental interactions, evolution is best conceived as an unending process that can be studied and used in our management policy of the GYE.

People share YNP, GTNP, and the surrounding region with thousands of bacteria, plants, and animals, all of which were there long before modern humans entered the scene. Typically, we pay little, if any, attention to most species, as the vast majority of living beings are small-bodied. So how much life is there in YNP and throughout the GYE? Little is known about many species of bacteria and of lower plants, including mosses and lichens, and invertebrates. Simply identifying some of these less charismatic life forms is a challenge, much less cataloging them. Without records, we can only estimate that there are likely many thousands of species in the GYE, varying greatly in size, life requirements, and adaptability. They congregate in various assemblages or biotic communities named after their setting or dominant overstory species (e.g., marshes, grassland, sagebrush, willows, aspen, pines, firs, spruce, alpine, bogs, lakes, ponds, riparian aggregations). Some communities are widespread, such as lodgepole pine forests, while others are more restricted, such as bogs and alpine areas.

Plants

More than 1,500 species of vascular plants (plants with specialized tissues for conducting water and nutrients) live within the GYE, in addition to the unrecorded thousands of nonvascular species. Most plants conduct photosynthesis, a biotic process of converting the sun's light energy and soil nutrients into carbohydrates to fuel their life activities in primary production. Thus, photosynthetic plants connect all life to the sun, light, and energy. Animals and other heterotrophs (species that cannot manufacture their own energy) then eat plants or other animals that eat plants, thus gaining both the stored energy and nutrients. Therefore, nearly all life that exists, and has ever existed on Earth, is a product of solar energy, which is essential to life as we know it. The few exceptions include species living in hot springs and thermal pools, at the bottom of lakes and oceans, and in steam vents. These life forms often aggregate in relation to one another in coevolved communities. Some of these rare species can be found in YNP.

Botanists recognize seven conifer communities, commonly referred to as evergreens (lodgepole pine, Engelmann spruce, subalpine fir, limber pine, whitebark pine, Douglas-fir, and Rocky Mountain juniper) in YNP. Other recognized communities based on their plant inhabitants include sagebrush-steppe, wetlands, and hydrothermal communities. YNP is also home to some extremely rare plants, such as Ross's bentgrass found in geyser basins, sand verbena found at high elevations, and sulfur buckwheat found on geothermally influenced barren ground.

A major problem for plants in the GYE today is the encroachment of invasive plant species. Approximately 220 invasive species are currently known in YNP. Generally speaking, invasive species are those that crowd out native species, change plant communities, and affect fire frequency and intensity. Wildlife often do not target invasive species to eat, so natural predation or herbivory cannot control their populations. While some invasive species can be eradicated if found early enough when their numbers and geographic spread are still small, others are widespread and common, so little can be done about them (e.g., spotted knapweed). The surrounding states of Wyoming, Montana, and Idaho list about forty species as "noxious weeds" harmful to agriculture, fish, wildlife, aquatic navigation, or public health. Invasive species remain a significant concern in the GYE today.

Animals

The GYE is teeming with animal life, although not with anything like the great diversity we see in tropical ecosystems or in the plains of Africa. The GYE has about 460 species of vertebrates, in addition to many thousands of insect species, 11 amphibians, 10 reptiles, 337 birds, and 81 mammals.[2] Although amphibians are receiving some attention recently, there is still a significant need for long-term monitoring of most species.

Growing problems today often revolve around various human–wildlife interactions, including highly visible road kills and increasing disease due to human behavior, which along with seemingly natural new diseases are potentially catastrophic for some animal species. Harmful interactions include people's dogs harassing wildlife, skiers on or near critical winter range stressing animals, and poaching, to name a very few problems. Many newspapers and technical literature give other examples of people harming wildlife, covering the range of direct and indirect interactions with humans. Much of the conflict is indirect, such as displacement, increased stress, and habitat interference. Other interactions are direct, such as domesticated dogs chasing and killing wildlife or vehicle collisions with animals. In direct conflict situations, such as a mountain lion in a backyard, wildlife is often forcibly removed or killed.

Much conflict comes from unintended consequences of human action (e.g., building or recreating in wildlife habitat). Finding adaptable formulas for human–wildlife coexistence, rather than focusing only on conflict situations, remains one of the most important long-term challenges we face. However, many problems are invisible to the casual observer, recreationist, or tourist, and even officials.

Invertebrates

Comprising 94 percent of the world's animal kingdom, invertebrates (animals without backbones) are the single largest group of wildlife. In their diversity and abundance, invertebrates lie at the heart of healthy ecosystems.[3] Butterflies are some of the most visible and colorful.[4] The services they provide range from pollinating and dispersing seeds to becoming food for other wildlife, recycling nutrients, cleaning water, and building soil. Invertebrates perform functions critical to all life on our planet. Without them, whole ecosystems would collapse. Unfortunately, invertebrates are often the first to be imperiled by human activities. Compounding this problem, these essential and diverse creatures are often not considered when decisions are made about environment management policy.

No full inventory of invertebrates in the GYE has ever been attempted. It would be a nearly impossible job. Thus, the importance of these animals in the functioning of ecosystems is little appreciated. But conservation is a "time-urgent" activity. Because so little is known about most invertebrates, and the extent and rate of change caused by humans is so pervasive, there have been many extinctions already. Some extinctions have been recorded, but most are unknown. We need to develop new technologies to identify samples of invertebrates. In the meantime, the best conservation strategy for invertebrates is "ecosystem-level conservation."

Carnivores and Predation

Some of the most captivating species in YNP, and more broadly around the world, are the large carnivores, including bears (*Ursidae*), wolves (*Canidae*), mountain lions or cougars (*Felidae*), and wolverines (*Mustelidae*). These species are the largest-bodied in their respective families; and they partake in the ecological process of predation, as is characteristic of species in the order *Carnivora*. While carnivores are fascinating in their own right, they also raise many ecological questions, such as why large, fierce animals are so rare, as opposed to their prey, which are more abundant?[5]

Carnivores

Few ecosystems in the contiguous United States are large enough to house carnivorous animals and a robust prey base. When there are sufficient resources for carnivores and the subsequent predation that occurs, there is often significant controversy over the management of such species.[6] Historically, predators are the category of wildlife with which humans most often compete. In 1999, along with colleagues Peyton Curlee, Paul Paquet, and Steve Minta, I worked to bring together, for the first time, the status of our knowledge of GYE's carnivores.[7] This project explored the distribution, abundance, and conservation of large carnivores in the Rocky Mountains of the United States and Canada.[8] We reiterated that the GYE is home to the large species referenced above, as well as many mid-sized carnivores, such as lynx, coyotes, badgers, and otters, and smaller carnivores, such as weasels and martens. This total predator guild plays a significant and visible role that is essential for a healthy functioning of the GYE.

Animals come in all sizes and abundances, but there are basic ecological principles that can guide our understanding of species prevalence. For instance, small species tend to be more numerous than larger species. The few large carnivorous species in the GYE have populations in the hundreds. In comparison, there are many millions of voles and mice. Relatively speaking, large-bodied predatory animals are rare. Why is this so? Animals are usually bigger than the things they eat, but there are exceptions. Wolves often eat mid-sized animals but can also target bison and elk, which are significantly larger. To hunt such large prey, wolves evolved in packs to employ cooperative hunting tactics. Cougars hunt with an ambush strategy, combined with powerful biting teeth and strong front limbs. Grizzly bears sport large claws for digging and to hold prey. Most of their food is small, but they are fully capable of hunting large animals. Body size, predatory strategies, and prey relations are active areas of study since predation is a significant ecological process. In the last few decades, ecological scientists have intensively studied how carnivores influence ecosystem processes. This process is complex ecologically, and to make it even more complex, it now involves human's social, cultural, economic, and administrative dimensions (i.e., management and policy).[9] However, traditional wildlife science will continue to play a central role in understanding the ecological process.

Carnivores might be best managed by focusing on how they affect people's welfare outside of protected areas. This could lead to increased tolerance for wildlife and opportunities for successful coexistence.[10] The focus of attention should be on local and regional scales, while keeping the overall context clearly in mind. Addressing local problematic interactions is important. We must establish a sliding conservation gradient flowing from protected areas where carnivores are given priority to agricultural landscapes where people are given priority, in order to develop flexible, coexistence management. Zoning and buffer zones will be needed. Large carnivores such as bears, wolves, cougars, and wolverines often serve as a "referent" for nature or "untamed wildness" in the GYE (Figure 1.2). These animals are a main attraction of people, but, as Jason Sunder of the *Jackson Hole News & Guide* notes, they are also a potential threat to unprepared residents and visitors alike.[11]

Bears

Black bears (*Ursus americana*) are common across the United States. They are small but in comparison to other bear species can differ in color, and are wide-ranging. The grizzly bear (*Ursus arctos*) is much larger in size, has a different disposition, and once ranged across much of what is now North America. Today, however, grizzlies in the contiguous United States are found only in the GYE and a few other places in the northwest (e.g., Glacier National Park, Montana). As seen in Figure 2.2, grizzly bears in the GYE occupy a larger area today than they did twenty years ago, but how this relates to their population growth is controversial.[12]

In 1975, grizzly bears were listed as a "threatened" species under the 1973 Endangered Species Act (ESA). At that time, there were no more than 250 individuals remaining. The GYE population of grizzly bears was delisted by the US Fish and Wildlife Service on July 31, 2017. The associated *Conservation Strategy for the Grizzly Bear in the Greater Yellowstone Ecosystem* is intended to ensure sufficient habitat and population protection measures will be implemented by federal, state, tribal, and county agencies into the future, to sustain a healthy population of many hundreds of bears widely distributed across suitable habitat within the region. Official estimates suggest there may be 700 grizzlies in existence, but other analysis indicates the population may be around 500 and declining.[13] The actual number of GYE grizzly bears is likely higher than ~700 because that estimate was derived using the Chao2 population model based on counts of unique females with cubs. That model has been shown to underestimate true bear numbers by 40–50 percent at the higher population sizes.[14]

This carnivore remains highly controversial in some circles, as reported in the media.[15] Newspaper headlines capture some of the conflict. For example, "2015 grizzly death toll tops all previous years," "Voodoo Science: Yellowstone Grizzly Count is a 'Flexible Fiction'," and "Upper Green stockmen navigate constant griz threat." For some, the "grizzlies of Yellowstone are emblematic of [...] divergent concepts of wilderness.

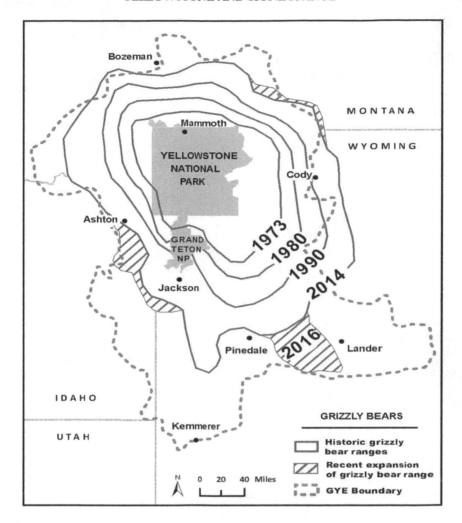

Figure 2.2 Grizzly bear distribution showing a range extension in the Greater Yellowstone Ecosystem from 1973 to 2016.

US Geological Survey, "Grizzly Bear Boundary Layers" [shapefile], 2016, https://www.sciencebase.gov/catalog/folder/52fe7f75e4b0354fef6de4f0 (February 1, 2017); US Department of the Interior, US Geological Survey, *Yellowstone Grizzly Bear Investigations: Annual Report of the Interagency Grizzly Bear Study Team*, ed. F. T. Van Manen, M. A. Haroldson, and B. E. Karabensh (Bozeman, MT, 2016).

Although the bears are deeply bound to a specific ecosystem, the tension between the desire to protect and the demand to hunt them represents a larger dynamic in the nation."[16]

The grizzly bear has been intensely studied since 1959, first by the famous John and Frank Craighead brothers and now largely by federal, state, and tribal governments as part of the Interagency Grizzly Bear Study Team led by the US Geological Survey (USGS). This speaks to the bear's symbolic importance as a key referent of nature and wildness. No single history has been written about the conservation efforts of the grizzlies, although there are thousands of articles, books, and blogs about specific periods in this convoluted history. On their official site, the Interagency Grizzly Bear Study Team offers one view of grizzly conservation that can be compared to the many other conservation organizations' assessments of conservation adequacy.[17] Currently, humans' relationship to grizzly bears remains highly complex and contentious.[18] This dynamic reflects our ambivalence about what we see nature to be and our relationship to it.

Perhaps the most significant study of our ambivalence was that by Dr. Christina Cromely, who examined the killing of grizzly bear #209 in GTNP in 1996, a bear at the center of high-profile incident.[19] She identified differing norms or prescriptions for bear management policy at play in this case, each was based on different notions of the bear's role in the park and nature. For example, when is it OK to remove a bear in a national park? Is it OK to do so when private cattle are grazing in the park under a special use permit? And, what is the role of public views in making decisions? Answers to these questions, and many more, were in conflict in the administration of management policy at the time.

Yet, nearly all visitors to YNP want and expect to see a bear and wolf, but about two-thirds actually do. There is worldwide support for the conservation of grizzly bears in the GYE because many people consider it YNP their home, due to past visitation experiences, television shows such as Yogi Bear and Craighead documentaries, and magazine articles (e.g., *National Geographic Magazine*). Other people want to remove special protection of these bears as a result of their spread outside preserves and wilderness. Bears continue to be an extremely contentious issue among agencies and citizens, resulting in litigation.

Wolves

Another iconic and contentious species is the wolf (*Canis lupus*).[20] As my coauthor, Denise Casey, and I said in our 1996 book on wolves:

> The gray wolf that haunts the imagination of North Americans does not travel lightly. Wherever he goes, what-ever [*sic*] he does, he is burdened with a heavy load that we have laid on him—all of our images of him, our dreams, our fears, our stories. They have accumulated over the centuries, carried from many lands in the Old World, dredged from the ancient past of North America's own people, fashioned anew in the New World's peculiar geography, history, and society.[21]

Since wolves were reintroduced to YNP in 1995, Douglas Smith, YNP wolf biologist, has led the NPS research efforts. He, along with several dozen colleagues, studies the species' behavior and predator–prey relations. In over 20 years of research, we know a lot about wolves' ecological impact. An entire issue of *Yellowstone Science* was devoted to celebrating 20 years of wolves in Yellowstone (Figure 2.3).[22]

The wolf is a lightning rod for major controversy, perhaps more than any other species.[23] Historically, the wolf has been a target for extermination, and more recently for reintroduction, recovery, and protection. In an earlier book I coauthored with Denise

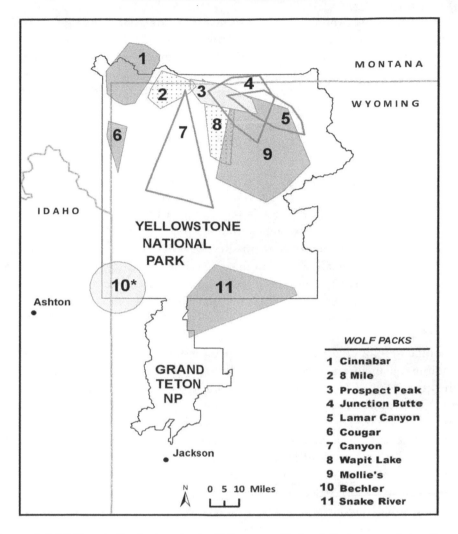

Figure 2.3 Wolf pack distributions in the Yellowstone National Park, part of the Greater Yellowstone Ecosystem, 2014.

US Department of the Interior, *Yellowstone National Park Wolf Project Annual Report 2014*, ed., Douglas Smith, Dan Stahler, E. Stahler, M. Metz, K. Cassidy, and Rick McIntyre (Yellowstone National Park, WY, 2015).

Casey, *Tales of the Wolf: Fifty-One Stories of Wolf Encounters in the Wild* (1996), five historical periods in the human relationship with wolves are outlined:[24]

1. Tales from eight Native American tribes
2. Early Explores and Naturalists (1748–1835)
3. Frontier Encounters, Old World Attitudes (1750–1876)
4. Serious Killing (1876–1942)
5. Contrary Point of View—Protection (1892–1944)

Today the latter two worldviews (serious killing and protection) are in direct conflict. Some local and regional interest groups support more killing; whereas others, from local to international, support more conservation. The seemingly endless, highly emotional conflict over wolves speaks to how important the wolf is as a referent for nature and wildness in the GYE. Just as with grizzly bears, newspaper headlines capture some of the conflict: "New study questions wolf hunting policies," "Mont. Takes aim at wolves," and "Out of bounds: The death of 832F, Yellowstone's most famous wolf." We can expect this dynamic to continue.

Cougars

Our third large carnivore to assess is the mountain lion or cougar (*Puma concolor*). Lions exist throughout the ecosystem and well beyond. Because they are secretive and solitary creatures, few people see them. Receiving decades of intermittent research in the West,[25] they, too, are a referent for the GYE and nature.[26] Historically, as with many large carnivores, mountain lions were almost exterminated and are still hunted today.[27] Headlines capture the drama: "Wyoming's lions escape trapping plan," "Commission locks in higher cougar quotas," and "Stop killing mountain lions for sport."[28]

Cougars' ecology and behavior are quite interesting as they make their secretive life sometimes quite close to people. Overall, cougars function and interact dynamically with other species in the ecosystem. Two studies of mountain lions are ongoing in the GYE (Figure 2.4). One study site includes the northern portion of YNP from Gardiner to Cooke City, Montana, through the Lamar Valley region.[29] The second site is in Jackson Hole in the Gros Ventre drainage.[30] These studies, under Dan Stahler, a Yellowstone biologist (and his predecessor, Tony Ruth), and Mark Elbroch, with Panthera in Jackson Hole, have yielded much new information on cougars, especially their social behavior. Videos of cougars interacting in the wild show previously unknown social interactions between individuals.[31] Videos show that cougars are much more tolerant of each other and social than once thought. Monitoring their population size and behavior is difficult; and while radio-collaring provides much new information, it is labor intensive and extremely expensive. Further, genetic sampling methods and molecular technology have recently provided useful data on cougar biology.

Cougar research occurred in two phases in YNP with the first running from 1987 to 2006. This study yielded a broad understanding of cougar ecology, predation, and population dynamics before and after the wolf reintroduction. Both radiotelemetry and snow tracking were employed. Prior to wolf reintroduction between 1987 and 1993, the study area was occupied by 15–22 cougars, which included adults, subadults, and kittens. After wolf establishment (1998–2005), there were an estimated 26–42 cougars. The second phase of research began in 2014 and focuses on population size estimates and kill rates of prey. These data provide insight into cougar relations with wolves, bears, and other cougars. YNP's northern range is a valuable resource for cougars that are dispersed throughout the region.

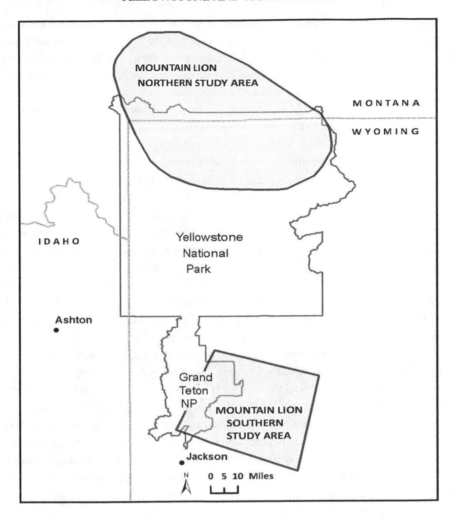

Figure 2.4 Map showing two cougar study areas, one in northern Yellowstone and the other in Jackson Hole region.

"Primary Study Area" [map] from Mark Elbroch, personal communication, October 18, 2017; Tony Ruth, The Yellowstone Cougar Project study area [map], "Ghost of the Rockies—the Yellowstone Cougar Project," *Yellowstone Science*, 12 (2004).

Wolverines

Our fourth carnivore to consider is the wolverine (*Gulo gulo*). As with the previously discussed carnivores, the wolverine is a key referent for nature, though less visible than grizzly bears and wolves. The Wolverine Foundation, under the directorship of Rebecca Watters, updates matters regarding wolverines on a regular basis.[32] Biologists estimate that fewer than three hundred wolverines live in the lower 48 states. Wolverines are geographically constrained by a requirement for cold climates, including deep spring snow and cool summer temperatures; thus, in the lower 48, they are confined to high-elevation regions that provide these conditions, such as Glacier National Park.[33] Although they are carnivores and they do hunt, a large part of the wolverine diet consists of scavenged ungulate carcasses.[34] Wolverines require extensive and exclusive territories, with males defending areas of up to 500 sq. mi. and females up to 300 sq. mi. In combination with their low reproductive rates, these territory requirements help to explain why wolverines are naturally rare and sparsely distributed on the landscape.

At the time of European settlement, wolverines were present throughout the mountainous regions of the GYE and in ranges as far south as New Mexico and as far west as California's Sierra Nevada.[35] They quickly fell prey, however, to trapping and poisoned bait set out as part of programs to control other predators. Wolverines were extirpated from a large part of their historic range by the early twentieth century, and began to slowly recolonize the GYE through the second half of the twentieth century. The state of Montana maintained an unlimited trapping season on wolverines, until they began to restrict the season in response to conservation concerns. In 2013, the trapping season was suspended on court order, pending the outcome of the ESA listing process. However, during court proceedings in February 2016, Montana asserted that their management of wolverines was sound and that trapping is viable into the future. Trapping continued.

Wolverines remained largely ignored by conservation science in the GYE until the 1980s, when early studies, such as those conducted by Maurice Hornocker, began focusing on the rare carnivores.[36] Throughout the 1990s and early 2000s, wolverines were collared and studied by state and federal projects in Grand Teton, Yellowstone, and Glacier national parks. The Wildlife Conservation Society began a large-scale study of wolverines in the GYE in 2001. The prospect of ESA listing combined with advances in technology, such as camera trapping and noninvasive genetic monitoring techniques, have led to an explosion in research efforts among both agencies (including the Forest Service and state Fish and Game departments) and nongovernmental organizations over the past several years. The wolverine's profile has increased substantially among the wildlife-interested public, due to the increase in research programs, the public debate about ESA listing, and the efforts of both scientists and advocates to educate the public (sometimes with competing messages). Wolverine tracking in the GYE and American West shows a few long-distance movements by individual wolverines. For the most part, it is the younger wolverines that are dispersing. The current range and management area of wolverines are shown in Figure 2.5.

Today, the fate of wolverines in the GYE and American West is uncertain and in jeopardy. This large carnivore, a member of the weasel family, requires deep snowpack to den and raise young. Most of the needed habitat will disappear with climate change in the next few decades. For this reason, conservation groups requested the US Fish and Wildlife Service (USFWS) to list wolverines under the ESA. In 2013, the USFWS issued a proposal to classify wolverines as threatened; but in 2014, a regional director reversed that recommendation, citing "uncertainty" in the climate data. Environmental groups sued for an appeal. On April 4, 2016, a district judge ruled that the USFWS refusal to protect wolverines was "arbitrary and capricious," noting "immense political pressure" from Wyoming, Idaho, and Montana. He wrote an 85-page directive telling USFWS, "If ever there was a species for which conservation depends on foregoing absolute certainty, it is the wolverine."[37] This decision has been a subject of significant controversy ever since.[38]

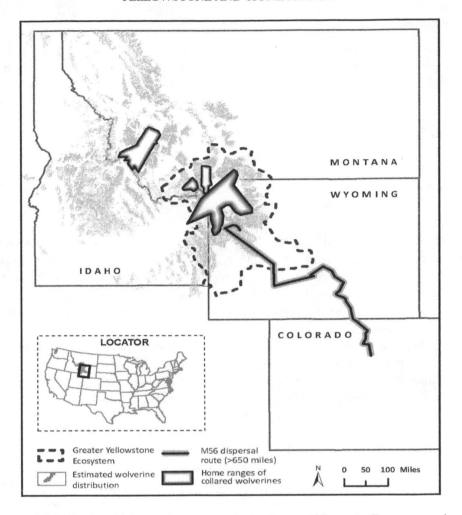

Figure 2.5 Wolverine sightings and movements in the Greater Yellowstone Ecosystem and well beyond the system. Note that there are very few resident animals and only an occasion transient. US Geological Survey, *Gulo gulo wolvx Raster* [Raster dataset], 2013, http://gapanalysis.usgs.gov/species/data/download/ (April 1, 2017); Inman, *Greater Yellowstone Wolverine Program*.

While existing literature supports the idea that wolverines prefer persistent spring snowpack for successful reproduction, part of the disagreement is the question of how much snow (depth) and how long that snow needs to persist is debatable. One set of researchers looked at the obligate relationship of female wolverines and snow depth in Alberta, Canada, and found no difference in occurrence in trapping captures in areas with and without an early spring.[39] Also, evidence of reproduction in areas with relatively low amounts of spring snow cover (e.g., Jardine, Montana, area) suggests wolverines may be more flexible than previously thought.

These four large carnivores—bears, wolves, cougars, wolverines—are fascinating animals in their own right. All are powerful referents of nature and the wild. They are the most iconic species in the GYE and American West for most people.

Predation

Large carnivores show less species diversity and have much smaller populations than animals of lower trophic levels, hence ecologists often refer to a trophic "pyramid." Animal population sizes are ultimately the result of energy flows. As mentioned previously, all life and energy on earth comes from the sun. Plants use light energy to sustain their life systems. In turn, herbivores eat the plants and take up some of this energy for their own life functions. However, as energy is transferred from one trophic level to the next, the vast majority of it is lost. Consequently, less energy is available to animals that eat herbivores. Carnivores, especially large ones, have access to limited energy. There is not enough energy at higher trophic levels (given body sizes) to support any large carnivores that could prey on grizzly bears, wolves, and mountain lions today.

What regulates carnivore abundance and distribution besides human-caused factors? All plants and animals reproduce as fast as they physically can in order to maintain their population size. Over the past 30 years, bear and wolf numbers have increased substantially. However, there is some variation in the abundance and distribution of elk, bison, and large carnivores in recent years. Much of that change is due to human actions, both directly and indirectly.

Another ecological question is how carnivores figure into ecosystem dynamics through predation, regulating prey species, and indirectly affecting plant communities. A vigorous debate is underway to answer that question, and the process of "trophic cascades" is under research. In this predatory–prey process, carnivores indirectly control plant diversity, biomass, productivity, and nutrient cycling by affecting where and when prey graze and browse.[40] In YNP, human influence is relatively minimal, so this ecological theory can be tested. Some studies are consistent with this theory. Researchers have shown that the reintroduction of wolves in 1995 restored the ecosystem's functioning by reducing herbivory on plants (e.g., eating of aspen and willows). That conclusion captivated the public's attention and stimulated a debate in ecological circles. The key question being asked is, "Are results being interpreted adequately?" For example, newer studies suggest that there is only a weak change in elk behavior and grazing in response to wolves and competing predators.[41] Despite the outcome of ecological studies, large carnivore status and management policy will turn on site-specific human values and behavior, especially from agriculturalists and the public. So how can or should large carnivores be conserved?

Migrations and Movements

Another major feature of the GYE is migrating animals.[42] The lead article in a 2010 special issue of the *National Geographic Magazine* asks, "What is it that makes annual migrations such a magnificent spectacle for the eye and mind?"[43] These large animals, such as mule deer, elk, and pronghorns are especially wide-ranging ones. These are important species in the public's mind. Not knowing of the human administrative boundaries that we superimposed on the GYE, these animals move over the landscapes as they historically have done. All migration processes are under growing stress from humans, and there is significant controversy over conserving these migrations routes. Ecologically, migration is one adaptation to a sharp change in seasonal environments, where food and cover vary across landscapes.

Much of what we now know about migrations in the GYE and the region is summarized in the book *Wild Migrations*.[44] This very accessible publication is data rich with maps, graphs, and text by experts studying animal movements. It describes what migration is as a biological phenomenon. As well, it talks about migrations at global and regional levels. This book gives many maps of movements for each species in the GYE.

Threats to these historical migrations include clear-cuts, roads, and development, especially oil and gas. Additionally, recreation on public lands and other traditional land uses are a threat. Loss of migrations has consequences for ecosystem functioning.[45] Let's explore some of the most visible migrations in the GYE—elk, pronghorn, and mule deer.

Elk and other large ungulates

The public is keenly interested in big ungulates (hoofed animals). These species show the most spectacular and important migrations in the GYE. Bighorn sheep, moose, deer, and bison do migrate, but on a much smaller scale than birds. In the GYE, it is estimated that 75 percent of traditional migration routes have changed or been lost in the past 100 years.[46]

Elk

For the first time, all elk movement data in the GYE are being aggregated. Arthur Middleton, professor at the University of California-Berkeley, and his colleagues are creating a comprehensive map of all elk movements. So far, they mapped about 2.7 million elk data points over thirteen years on 271 individuals (Figure 2.6). These data show nine migrating herds ranging over three states. Some elk move seasonally 85–100 miles or more. The researchers are adding to the data set with ongoing field studies to refine these major routes and, perhaps, find new routes. Middleton notes that elk migrate out from several core areas of YNP like spokes on a wheel. They leave the high elevations in the park in winter and migrate to low-elevation valleys in Wyoming, Montana, and Idaho. Elk leave the park crossing forest, rangelands, sagebrush, and ranches. The total area occupied by elk annually is about five times the size of YNP (2.2 million acres).

As Professor Middleton notes, it is difficult to estimate numbers, but estimates suggest over twenty thousand elk are present in the GYE. This equates to about five thousand tons of elk pulsing in and out of YNP seasonally. These migrations can be used as a metaphor to better understand the importance of migrations in the GYE. Middleton says that YNP is like a beating heart, with the migrations as veins and arteries and the animals as the blood that hold the system together. This "circulatory system" is almost ten million acres for elk alone. It is even larger when we include other species such as

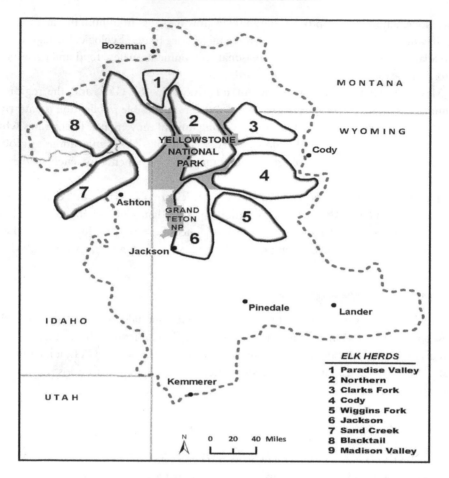

Figure 2.6 Map showing nine major elk migrations into and out of Yellowstone National Park or nearby, as the seasons change.

Chuck Preston, Invisible Boundaries: Exploring Yellowstone's Great Animal Migrations (Cody, WY: Buffalo Bill Historic Center, 2016); Wyoming Migration Initiative [map], "Elk Migrations of the Greater Yellowstone," 2015, https://migrationinitiative.org/content/elk-migrations-greater-yellowstone.

mule deer, and even more expansive when we include migrating birds. Thus, the GYE really has no boundaries.

Elk management in Wyoming is highly contentious, including GTNP, the only park that allows hunting.[47] For example, newspaper headlines note the following: "Closing feedgrounds for CWD [chronic wasting disease] a long shot," "New feedground plan bans anti-wolf efforts," and "Park's 'elk reduction' a travesty of hunting." Scientific papers also speak to the conflict over elk management, as seen in the following: "Addressing a Persistent Policy Problem: The Elk Hunt in Grand Teton National Park, Wyoming" and "Discourses of Elk Hunting and Grizzly Bear Incidents in Grand Teton National Park, Wyoming." Conflicting views commonly find their way into letters to the editor in local and regional newspapers.

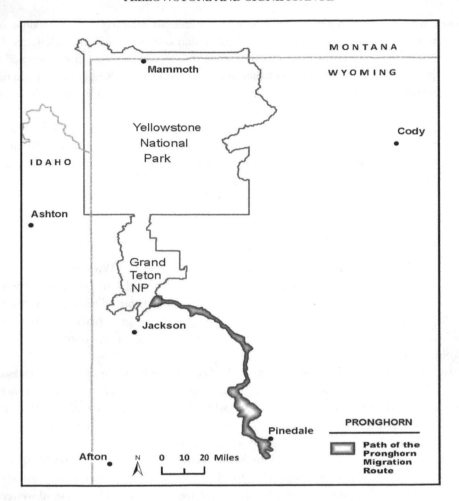

Figure 2.7 Migration of pronghorn from Grand Teton National Park in summer south to Sublette County, Wyoming, in the southern part of the Greater Yellowstone Ecosystem.

Wildlife Conservation Society, Path of the Pronghorn [map], "Pronghorn Migration on the Path of the Pronghorn," 2014, https://programs.wcs.org/northamerica/Wild-Places/Yellowstone-and-Northern-Rockies/Pronghorn-Field-Program/Pronghorn-Migration-Path.aspx.

Pronghorns

In the 1970s, I wanted to study Jackson Hole's migrating pronghorn; but my efforts came to naught, as I could not find an agency or conservation organization partner (Figure 2.7). At the time, there was simply no interest in such a project. However, by the 1990s, pronghorn migrations in and out of Jackson Hole became a big issue. Joel Berger, then of the Wildlife Conservation Society, took on a major study,[48] making pronghorn a very high-profile conservation matter. As it turned out, my colleague David Cherney, then with Yale University, studied pronghorn migrations from Jackson Hole to Sublette County to the south, a distance of up to 170 miles.[49] We were mostly interested in how conservation decision-making played out as the animals spanned the jurisdictions of three federal agencies, three Wyoming counties, and over forty private landowners. In addition, there were over ten nongovernmental conservation organizations, two major state agencies, Wyoming's executive office, and many citizens involved in the issue. In the end, the pronghorn migration route was largely protected, but remains under threat on private lands.

As I drove from Rock Spring to Jackson on an early spring day, I saw thousands of pronghorns migrating, traveling north on both sides of the highway for nearly 80 miles. Mile after mile of pronghorn pushed north in family groups of hundreds. Some paused to forage, most just walked, a few ran, heads up and alert. At the time, I estimated seeing 5,000 animals. It was a high for me in my wildlife work over the decades!

Mule deer

Next, we look into mule deer migrations (Figure 2.8) in the Wyoming part of the GYE.[50] Deer are common throughout the ecosystem and can be easily viewed. I became interested in deer in Jackson Hole shortly after I started ecological research in the region in the late 1960s.[51] This interest, in part, sprung from my master's work on mule deer in the Oklahoma panhandle, in Black Mesa country along the Cimarron River.

In 1978, I started a mule deer study on the Gros Ventre Buttes in the floor of Jackson Hole. Along with colleague Tom Campbell, I radio-collared deer and observed them over the next dozen years. We documented winter habitat use, migration corridors, and even one long migratory movement (35+ miles). Our work was the first deer study in Jackson Hole.[52] Since then, there have been many other studies.

Mule deer, always important regionally, have been catapulted to a high-profile species in the last decade. This is largely due to Hall Sawyer and colleagues "accidentally" finding that deer migrate 150 miles from south of Jackson down to I-80 in southwestern Wyoming (Figure 2.8, herds 5 and 6).[53] Deer have apparently been making these movements for eons, but only recently was this realized through sound scientific studies. The Wyoming Migration Initiative is focused on learning more about deer movements and habitat relations throughout western Wyoming.[54]

Bison

Bison are the American icon, the national mammal. They once roamed the Great Plains numbering up to 65 million. Nearly slaughtered to extinction by the mid-1880s, only about 25 animals held out in YNP. These few animals, augmented with 21 bison from northwest Montana and Texas, led to recovery of the species today. Presently, bison are one of the GYE's most spectacular ecosystem stories (Figure 2.9). They can be seen in herds of several hundred in most big valleys of the park (Lamar, Hayden, Pelican) and a herd exists in Jackson Hole as well. Bison range widely and tend to move out of YNP in the winter that brings them into conflict with surrounding states. These states fear that

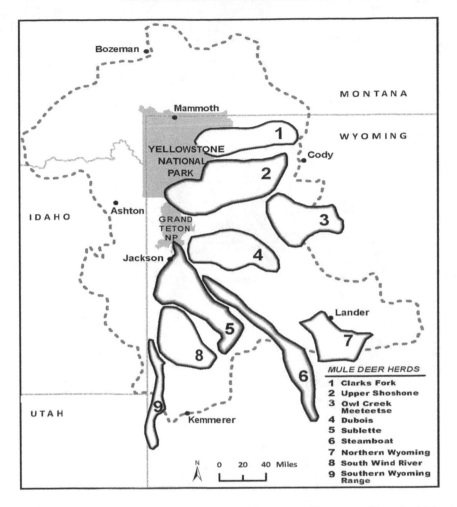

Figure 2.8 Migrations of mule deer in the Greater Yellowstone Ecosystem. Data for Idaho and Montana not available.

Wyoming Migration Initiative, Tracking Mule Deer Migrations of the Eastern Greater Yellowstone Ecosystem [map], "Eastern Greater Yellowstone Mule Deer Project," 2016, https:// migrationinitiative.org/content/eastern-greater-yellowstone-mule-deer-project.

bison carry brucellosis, which can be transmitted to cows, so the states push for stopping bison at the border of YNP.

As iconic as they are, it is not surprising that bison are targets for conflict.[55] Bison management is one of the most complex wildlife issues in the GYE. Take these headlines as examples: "Decades in, bison fate still an American shame," "Yellowstone chief: Bison slaughter to continue for now," "Montana defies logic in killing bull bison," and "Y'stone proposes culling roughly 900 park bison."[56]

I became involved in bison conservation in the mid-1980s in Jackson Hole when there were only a hundred animals. The Wyoming Game and Fish Department proposed killing them back to 50 individuals, but the rationale for doing so was highly questionable. Decades

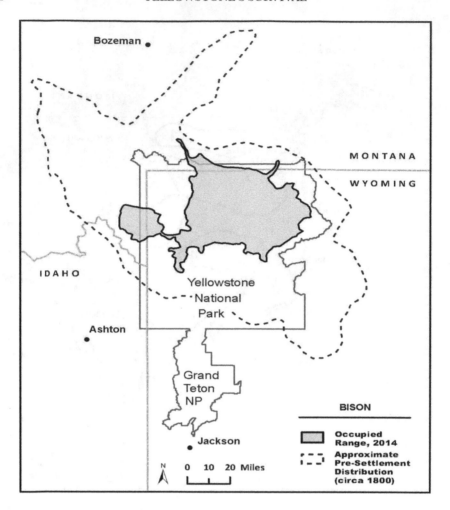

Figure 2.9 Migration of bison in the Greater Yellowstone Ecosystem showing major routes bison use to exit the Yellowstone National Park in winter. A few bison leave the park to the east; fewer still leave to the south.

of public involvement at that point forced the state to modify its plans. Today there are fewer than nine hundred bison in Jackson Hole. Bison conservation remains a salient issue that typically results in the killing of bison leaving YNP and Jackson Hole through hunts.

Bighorn sheep

Bighorn sheep is another species whose numbers and range have been hugely reduced in North America and in the GYE. The herd in GTNP is the most threatened today. Today, sheep exist only in scattered, isolated herds, some of which can be viewed up close on the National Elk Refuge in Wyoming from the road in winter.[57] They are susceptible to disturbance, stress from humans, diseases from domestic sheep, and competition with introduced goats.

Loss of bighorn sheep winter range near the Teton Village ski resort was the focus of a six-year study that measured how skiing and snowboarding are influencing Teton bighorn. "Sheep are being displaced from both regularly and infrequently skied habitat in the range," says Aly Courtemanch of the Wyoming Game and Fish Department.[58] She went on to note that the population is small and highly vulnerable. Recreationists, Teton Village managers, and the US Forest Service (USFS) all need to be very attentive. The precarious status of sheep is reflected in news articles. Headlines include "Skiers disturb Teton [bighorn] Sheep" and "Dead bighorns blamed on livestock disease."[59] Mountain goats compete with sheep for limited land and resources, and in the end, sheep lose out.[60]

Moose

Moose are the largest member of the deer family. They are popular with tourists. They can be seen along the rivers, streams, lakes, and in beaver ponds. They are typically alone as solitary animals or a mother and calf. Sometimes bulls hang out in small herds. They eat willows and other woody as well as aquatic vegetation.

Moose numbers have decreased drastically in the GYE since the 1990s. One day in the late 1980s I saw 54 moose in Jackson Hole and I wasn't even trying to see how many I could find. Hunters often blame wolves as the cause, but this does not hold up under careful examination. Moose are still visible in certain locations, such as around Blacktail Butte in Jackson Hole; but they are killed too frequently where the highway cuts the riparian corridor near Wilson, Wyoming. Take these few newspaper headlines that frame the story: "1 more moose killed, but overall toll falls," "Planners OK new park, want moose protected," "See moose triumvirate at wildlife party Tuesday," and "Bilers [snowmobilers] crossed moose habitat."[61] Moose are definitely a popular animal and in the public's mind.

Birds, bats, butterflies

Many animals migrate, including insects, hummingbirds, sandhill cranes, and bats, in addition to large mammals and other birds (Figure 2.10).[62] About a third of the 160 or so bird species that summer in the GYE stay through the winter. The remaining bird species migrate to more favorable environments to overwinter. We have all seen flocks of waterfowl massing in the fall to go south. We look forward to the spring arrival of red-winged and other black birds. Like all migrating animals, there are often population segments with some segments migrating before others. Not all birds take the exact same route between summer and winter ranges. Some birds migrate south in winter to the GYE (e.g., snowy owls), while others travel distances to the Caribbean, southern United States, and Central and South America. For example, the western tanager may travel over 3,100 miles south of YNP to Costa Rica. Swanson's hawk migrates up to 6,200 miles to Brazil, Uruguay, and Argentina, flying down the Isthmus of Panama. Many shorebirds also migrate, including the black-bellied plover, semipalmated sandpiper,

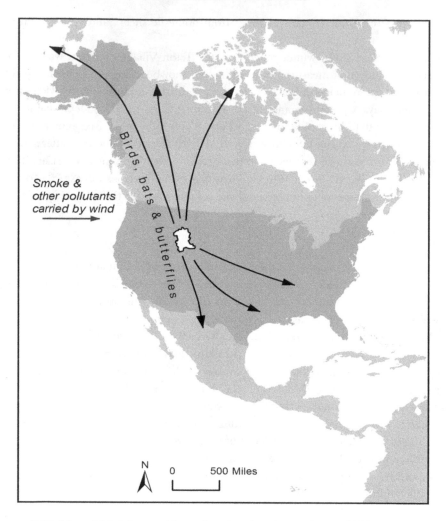

Figure 2.10 Map of bird, bat, and butterfly migrations into and out of Yellowstone National Park or nearby, as the seasons change.

black-necked stilt, and American avocet. For these shorebirds, the GYE is a stopover in their annual ecology where they rest and restock their energy levels.

Insects migrate, too. A key food, and perhaps the most famous, example is the army cutworm moth. It is a key food for many grizzly bears in the eastern portion of the GYE. This moth lays its eggs in the agricultural fields of the Great Plains in late summer. Caterpillars hibernate in the plains of Nebraska, Oklahoma, Kansas, and eastern Colorado. In spring, they emerge and eat crops. Many individuals are killed by pesticides. Survivors pupate in the soil, emerge, and migrate to the Rocky Mountains, many to the Absaroka Mountains on the east side of YNP. These moths feed on alpine plants and then become high-fat food for grizzly bears that move to high elevations to exploit the massive concentrations of cutworm moths. A single bear may eat thousands of moths each day. By late summer, surviving moths migrate back to the Great Plains; and the cycle repeats.

There are 14 species of bats that summer in the ecosystem; several migrate and overwinter in places outside of the GYE. For example, the horary bat migrates long distances to spend winter in southern Mexico (1,600 miles away). Overall, little is known of the GYE's bat species, their ecology, or their movements.

Overview of migrating species

Overall, the changing land-use patterns, urban sprawl, and mushrooming front and backcountry use (especially adventure sports and hunting in the fall and winter) are causing problems for all migrating species and large predators. Challenges include introduced and invasive species (i.e., mountain goats), road kills (e.g., in Teton County, on average 114 mule deer, 35 elk, and 15 moose are killed on roads each year), and pets, especially dogs, which harass and kill wildlife.[63] Disease is also a significant threat.[64] Take these news articles: "19% of deer herd a year dies of chronic wasting disease," "Wasting disease nears park," "CWD study feeds debate," "Supervisor says CWD worse in feedgrounds," "CWD study predicts hunting will suffer," and "Wasting disease crept further west this year."[65] Identifying problems and protecting these ecological processes should be the highest management policy priority.

Conclusion

Most conservation and controversy in the GYE presently focus on wildlife and the human-to-human conflict wherein people argue with one another over what is the status of the animals, is management adequate to ensure species populations, and who is right and who is wrong in their positions on these complicated issues. The two big conservation issues in the ecosystem currently are about large carnivores and migration of large mammals. Both matters, when viewed from a large ecological standpoint, are dramatic evolutionary stories.

Ecologists study these ecological phenomena, as part of the "ecological theater and evolutionary play" in the GYE, and present that story to us. Predator–prey dynamics and migrations are windows into the great evolutionary drama unfolding before our eyes within the GYE's communities and ecosystems, giving us insight into how the ecosystem is structured and functions.[66] Migrations tie the whole system together, as thousands of individuals pulse in and out of the core of the GYE each year to winter range, some species moving over 150 miles. Carnivores, in part, help regulate abundance and distribution of all other animals and plants. Thus, predation and migrations make up a system of interconnections and functions.

The typical response to perceived conflict, which is often misdiagnosed as "conflict with wildlife," is to take concrete, engineering-like actions by building wildlife overpasses, erecting signs, printing brochures, implementing vehicle speed restrictions, and closing down bike paths and trails, maybe temporarily. More permanent remedies that deal with the real conflict, which is human-to-human in origin, are complex and require foresight.

In short, remedies include a more knowledgeable public, more accessible cooperative management process, and farsighted leadership in all organizations—governmental and nongovernmental. As the human population increases, the task of education about responsible development and recreation will be endless.

Although we spend a lot of time and energy talking about grizzly bears, elk, or other species, basic and foundational issues go unaddressed. It seems as if we cannot ask and dialogue about the foundational and functional matters directly, systematically, or consciously. Thus, the public dialogue about our relationship to wildlife is limited to only conventional perspectives. What we are seeing is a public contest over whose view of nature, wildness, and which referents will dominate in the region. In the meantime, wildlife pays the price for our lack of clarity, concern, and action.

Chapter 3

GREATER YELLOWSTONE AS A SYSTEM

It is clear [...] that the fate of humankind [and nature] increasingly depend
on what humans do, and in turn, it depend[s] on their
images of the future—their visions of preferable, possible, and probable
futures (and today I would add preventable futures), as well as their ability
to design effective actions to achieve their preferable future
and to avoid undesirable futures.[1]

The US Congress passed the Wilderness Act in 1964 committing us to the value of preserving areas "of the earth and its community of life [that are] untrammeled by man, where man himself is a visitor who does not remain." Designated wilderness areas were to be managed in such a way to leave them unimpaired as wilderness for the American people. Although little of Greater Yellowstone is formally designated Wilderness, much of it is of a wilderness quality. The Wilderness Act stated clearly that they are to be preserved as wilderness and for the gathering and dissemination of information regarding their use and enjoyment as wilderness. There are several official wildernesses in the GYE (e.g., Jediahiah Smith, Gros Ventre, Teton, Washakie, Absaroka). They are managed by the National Park Service, Forest Service, and Bureau of Land Management and other government agencies depending on where they are. Importantly, the Act laid out a process for the long-term study and additional official wilderness designations. Two such wilderness study areas (WSAs) are south of Jackson, Wyoming—the Palisades and Shoal Creek Wilderness Study Areas.

The fate of these two WSAs, each several tens of thousands of acres, was deliberated for two years by citizens and officials in Teton County, Wyoming.[2] By the fall of 2018, there was little agreement on what to do despite formally designed and facilitated workshops and the investment of thousands of person-hours and tens of thousands of taxpayer dollars.[3] Finally, the news headlines said it all: "County gives up on land initiative: No recommendation coming from commissioners on new wilderness."[4] Perhaps this effort failed to secure an integrated outcome in the best interests of people involved because the wrong "problem-solving" design and facilitation process was used. Perhaps it failed because the interests, knowledge, and skills of people involved were not up to the task asked of them. Or perhaps, other factors caused it to fail. Regardless, the fate of the WSAs and hundreds of thousands of vital natural acres and the wildlife within them remains in limbo. One clear lesson is that in the future, we need to do a much better job addressing our social and environmental challenges, as this example shows. After all, the WSA initiative was about clarifying our relationship to nature, wildness, and wildlife and how we want to live in the region. Our answer in that case has affected our communities, the wildness of the region itself, and the future for its inhabitants (human and nonhuman wildlife).

In this chapter, I will first consider the idea of ecosystems and greater ecosystems. Second, I look at "systems thinking" and the GYE as a big system. Third, I use complex adaptive systems (CASs) science, a recent branch of science, to explore the growing human footprint in the region. And finally, I survey our present goals for the GYE, our past commitments, our concerns for the future, and our present challenges. In doing so, I consider how life and ecosystems function as CASs and how humans operate within this system. As I go, I apply the story outline presented earlier, a heuristic for *people* finding *meaning (values)* through *society (institutions)*, using and affecting the *environment (resources*, both natural and cultural).

Life and Ecosystems

The GYE has many different kinds of plants and animals. We need to take them into consideration as we make decision about the future. All these life forms are constantly interacting with one another and their physical environment through ecological and behavioral systems, and so are we as humans. What is the GYE's environmental system or what we call today—the Greater Yellowstone Ecosystem? Put simply, it is a complex of land, water, air, life, and energy—the living and nonliving units of an environment. This system is open, with many inputs (e.g., weather) and outputs (e.g., water flows). It is connected to the larger global system or biosphere—the whole planet—through complex living biotic and abiotic interactions (e.g., migrations, water cycle, climate). The present political or agency boundaries of the GYE are not absolute, because the concept of an ecosystem is a scientific, functional construct not an inherent property of the place itself or a clearly bounded political or agency-managed landscape. Nevertheless, the concept of an ecosystem does serve as a realistic unit of management and policy.

Ecosystems

The two national parks within the GYE—Yellowstone and Grand Teton—are too small to maintain whole ecosystem function for all biotic and abiotic processes and cycles to occur in their natural state. This is particularly true of biodiversity conservation (e.g., elk, grizzly bears, bald eagles). The GYE is not one big "machine-like" entity regulated by a "balance of nature" or a pristine wild system, untouched by human influence. We are a part of the GYE, and we influence the landscape both directly and indirectly, short term and long term. Today, we know that the ecosystem is highly dynamic, ever-changing, and open to the world. It is a CAS. At the fundamental level, ecology is the study of CASs and therefore we are called to examine the whole system and to consider its internal and external relationships, including ourselves, in an integrated way.

Greater ecosystems

In the late 1970s, the term "Greater Yellowstone Ecosystem" was first used to describe the region.[5] As ecologist Frank Golley of the University of Georgia noted that ecologists (at the time) do not share a common understanding of the concept of an ecosystem.[6] Professor Golley looked at an ecosystem as a "complex organism," a biophysical concept, and offered terms to help us think about how one works. He laid out the historic and ecological context of the word "ecosystem," first used in 1935.[7] An ecosystem is a human abstraction, conception, or construct of the patterns we see in natural communities and their environments. An ecosystem's processes include the interactions of the biophysical environment and the living organism that inhabit it.

In thinking about the GYE's system boundaries, remember that humans affect the ecosystem in many ways—both direct and indirect. The rising number of tourists, development, and backcountry recreational uses threaten the system's integrity, just as air pollution from Asia is brought to the GYE by global air circulation. Impacts are sometimes directly harmful (vehicle road kills) and at other times subtly damaging (acid rain). The ecosystem is not a closed vessel but rather a porous membrane that allows the constant flux of materials and energy (resources) in and out. This brings up the issue of boundaries and our felt need to superimpose lines on unbounded systems for our uses and policy attention.[8] We typically set boundaries of the system for management purposes, focusing on government authorities largely at national to county levels, and other group interests, not on actual ecological structures and processes. Grizzly bear conservation is one very good, clear example of this setting of boundaries.

We now know that YNP is much too small to sustain many of the megafaunal species within its boundaries, and in fact we have known this since the 1880s when General Phillip Sheridan called on Congress to expand the park. But even the entire GYE is not enough to conserve ecological processes unharmed by human influences. The breadth of human influence on the natural world is staggering and it is rapidly growing. As humans, we are capable of changing whole ecosystems at the global scale, for example through rapid climate change, and in fact we are doing so to our harm.

Ecosystems everywhere are under siege, and the GYE is no exception. The GYE is rapidly becoming a geographic island thanks to encroaching physical development of houses and roads and increasing human presence almost everywhere, and through mushrooming recreation demands.[9] The human population surrounding the region is rapidly increasing, and the population shows diversifying special interests, each interest wants to use the GYE more intensely for its benefit, in old and new ways, often without considering the overall consequences. All uses have some harmful consequences on the wildlife. Far too often, special interest groups make demands for using the ecosystem without much knowledge of its history, biology, or their own potential impacts. This brings up basic questions about ethics and responsibilities (Figure 1.3).

For now, in this book, we are left with the questions: What is the spatial extent of the GYE necessary for functional ecosystem processes to continue and remain adaptable into the future? Are we knowledgeable and organized to think and manage in terms of open ecosystems, changing scales, and an uncertain future, all in the long-term best interest for all involved (human and wildlife)? Are we self-aware and knowledgeable enough to take on these questions in the first place? The GYE is evolving as a biophysical entity, and we are too, individually and collectively. If we wish to secure the common interest for all concerned—locally, nationally, and globally—we must develop resilient and adaptable democratic institutions that are prepared to address such challenges.

The common interest is our goal in democratic societies. The common interest for the GYE is already there in legislation establishing the national parks, forests, and wilderness areas, as well as in the Endangered Species Act, and much more. The goal in the Teton County, Wyoming, and other county land-use plans, is in the common interest. We all know that interests of all kinds (wealth, recreation, knowledge, education, and more) are at the heart of the GYE conservation debate.

Most people are versed in the concepts of special and common interests. Interests are patterns of demands (values) supported by people's beliefs that those demands are justified. Common interests are those interests that are widely shared with a community—local to global. These are demanded on behalf of the whole community. Safe drinking water is a common interest and is demanded by people and communities everywhere. It is reasonable and justified to demand a healthy environment. The GYE is one such place where demands, old and new special interests, conflict over its current and future use. Many of the new uses that are being demanded depart from the long-term historic common interest! Incompatible, unregulated recreation is but one example.

Evolution

"Nothing in biology makes sense except in light of evolution," noted Theodosius Dobzhansky, the geneticist and synthesizer of evolution for modern understanding.[10] The earth, its present form and its history, can only be understood in terms of its evolution—change. This is true for the GYE, too. The modern synthesis of evolution is about understanding observed facts of the structure that we see in plants and animals, and in their interactions over time. The GYE is an ecological theater and an evolutionary play.[11]

In other words, if we want to understand the GYE and our place in the ecosystem, we need to drill into how evolution works. It works, given our modern evolutionary synthesis, in six foundational ways:

1. Species tend to over-reproduce.
2. There is a struggle for existence between organisms.
3. There is variation among individuals within a species.
4. The fittest survive.
5. The variations in individuals are hereditary.
6. Through the favorable accumulation over time of hereditary traits, new species eventually develop.

This understanding of life and change is evident everywhere in the world, including in human ecology.

What makes some plants and animals common and others rare? Why do the common stay common and the rare stay rare? The answers to these questions lie in differences between the resources used and the function played by different species in the environment. Each species occupies its own ecological niche, the full set of biotic and abiotic interactions in which a species participates. Species tend to differentiate their niches through the process of niche partitioning in order to minimize competition for critical resources. Today, we know that all plants and animals have slowly changed to their present forms and ecologies through evolution. Their bodies and behavior are adapted to sustain their lives within the niche they occupy. All animals have evolved to meet their own unique life requirements—finding food, surviving hazards, and reproducing—in the context of their dynamic environment.

There is a continual feedback between the environment and evolution, with each always acting upon the other. As the environment changes, species adapt to the new conditions through evolution, but as they evolve, they also begin to affect the environment in new ways. This is known as an eco-evolutionary feedback system. Traditionally, ecology and evolution have been considered separate and discrete fields. Ecologists ask questions such as, "What function does the pronghorn antelope play in the GYE?" Evolutionary biologists ask, "What conditions in the GYE shaped the modern form of the pronghorn antelope?" Obviously, the biotic functions of an ecosystem are the product of natural selection and evolution. Species evolve in relation to one another over long periods of time, sometimes for millions of years.

One example of these coevolutionary ecological processes is predator–prey dynamics. Predator and prey species are constantly changing and optimizing through ongoing evolution. Prey species adapt defenses to avoid being captured by predators while predator species simultaneously evolve to become more skilled and more efficient hunters. Let us consider again the pronghorn antelope. The pronghorn antelope is one of the fastest land mammals capable of reaching speeds over 50 miles an hour. However, across its habitat, including the GYE, it has no predators today capable of reaching such high speeds. An evolutionary biologist may ask then, why did the pronghorn adapt to be such a powerful runner? The answer lies in the fossil record. Thousands of years ago, the pronghorn antelope occupied the same region as the American cheetah, a swift predator species that died around twelve thousand years ago. In this case, the ongoing biotic interaction between the cheetah and the pronghorn resulted in coevolved, highly efficient runners. The GYE is only one arena in which these processes and products are being studied today. In the end, evolution is the mechanism by which changes are produced in species and their interactions.[12]

Yet, the future of the GYE, antelope, and predators and prey (and all other eco-evolutionary processes) has personal, political, and philosophic dimensions. Clearly, these dimensions cannot be understood by ecology alone, or conventional, everyday thinking. What would a psychologist, sociologist or a rancher, tribal elder, or a tourist skiing at Teton Village, Teton Pass, or in the WSAs say about the GYE or wildlife? Recognizing these many human dimensions (human mental level) is why this book considers many different people, values, and demands and also the fields of ecology, geography, anthropology, psychology, sociology, policy sciences, pragmatics, and more. Consequently, this book is my effort to recognize these dimensions and take an integrated, common interest-oriented look at all of them together.

Ecology and ecologists

Ecologists study the interactions of plants, animals, and their environments. The word "ecology" was coined in 1869 and "ecosystems" in 1935. Since those early days of ecology, the field has exploded as a major study.[13] Ecologists study the workings and pattern of living systems, wherever they occur (Arctic and Antarctic to tropics, mountain tops to bottom of the oceans).

Ecology is a diverse field, and the importance of scale and context is becoming increasingly important in our understanding of Earth's biotic and abiotic systems. These different levels of organization follow a hieratical order from the smallest unit, the individual, to the largest unit, the biosphere. With each level, there is an increase in the complexity and intensity of interactions among various organisms and their environments. Organismal ecologists study the response of individuals to small-scale changes in the environment. Population ecologists consider these individuals in relation to one another, including biotic interactions such as reproduction and competition. At the community scale, ecologists consider the types of interactions that emerge between different species, such as predation and symbiosis. Not until the ecosystem's abiotic interactions are considered do we look at the ways that species respond to changes in energy or nutrient cycling or to disturbance events.

Today, the GYE is being studied at many levels, from individual animals to biome affects, each a series of ecosystems that extends over vast geographic areas. The NPS employs dozens of ecologists in YNP and GTNP, and many more scholars work there from universities around the world. These people have generated a picture of what is happening in the GYE ecologically, which is part and parcel of a larger global situation. From studies of the GYE through time, we understand that humans have significantly affected the ecological systems, as they have everywhere. Our technology, especially our exploitation of fossil fuels, has brought on a huge transition in the way we live, a trend that is still underway. The trend has tremendous environmental consequences (e.g., climate change, species extinctions, pollution, and soil loss). Moreover, our exploding population, not only in the GYE but also globally, puts a very heavy pressure on earth's ecological systems.

The overall theme of our stories about nature, wildness, and our living is that we are the central character in all of them. There exists a dualism between the natural world and us on our stories and we therefore see ourselves occupying a place somewhat separate and removed from the ongoing ecological world. That story, our dominant story, is highly problematic! Many growing destructive consequences flow directly from that old story. We need a new story of conservation for the GYE and the world.

Harmful ecological trends and increasing awareness of the flaws in our cultural narratives or stories about our relationship to the natural world bring out diverse philosophic, practical, and policy concerns, and conflict. What we have learned about ecology, especially our own ecology in the GYE, is forcing us to reconsider much of what we think about our role in an ecological world, and also to reconsider our past harmful conventional thoughtlessness. We have been thinking and behaving too small. We are now being forced to confront major environmental changes that are entrained by our society, such as climate change and even aspects of the destructive recreational businesses and uses of nature now coming to the forefront in the GYE. No doubt there are many recreationalists who seek a sustainable human–nature relationship and want to join with other people working for the overall common interest. Yet, it is clear we have a long way to go to clarify and secure that common interest.

Overall, it is clear that humans are overwhelmingly the dominant force of the planet's ecology. This is true in the GYE and growing more so with the mushrooming new residents and visitors and their new and intensified uses of the GYE.[14] As population and consumption trends continue to mushroom, we need to hasten our democratic deliberations on these pressing issues because time is running out.

Living in a World of Systems

We are living in a world of embedded systems that can be considered at a number of scales. Humans are living in nature, within complex regional and global systems. We need to learn about our heavy footprint on nature and in these systems. In short, we need a systems "lens" on the world to frame our understanding of all this. We must think of this systems view as a window that includes ourselves. As historian James A. Pritchard then with the University of Kansas said,

> Most visitors to the parks are unaware of scholarly debates over definitions of what is natural [...] In the real world, "natural" is simply a matter of degree. We know nature when we see it. [...] The purpose of the national parks remains one of the most important cultural considerations regarding the management of our public lands.[15]

We need to delve into systems thinking and what today is called "complex adaptive systems," if we are to better understand the GYE and our role in it. The major feature of systems is that they are dynamic, constantly changing. Such change demands new thinking and behavior, but currently we lack an adaptive framework that would allow us to readily adjust our viewpoints to meet the realities of the changing world. We are currently constrained by our conventional, static, and self-interested thinking. We seem locked into old hierarchical, bureaucratic ways of thinking and deciding.

We take conventional thinking such as the human versus nature divide as the way it has always been and therefore cannot be changed. And we take our immediate self-interests first, too often without considering what we are doing to wildlife, for example. Overall, and individually, we must understand how to recognize when we fall into these conventional, self-interested patterns of thought and, most importantly, how we can learn and work our way out of them. These conventional traps, once we fall into them, become culturally rigidified and institutionalized and block our potential for learning and adaptation.

Systems thinking

The key to systems thinking is to see the relationship between the parts and the processes. What is the relationship between the elk we see along the road in the spring and fall and the process of large migrations of thousands of animals? There are many similar examples where we see the parts but not the processes. We must be able to scale our thinking to consider each individual component of the system, as well as the ways that that view informs or prescribes the function of the system itself.[16] If we learn to see relationships between the part and the whole, perhaps only a few of them at first, we will begin to understand how systems work.

Ideally, we will then be able to shift toward behavior that works better for us all, given our goals as described below. As Donella Meadows, the famous scientist from Dartmouth, notes in *Thinking in Systems*, "As our world continues to change rapidly and become more complex, systems thinking will help us manage, adapt, and see the wide range of choices we have before us."[17] Systems thinking requires us to address the basic and foundational questions that were introduced in the first chapter in Figure 1.3. Systems thinking can transform how we see and think about the GYE and ourselves in it.

Systems thinking came out of mid-twentieth-century work, largely from a critique of scientific reductionism (positivism), which still dominates most thinking in the GYE today. The positivistic belief assumes that the ecosystem is simply an emergent property of the pieces. Positivistic belief in the public, the agencies, officials, and nongovernmental groups is a major, unrecognized barrier to our moving on to a new story about human nature and wildlife coexistence that we can live out. Positivism assumes that there is an objective reality, that our senses and beliefs give us a realistic grasp of the world and ourselves in it. From that, we assume we know what to do to lead ourselves to a healthy life, relative to nature.

Positivism is reductionistic (it breaks things down into parts and studies the parts thinking this is way to understand whole systems). This approach produces knowledge about the pieces, the constituent parts of complex systems, not about the complex systems. It is assumed that by studying these (few) simple parts or elements, we can understand the overall phenomena (e.g., the greater ecosystem or the biosphere). We live in a civilization that glorifies the power of this kind of positivistic reasoning. The name we give positivism is science, in the narrow sense of the concept. It is a view that is accepted widely without question on a scale today unprecedented in all of human history. Some people see that it overlooks the fact that individuals and society organize themselves around principles of morality and social and political life (the human mental level). Positivism seems to rule out the need to find reasoned answers to how we might learn to live sustainably with nature and wildlife in moral and political ways. And, it rules out how we might discriminate between necessary and unnecessary desires and beliefs and come to the goal of right living in the common interest. For human beings, as I see it, we must find a way of knowing and living that is compatible with our commitment to individuality, democracy, and living in a healthy sustainable environment with nature, rich in wildlife. However, our current thinking is not system thinking in the full sense of the concept that I am calling for. Perhaps the new thinking and story is our greatest system challenge.

Emergence and interaction are the two fundamental ideas of systems thinking. The notion of emergence is popularly understood as follows: "the whole is greater than the parts," and with the parts; it is assumed the parts are working in synergy. Conventionally, it is argued that valid knowledge and meaningful understanding come from adding up parts to see the whole picture. We now know this is not true.

We are not adequately employing systems thinking in the GYE today. Instead, we are lurching from case to case, incident to incident, and individual to population issue, all in isolation from the other cases, incidents, and issues. We are not seeing the bigger picture or the underlying foundational reasons for our shortsightedness. As a consequence, we are offering technical solutions to discrete events and getting caught up in the web of minutia of everyday events. In doing so, we are creating harmful and unintended consequences that are accumulating.[18]

One example of these unintended consequences is the outbreak of brucellosis in elk on the Wyoming feed-grounds. The state of Wyoming began concentrating the elk on winter feed-grounds to maintain high populations for hunting. However, this incidentally created a reservoir of brucellosis. Recent studies show that brucellosis was introduced into wildlife in the ecosystem at least five times from cattle,[19] and the feed-grounds concentrated elk and contributed to the rapid spread of the disease. This prompted fear about the continual spread of brucellosis back into livestock populations. To address brucellosis transmission, officials kill large numbers of bison at the borders of YNP to try and "manage" these animals, while they continue to concentrate elk on feed-grounds. Clearly, we need more farsighted systems thinking from our leadership and managers.

There is resistance to systems thinking for several reasons. We have been socialized to think in simple linear cause-and-effect terms. If condition A exists, then there must be a condition B that caused it. However, the real world is not necessarily that simple. For example, it is generally accepted that if I have a cold, then a virus caused it. But perhaps it is caused because I did not sleep enough the night before. Or perhaps the cause is my own immune response. Causality is only provable if event A always happens following event B in certain circumstances. In this case, however, we know that our bodies are always infected with viruses, but we do not always exhibit symptoms of sickness. This kind of linear thinking gives us a false sense that we can control what matters and that if we just understand enough, we will be able to make a singular informed choice that will achieve our intended goal. Over evolutionary time and through cultural experience, we have amassed a set of prescribed, conventional truths about the world that are not generally questioned (i.e., convention). One example of these conventionally accepted truths is the false nature–human dualism that we now hold. This is where systems thinking would be beneficial in allowing us to see both the forest and the trees, and the foundation for our knowing about the world, and ourselves in it. However, there is significant resistance to complex systems thinking that results in delays and overshoots to understanding and addressing problems.

Fortunately, we do have some innate understanding of systems features. The phrase, *a stitch in time saves nine*, for example, captures the idea of feedback delays in complex systems. Another phrase, *don't put all your eggs in one basket*, captures the features of multiple pathways and redundancies, recognizing that systems with these features are less vulnerable to disruption than systems without them. Since the Enlightenment Period and the Industrial Revolution, we have come to rely on positivistic science, reductionism, and experimentation. This philosophy of knowing typically puts the cause of events "out there," rather than inside our own thinking or "in here, in our heads." Consequently, we typically blame, externalize, or assign the cause of something (wolf predation, bison movements) to something else, even other people, some ecological process, or an animal "out there." This comfortably shifts any responsibility from us and our observations and thinking to an external objectified target.

As a result, we focus on finding the control dial, the technical fix (e.g., the overpass for wildlife) that will make things right somewhere "out there" and ignore our own thinking as a variable. Our focus of attention comes to rest on external entities and agents that we can manipulate, engineer, or think that we can. This is an oversight, a weakness in our thinking, a blinding mistake behind the many problems we now confront.

Systems basics

To understand the GYE, we need to understand the basics of systems. Dr. Meadows defines a system as "an interconnected set of elements that is coherently organized in a way that achieves something."[20] Systems involve elements, interconnections, and a function or purpose. As noted above, predator–prey dynamics and migrations are but two (sub)systems in the GYE, each with its own elements, connections, and coherent organization. These subsystems are all connected with one another, making up the entire GYE, which is, in turn, an open system that interacts with the global biosphere.

There are many human systems operating in the GYE, but they often function in fragmented isolation from one another. Many important human systems like healthy individual, psychological, and sociological systems, as well as democratic systems, are often overlooked because a disproportionate emphasis is placed on the economic system. However, all of these human systems overlap and affect one another. For example, the question arises, should the Wyoming Tourist Board and the Jackson Hole Chamber of Commerce, as do many communities in the GYE, advertise widely for tourists to come to the region and spend money? They say they are conserving the "wild" in their promotional materials, but in fact they are helping to erode it. However, local accommodations, roads, and our public lands are already overloaded with visitors now. This overcrowding causes many local problems that seem beyond full understanding, fixing or managing, under present arrangements and conventional thinking. Because we are predominantly driven by economic incentives, these effects are often overlooked, but they are no less important to understanding our relationship to the GYE as humans.

What is not a system? Any conglomeration or collection of things and processes without any particular function or interconnections is not a system. The thousands of randomly scattered boulders in Jackson Hole are not a system. If you take away wolves, as we did decades ago, or destroy migration corridors, as we are now doing, we lose critical functions and connections and therefore cease to have a complete functioning ecosystem.

We lose "system-ness" when humans create disease reservoirs, overkill predators, displace and stress wildlife, and disrupt migrations.

But system-ness is not a discrete condition. Systems can change; they are not static. It is therefore the "goal" of systems to be dynamic and adaptive. Redundancies in function and in connection are one way that ecosystems can achieve this resilience. The GYE exhibits significant systems integrity, a wholeness and coherence that functions as it does because of a particular set of biophysical mechanisms working together to achieve a function. Unfortunately, the hard truth is that we are eroding the integrity of the GYE by eliminating and disrupting critical connections within the system through our own activities at present. Over-visitation, urban sprawl, and increased recreation all threaten the persistence of this beloved ecosystem.

A major barrier to the needed systems thinking is that people tend to conventionally focus on those literal, concrete parts of the GYE that are easiest to see: elk, trees, lakes, bears, wolves, and bison, among others. However, if you look closer from a systems view, you can see that these species and things are part of abstract systems of food webs, material cycling, and energy flows. They make up systems of herbivory, predator–prey relations, competition, migrations, and more. What we are doing now in the GYE is focusing too much of our attention on individuals, cases, and incidents that come and go in the public's attention, and by doing so, we overlook the greater patterns and processes at play, including those in our own thinking. In short, we will never be able to see the forest, if we only look at the leaves, as we are mostly doing now in the GYE. This is not to say that individuals, cases, and specific incidents are not important as in fact they are real, and highly symbolic and distracting. They are the product, not the cause, of our problems. When we treat these cases as the source of our problems and attempt to prescribe technical solutions to them, we fail to adequately address the challenges facing the GYE and to secure our overriding goals. Instead, we must learn to synthesize these discrete cases within the broader context of the ecosystem. To do this, we must apply systems thinking.

To move forward, to progress toward human–nature coexistence, we need to see systems interconnections. We need to look beyond the actors, the players, the species, and individuals, and instead focus on the rules of the "ecological-evolutionary play" in the GYE. For example, we can easily see the large mammal migrations that traverse the landscape each year, but this phenomenon is not understood unless we consider the broader seasonal changes that drive them. The whole ecosystem is constantly in flux. Large-scale biotic and abiotic patterns interact as a vibrant, functioning system. To better address the challenges that we are facing in the GYE, we need to start looking more deeply at interconnections—at the fires and plant communities, climate change and diseases, predation, hunting, tourism, and migrations. And, importantly, we need to look at our own role and compounding disruption in the GYE.

To understand the notion of function or "purpose" in systems, consider a thermostat, the system that keeps a room warm or cold, depending on the setting. This thermostat, like the GYE, is a dynamic regulating system. Simplistically, the various components of the GYE function as a thermostat. Predators regulate prey populations; climatic conditions regulate plant reproduction; and vegetation distribution regulates migration patterns. Each part of the system is simultaneously a cause and an effect. However, the most important function of each species is to operate in a way to ensure its own survival. Natural systems behave in ways that are surprising, and the collective interaction of many elements can produce unexpected or unintended outcomes. Interactions and feedbacks are not a simple thermostat-like process.

Complex systems are endlessly dynamic processes. Understanding systems requires a longtime, significant observation, and a combined focus on elements, interconnection, and functions. We are falling short of this in the GYE today. Instead, the public and management in the GYE argument are about details—the visible, concrete, and conventional elements of the system, rather than the function of the ecosystem itself. This focus of attention is an artifact of us as human beings, our system of knowing and thought, and the meaning we attach to it—our stories.

Scales

Geography and scaling are key concepts to understanding systems and the GYE. Scale refers to the spatial and temporal variation in our lives and the environment. Everything occurs at different scales of space, time, and complexity. An ecosystem or greater ecosystem is one variation on scale. Scale is a highly useful notion as an organizing principle for land-use management policy.[21] Differences in scale are evident through variations in climate, topography, soil types, water temperatures, biodiversity, and human communities, among others.[22]

Geographic

The concept of scales is a rapidly evolving field of study that has mushroomed since the 1980s. As noted by Canadian scientist David Schneider in 2001, our "skill and rigor in the design of field experiments, the growth of mathematically based theory, computer-mediated advances in analytic capacity, and the accelerating capacity to acquire and retrieve data" has resulted in huge gains in our understanding of scale dynamics.[23] These findings have placed greater emphasis on the need to understand scaling overall and in particular locations such as the GYE (Figure 3.1). Many major problems in ecology often unfold at larger and longer scales in intact ecosystems, such as vegetation patterns and climate change in the GYE.

Humans living at more immediate spatial and temporal scales can easily miss these long-term phenomena. Most variables, especially rates of change, can only be measured at specific sites. Patterns observed at small scales do not necessarily hold true at larger scales and vice versa, but we often use a collection of small-scale findings to extrapolate to the whole. In the GYE, we cannot address expansive and long-term problems using small-scale data, observations, and management policy.[24]

Space

In many cases, spatial and temporal scales are positively correlated. For example, environmental disturbances, biotic responses, and vegetation patterns are correlated

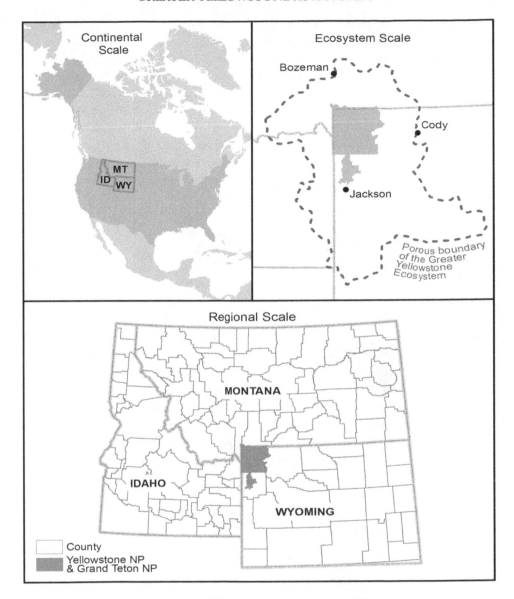

Figure 3.1 Three different scales for the Greater Yellowstone Ecosystem (continental, the Greater Yellowstone Ecosystem, and Yellowstone and Grand Teton National Parks).

in the context of space-time domains. As the spatial extent of geographic phenomena increases, so does the temporal extent over which the dynamics can be observed. In the GYE, for example, a single predation event in the Lamar Valley is spatially and temporally distinct from the broader boom-and-bust cycles that result from fluctuations in predator–prey populations. This concept of the space-time correlation can be applied at even larger scales (e.g., global climate change, disease ecology, tourism).

In the GYE, a large portion of the spatial variation is related to geology and climate, among other things. The GYE ranges from high plateaus and massive mountains to sagebrush steppes and deserts. These distinct spatial units vary in the types of species and processes that they support—the vegetation in the high alpine differs drastically from that of the riparian zone. This kind of spatial variation is easy to see across the landscape. Just hop in your car and drive from Bozeman, Montana, to Kemmerer, Wyoming, a distance that covers almost the entire latitudinal extent of the GYE. Similarly, drive from Idaho Falls, Idaho, to Cody, Wyoming, to traverse the entire longitudinal extent. Along these axes, you will see great geographic and ecological changes across the landscape. Such spatial variation also yields differences in how dynamic geographic systems evolve over time.

Time

The temporal variation is defined as change in the environment over time. In many cases, temporal variation is related to spatial variation. For example, growing seasons are shorter on mountain tops than in valleys (Figure 3.2). We experience many forms of temporal variation. Daily solar cycles and annual seasonality are two of many examples. Daily, we see the change from day to night. Annually, we experience the solar variation from summer to winter. There is significant variation in temperatures, precipitation, drought, and fire across time.

As part of each species' ecological niche, different life forms have different tolerances for changes in the environment over time. Some species, particularly those with broad niches, are better able to sustain rapid, significant, and sustained change. Global climate change is one such change. So are fires and other disturbances, like floods. The environmental conditions that are tolerable to a given species partly determine the distribution of that species on the landscape. Thus, as the environmental conditions change, the distribution of plants and animals changes as well. Across the GYE, rapid changes are underfoot that will surely alter the spatial, temporal, and complexity patterns for most species and communities in the GYE, including humans.

Complexity

Complexity is a property of an object or system; some systems, ecological or otherwise, are relatively simple. Other systems are unimaginably complex (Figure 3.3). The GYE, as a CAS, is by definition high on the complexity scale. The GYE is geographically big, spatially large, time-wise a long-term entity, and when taken all together with all its processes and humans, it is a complex system. Its complexity is far beyond what we can ever fully know. The study of complex linkages at various scales in the GYE and beyond is the main goal of complex systems theory and sustainable conservation policy in the common interest, and is discussed below. What we need for our living in the GYE is a strategy for coexistence as introduced in Part 3 of this book.

Definitions of complexity often depend on the concept of a system, which is a set of parts or elements that are related to one another but differentiated from outside elements.

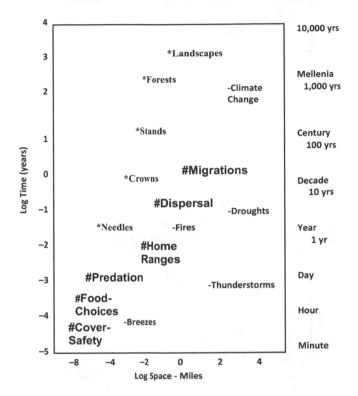

Figure 3.2 Three sets of relationship over time and space in the Greater Yellowstone Ecosystem: (1) * "needles to landscapes," (2) - "thunderstorms to climate change," and (3) bold for elk and carnivores "cover-safety to migrations."

Figure modified from C. S. Holling, "What Barriers? What Bridges?" in Barriers and Bridges to the Renewal of Ecosystems and Institutions, ed. Lance H. Gunderson, C. S. Holling, and Stephen S. Light (New York: Columbia University Press, 1995), 3–34.

Many definitions of complexity tend to postulate that complexity is a condition of numerous elements in a system and the numerous relationships among those elements. However, our definitions of complexity are relative and change through time. This is particularly true for CASs. CASs, like the GYE, are those for which the whole is greater than the parts. By that, I mean such systems are a dynamic set of networks (spider webs) of interactions and relationships among the physical environment and living species, all of them. And that the individual and collective behavior of all of them is constantly changing as a result of experience with changing contexts. The GYE is truly a complex dynamic system.

Stressors and opportunities

The world-famous GYE, as well as other large landscapes, is under more stress than ever before. The GYE and many of those other ecological systems are showing wholesale shifts in weather, vegetation, and wildlife movements, with more changes to come

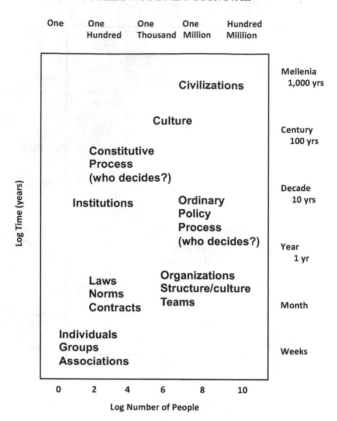

Figure 3.3 Scales (relationships) illustrated in the numbers of people over time ranging from individuals and their actions up to millions of people interacting over long times. Throughout, ask what is the relationship between humans and nature?
Figure modified from Holling, "What Barriers? What Bridges?" 3–34.

(e.g., extinctions, invasive species, disease). Much of the system-level changes we are seeing now are likely invisible to casual observers. These large-scale changes manifest in localized cases, which we understand, but the actual overall change is largely invisible to many. Visitors, outdoor adventure enthusiasts, and the business community all tell stories of a utopian Jackson Hole and the GYE. Most of these people see the GYE as a pristine system, with fresh air, clean water, and abundant wildlife. "What is the problem?" they ask as people move to this region, seek more outdoor adventures, or strive to maximize business profits. Many of these people are short-term residents of the system and, thus, have not been around long enough to experience the harmful change for them, let alone to other animals. This short-term experience is just one of the ways that we are trapped inside certain ways of conventional seeing, thinking, and behaving.

Given the diverse perspectives on the GYE, it is easy to see why some people argue over issues like just how many grizzly bears we want or need, where they should be tolerated, and the like. This is important to be sure, but I suspect that many people externalize much of their thinking about these matters out to the bears and never to simultaneously looking at their own perceptions, interests, and thinking. Consequently, we may be overlooking the significance of our individual and cultural stories that influence what we see and do.

If we do fall prey to bounded thinking, it can severely limit our understanding of the GYE and ourselves in it. Such a bounded outlook causes us to materialize, objectify, and instrumentalize nature, the GYE, and all its plants, animals, and systems processes. Because everyday science, business, and recreation are organized around these objectified things in this view, they are all used to further our human quest for happiness, entertainment, and possession, without considering consequences. Our cultural system of thought can block examination of our role in the GYE (the "internalize" view as in Figure 1.3). This point is what David Bohm, the theoretical physicist and philosopher, means when he writes about deep problems or invisible traps in our system of thought.[25]

As Donella Meadows concluded, "At a time when the world is more messy, more crowded, more interconnected, more interdependent, and more rapidly changing than ever before, the more ways of seeing, the better. The systems-thinking lens allows us to reclaim our intuition about whole systems."[26] The systems-thinking approach, if we use it, will (1) hone our abilities to understand the parts, (2) see interconnections, (3) ask "what-if" questions about possible futures, and (4) be creative and courageous about systems redesign and our own behavior and its consequences. I argue that we need to quickly shift gears from conventional thinking to systems thinking, greater reflectivity, and new action, if we want a healthy GYE in the future.

We need to learn how systems are structured and how they behave, while remembering that we are part of those systems. We know that natural systems work very well if we leave them alone. We also know that human intrusions can disrupt whole systems and, in the worst cases, cause systems to collapse. The loss of species through direct destruction can cause system collapse too. Our task, as an evolved ecological species living within these ecological systems, is to learn when to intervene and when to keep hands off. In the end, we must learn to live as active agents within a world of systems.

Complex Adaptive Systems

Systems ecologists have recently come to appreciate that human and natural systems are closely coupled in one integrated system. Philosophers, historians, and geographers have known this for generations (the integrative disciplines formulated clear connections of systems decades ago). Positivism and the political organization of science and management blocked the complex system view of the world from becoming main steam. It is our job currently to adopt a view of ourselves in the GYE as a CAS, to reconsider our impacts on nature and the way that we live. No work in the GYE on CASs has yet been done at a scale and duration needed, so CAS interactions are not well understood.

Systems concepts

There are two contrasting ways to think about systems and the GYE.[27] The first assumes that the GYE system, for example, and our knowledge of it are objective (the "hard system thinking" approach). In this approach, understanding the system requires a belief that there is an objective reality out there, apart from us as observers. And that we can identify it and improve it using positivism and manipulative management. This approach most strongly dominates in the worldviews of most people concerned with ecosystems as physical, technical things today, including representatives in government and nongovernment.

The second way of thinking about systems assumes that they are partly objective and partly subjective. This new, perhaps strange way of perception and thinking is likely novel to some people in the GYE and beyond. This way of seeing comes out of the landscape of thought over the twentieth century. There is underway now a proliferation of differing philosophies and names for how we perceive, think, and act today. Among the labels are the terms "modernism" and "postmodernism" or "deconstruction." It is beyond this book to dig deeply into these ways of thinking, and contrast them and their implications, but at some level I think we all experience that the time we are now living is rapidly changing, in large part due to the changing landscape of our social and political lives. We feel, or many of us do anyway, that something big is happening, even if we are not quite sure what that is or means. Different ways of knowing and thinking are ever more diverse these days and in direct competition with one another in the GYE, nationally and worldwide, thus the "cultural wars" that we are living out now in our society. All this makes us ask, what will the future be like?

This second way or approach understands reality, the truth, as a "creative construction" of human beings through our minds (thinking), culture (philosophy), and language (biosemiotics) (see Figure 1.3). Its sees that our social reality is merely the result of people's interpretation of their own experiences, which are guided by hard realities. We are not talking about biology, but rather about interpretative understanding and meaning-making process—individually and culturally. Consequently, the second approach works with an evolving appreciation of people's perspectives, values, and commitment. It is sometimes called the "configurative, comprehensive, integrated approach."[28]

The CAS term is used to label the loosely organized field in academia that is maturing around the study of ecosystems and landscapes. There are many complex systems in society: the stock market, the human brain, immune systems, communities, and civilization, among others. In nature, complex systems include ant colonies, embryo development, and landscapes. Much systems thinking is grounded in complex visualizations (box and arrow charts, flow and flux diagrams) and expressed in chemistry and physics, mathematics and computer simulations, and evolutionary and adaptation terms.

CAS is defined as a system of multiple interacting agents. John H. Holland, a CAS thinker, says, "CAS are systems that have a large number of components, often called agents, that interact and adapt or learn."[29] From this perspective and line of thinking, YNP and the GYE are a CAS. These systems show emergent properties that come out of the structure, complexity, and interactions of the system's agents. Another emergent property is that of our "mind" comes out of our physical body, the bones, muscle, lung, heart, and nerves and brain. Our consciousness and self-awareness are emergent properties of our bodies and evolution.

There are features such as self-organization, where prey and predator develop a dynamic relationship over time. Other key properties of CAS are spatial and temporal patterns, communication, cooperation, and specialization. Cooperation takes place at all levels, from agent (species) to system (ecosystem) level. Human and natural systems interact through complex feedback loops too. CAS features are not necessarily the things we think about as we go through our *daily* lives because most of us do not think in these concepts. We are more interested in where we go to get groceries and the like. Nevertheless, CAS features are real and affect all that we do, how we live, and what the future may hold in places such as the GYE.

Systems features

The GYE is indeed a CAS. Below, I will consider some of the ways in which the GYE exhibits principal features of CAS.[30] We should keep these in mind at all times, especially as we create policy and carry out management. Key features include the following:

1. CAS show *nonlinearity* (not 1 to 2 to 3 ...) *and thresholds* (steps at which things are different than the step below). For example, trees that fall into lakes and creeks provide critical fish habitat. For example, there are examples of a threshold—the point of transition between one state and another where things change, such as steam that gains volume and speed. At one point you can safely cross it and at the next point it is too swift and powerful to cross safely—and represents a common form of nonlinearity.
2. CAS show *surprises*, such as the 1988 Yellowstone fires and other natural disasters. It means that we simply do not know very much at present about systems, if we ever can.
3. CAS inherently have *legacy effects and time lags*. Sometime things happen because we did something much earlier. The cause and effect are separated in time, but nevertheless directly connected. Legacy of historically caused effects are impacts from earlier times in which human–nature interactions created conditions for later outcomes. The construction of towns and highways along critical migration corridors still impacts those patterns of migration today. Legacy effects typically manifest within a few years or a few centuries (e.g., as with climate change).
4. CASs have varying degrees of *resilience*, the capability to retain similar structures and functioning after disturbances. The resilience of a CAS is affected by many factors; currently, human intervention plays a major role in maintaining resilience in nature, although healthy ecosystems are capable of maintaining their own dynamic equilibrium. We technically manage wildlife through hunting permits and reintroductions instead of adopting a healthier concept of the human–nature relationship. This will require major changes in the future, if we want to be more sustainable.
5. CASs are *heterogeneous*. They vary across space, time, and organizational unit. In the GYE, the spatial heterogeneity is obvious. As there is land-use classification ranging

from wildernesses to dense human development. This includes intense agriculture, mineral extraction, gas fields, and more.

Because we know that the GYE is a CAS that exhibits nonlinearity and thresholds, surprises, legacy effects, and time lags, resiliency, and heterogeneity, we ought to transition from our present conventional, positivistic thinking and bureaucratic natural resource management organizations to an adaptive (rapid, co-learning) co-management system. This requires moving from agency experts making all the decisions to community involvement in democratic governance for adaptive comanagement of landscapes using appropriate knowledge and skills.

Overall, we are moving in the GYE toward greater hominization (a process of becoming more humanized, i.e., human dominated, and simplified, as we degrade systems components (e.g., migrations) given our limited understanding of the system and our role and responsibilities presently). This humanizing, dominating, and harmful trend is exploding as human population densities increase and land uses become more intensive and invasive. This will likely lead to more conflict among humans in the GYE over how we should use it, our responsibility to wildlife, and, in many other ways, if we continue on our present path. This path dependency or "fixed path" will only accelerate destruction of natural systems and be a loss for wildlife as the GYE continues to change as a complex dynamic system.

Learning about systems

Learning about and successful existence for all life in the GYE requires that we rearrange humans' current level of understanding and behavior of not only the ecosystem but also ourselves. The dynamics of human-natural systems are influenced by many factors, including people's behaviors, social institutions, government, policies, and leadership. It is critical for us to move well beyond how we currently interact with and intrude on natural systems in the GYE.[31]

In 1872, Congress drew a line on a map, somewhat arbitrarily, inside of which there was to be YNP and outside of which was undesignated public and private lands. With this Congressional Act, the GYE, as we call the region today, experienced its first act of administrative fragmentation. This initial act started a long process of dividing the larger ecosystem into various administrative units and private property. Over the following decades, people came and established homesteads, towns, and ranches. These boundaries that have been superimposed on the landscape seldom contain or delineate the natural processes at play in the region, particularly large-scale processes such as migration or the water cycle. Today, we are challenged to learn about this growing coupled human-natural system in new ways. Time is a-wasting. Human ecology may be the hardest for us to learn about.

Our Cultural Ecology

The GYE has a rich human ecology—mental level. The earliest human presence in the region was not recorded, so we know of it only from the artifacts that remain. Since those times, we moved from low-density, nomadic human populations to a vast, very dense population that settled permanently and built extensively over the past 150 years. That trend is accelerating with growing impacts from our numbers and technology.

People—us

Human presence in what is now the GYE has changed significantly over the past ten thousand years. More recently, change has been astounding. Over the past thirty-five years, we have gained a new systemic understanding of human ecology in the GYE, at least to a rudimentary degree, as well as knowledge of the status of some of the life in the region and beyond, and about the overall complex system. Much of this understanding is the basis for more modern stories of meaning making about humans in nature, as we live, visit, and play in the GYE. The full extent of that new knowledge has not yet been assimilated into our cultural perspective or policy.

Native Americans and Euro-American discovery

More than twenty-five Native American tribes used the current range of the GYE as their home, hunting grounds, and transportation routes, for food, clothing, weapons, decorative items, and unique stones. Diverse waves of people have occupied the current GYE landscape, using the ecosystem for over eleven thousand years.[32] Much of this occupancy is just now being discovered and written about. Recently, we have discovered more artifacts and records of pre-European settlements in the region, which challenge the perception that the GYE was a pristine wilderness never trodden by human footsteps until recent times. That view is incorrect, yet it is too often part of our modern, conventional story of the GYE and nature.

Many clashes developed between natives and settlers in the late 1800s and conflict continues to this day.[33] Efforts to create borders between the two groups led to the 1851 Fort Laramie Treaty, which granted the eastern third of what would become YNP to the Crow Nation. In 1868, another Fort Laramie Treaty granted land rights to Crow and Shoshone people in the area, but also included land cessions of what would become YNP. Other tribes were recognized in one way or another in these treaties, including the equestrian tribes such as the Blackfeet, Bannock, and Shoshone, all of whom hunted within what are now YNP and GYE.

However, by 1872, officials considered the land to be unused by Indians. The Bureau of Indian Affairs (BIA) expressed concern about clashes like tourists coming to the newly formed park. In turn, the BIA sought to limit off-reservation rights. In 1877, the Nez Perce War began. In 1880, the Crow ceded land to the government and were moved to a reservation, where they remain today. People from the Fort Hall and Lemhi, Idaho, and Wind River, Wyoming, reservations continued to use the GYE lands until about 1900. By the 1880s bison were eliminated. The effects of these treaties, loss of a food base, and forced removals are still felt today. Natives and non-natives alike continue to grapple with the question: How should Native Americans be involved in the ecosystem? This remains a complex policy issue, and federal and state authorities are working to find an accommodation, as are the tribes and various collations. Much work remains to address these complex matters in ways that genuinely address human dignity concerns for all.

Euro-American exploration and settlers

Euro-American explorers, trappers, traders, and wagon trains came through the region from 1805 to 1888. Later people were en route to Oregon and elsewhere. John Colter, the first recorded European American, visited in 1805. Mountain men followed until about 1840. Hunters, prospectors, miners, and eventually settlers (e.g., Virginia City, Montana) ushered in dramatic change for the influx of Euro-Americans. Forts were set up to ensure travel was uninterrupted. Wyoming, Montana, and Idaho were designated as states in the late 1800s.[34]

The inhabitants and visitors to the GYE have marveled in its natural beauty and considered what it means to them and the nation.[35] First, take the geologist Fritiolf Fryxel, who said of the Teton Mountains in 1938, "Irrespective of hour or season, whether on clear days or stormy, the Tetons are so surprisingly beautiful that one is likely to gaze silently upon them, conscious of the futility of speech."[36] Second, the conservationist, Olaus Murie, who said in 1940, "Jackson Hole is not merely a ski-piercing range of mountains [...] [I]t is a country with a spirit." Finally, the landscape photographer Ansel Adams in 1950 said, "A grand lift of the Tetons is more than a mechanical fold and faulting of the earth's crust; it becomes a primal gesture of the earth beneath a greater sky." The wonders of the GYE evoke sentiments and meaningful connections in many people. These feelings are embodied in our stories of meaning of the region and in terms of our individual and human cultural ecology.

Modern and postmodern humans

The Murie and Craighead families have become especially notable for their contribution to conservation of the region, and their accomplishments and legacies are hugely important. First, the Muries held a principled commitment to wildness and wilderness. Olaus Murie's vision for habitat preservation placed him at the forefront of wilderness conservation in America. Mardy Murie, his wife, was awarded the Presidential Medal of Freedom in 1998 for her lifetime service to conservation. Shortly after Olaus' death in 1963, the Wilderness Act was passed.

The Craighead family is also well known for their conservation accomplishments, especially their pioneering research. They pioneered a study of Yellowstone grizzly bears in the late 1950s and 1960s and radio-tracking of wildlife. Frank went on to write *Track of the Grizzly* in 1979 and, more importantly, introduced the concept of the GYE for grizzly bears. This is the first time the concept and word "ecosystem" was used in relation to the Yellowstone region. The Craighead Institute thrives today under the direction of Lance Craighead, and its goal is to sustain both people and wildlife in the region in healthy coexistence. The Craighead family continues advancing knowledge and education about the GYE.

The Rockefellow family was key to the expansion of Grand Teton National Park, Wyoming, and public lands in the region. Their initiative and leadership are exemplary.

The Muries, Craigheads, Rockefellows, and others strongly supported the Wilderness Act of 1964 and many other common interests.

Today, there are many other people doing good work for the GYE. Among them are Bert Raynes, a tireless conservationist and inspiration; Deb and Susan Patla, who study amphibians and swans, respectively; Lloyd Dorsey, with decades of advocacy work, strives for better wildlife conservation; Doug Smith, whose work with Yellowstone's wolves is cutting edge, also works for a better situation; Todd Wilkinson, writer and force behind *Mountain Journal*, is another leader; and Andrew Hansen, whose important "vital signs" publications have improved knowledge in the region.[37] Others work determinedly as citizens, in nongovernmental organizations, at universities, and also in the agencies, most of whom go unrecognized for their valuable work. All are committed to the idea of YNP, GTNP, wildlife refuges, wildness, the GYE, and the health of our biosphere.

Promises to keep

We humans have set goals over the years for the Yellowstone region. Goal setting is part of our human ecology and culture too. These are promises to ourselves and to future generations. Goals tell us about who we want to be and what we want for our children. By making commitments, we tell the world what we value and what we are willing to work toward. Some goals were set long ago. Basic, constitutive promises were made at the time of the birth of the United States about the kind of society we should be. Newer goals are still being clarified. How they are to be accomplished in our rapidly changing world is contentious. Goal setting at the national level is a major part of the story of collective meaning making of the GYE and nature. Many regional and local goals, given the diversity of people in the ecosystem today, make policy contentious.

Goals from history

The American public made a promise in 1872 and designated an area on a map where that promise would be met. This promise was one of self-restraint and conservation of the natural wonders in YNP. This promise was rooted in the cultural story of ourselves that prevailed at the time, and it continues to serve as a powerful story that helps ground our meaning and place in the world. In 1832, George Catlin described it well when he said, "What a beautiful and thrilling specimen for America to preserve and hold up to the view of her refined citizens and the world in future ages! A nation's Park, containing man and beast, in all the wild and freshness of their nature's beauty."[38]

We have reconfirmed this promise in YNP and in other areas around the country for nearly 140 years. However, this promise is threatened today by environmental degradation, anthropogenic ecosystem disturbance, and cultural change that suggests we may be abandoning those long-held goals. Take, George B. Hartzog Jr., director of the National Park Service in 1964–72, who observed:

> With the establishment of Yellowstone National Park, the Congress gave a new dimension to the conservation goals of our Nation. This clear dimension provided that portions of our

great natural heritage should be set aside and managed for other than immediate material gains and riches. This policy is now firmly imbedded in the public land management policy and programs of the Nation.[39]

This represents a hope for our future. But new challenges are emerging in addition to the growing issues of the past. We will have to overcome them, if we are to live up to past promises about democracy and the conservation of national parks, wilderness areas, and public lands in the common interest.

In 1872, the Congressional Act establishing YNP said the region was "dedicated and set apart as a public park and pleasuring-ground for the benefit and enjoyment of the people."[40] The park is open for all people and is a shared part of our cultural story. This story and the experience of YNP and GTNP compels people today to visit them on a quest for a supreme experience. Even today, 140 years after its founding, visitors are still able to experience the sublime and astounding features that made YNP a part of our story. Take the following, as noted by Newton B. Drury, director of the National Park Service from 1940 to 1951:

> The gleam of glaciers on a mighty mountain; the shimmering beauty of a lake indescribably blue, resting in the crater of an extinct volcano; the thunder and mist of water falling over sculptured granite cliffs; the colorful chapter in the Book of Time revealed by the strata of a mile-high canyon gashed by a rushing river; the sight of strange, new plants and animals living in natural adaptation to their environment and to each other; the roar of surf waging its eternal battle with the land; the silence that hangs over the ruins of the habitations of forgotten peoples; the lengthening shadows of the towering Sequoias—these and a thousand other vivid impressions are at the heart or the experience that national park visitors travel many miles to seek. All else that they do or that we do in the National Parks is incidental. If we can remember this, we can remain true to our high calling as trustees for the greater things of America.[41]

Perhaps, these goals are all utopian to some degree. When our nation's founders created the country called the United States, they presented an idealized view of the goals for society and for democracy far into the future. The Declaration of Independence and other documents (such as the Gettysburg Address by President Abraham Lincoln in 1863) shifted and reconfirmed our culture's basic beliefs—its doctrine and formula. These basic philosophic beliefs fundamentally speak to our sense of individualism, liberation, respect, freedom, individual and collective responsibility, and much more. These are themes that we struggle to maintain today in our country and in the GYE. Past generations left many public goods (e.g., parks, forests, lakes) that they created not only for the present generation but for future generations as well. We all enjoy these goods—hopefully our children will too.

Goals for the future

Speaking about his travels in what is now YNP, Ferdinand Vandeveer Hayden said in 1871, "We pass with rapid transition from one remarkable vision to another, each unique of its kind and surpassing all others in the known world. The intelligent American will one day point on the map to this remarkable district with the conscious pride that it has not its parallel on the face of the globe."[42] When Congress envisioned a future for the region, it made the promise to conserve it for future generations.

In turn, we have made promises to our children and all the people who will come after us. There is an obligation of our present generation to ensure that past promises are kept and that this important business is not left unattended. Many of us have children or may go on to be parents. We may have grandchildren. Even if we do not, many people will come after us. Thus, concern for present people implies a concern for future people. Humility ought to lead our present generation to act with prudence toward the well-being of future generations.[43]

President Lyndon B. Johnson said, "If future generations are to remember us with gratitude rather than contempt, we must leave them something more than the miracles of technology. We must leave them a glimpse of the world as it was in the beginning, not just after we got through with it." Perhaps Marston Bates, the famous conservationist, said it best:

> The world is full of interesting things. We live on 10 inches of topsoil. All plant life is nourished by it. Below is rock. If the topsoil blows away, all vegetables and trees will die. If they die, the animals will go. Including us. The earth then will consist of seas and deserts. In defying Nature, in destroying Nature, in building an arrogantly selfish, man-centered, artificial world, I do not see how man can gain peace or freedom or joy. I have faith in man's future, faith in the possibilities latent in the human experiment: but it is faith in man as a part of Nature [...] faith in man sharing life, not destroying it.[44]

Reconfiguring history

The historical account above is seen through the lens of Euro-American colonization of native people's lands, especially by younger generations today. This view of history, reality, and the value of individuals and indigenous people compels all citizens to see history in a fundamentally different way than history books have traditionally told us about it. This recent "critical theory" view of history sees Euro-Americans as genocidal, exploitative, and callous to native people's prior presence in the lands that we now call America. There are many documented accounts of indignity heaped upon native peoples by killing them, taking their lands, destroying their ecology and culture, and forcing them on to reservations and dependency. This history is well known.[45] It is appalling by any standards of human dignity that we hold today. This fairly recent colonization view of history has come to the forefront, for example, in many younger people's and some academics view of history and the human condition.

The colonization and decolonization narrative growing in popularity today raises questions about how we should conceive the relationship among people's and political communities, historically and presently. It raises deep ethical questions about how various humans draw on their philosophies and religious beliefs, including notions of obligation and progress—their mental level—to make meaning. We are clearly in a time of reevaluation of history, human rights, and justice. On many university campuses today, for example, there is much talk about diversity, critical theory, and revisionist history.

Diversity is becoming a placeholder for a new view of history and justice. It is used by some people as a yardstick by which history, people in other times, and present people are to be measured. This view raises questions. Among them, do human communities exist for the cultivation of diversity, however defined, as the highest and most valued outcome communities should perform? If this is the highest moral ambition of the state today, what should modern people do to address past and present wrongs to native peoples? Our current responses are clearly not adequate from the point of view of Indians and many Euro-Americans who hold the diversity worldview.

Goals for today

We share the planet with other life, but are we responsible for it, at least in part? Are we responsible for nature, the overall health of the environment? What story of meaning should we tell ourselves, in fact need for today and the foreseeable future, if we are to thrive as a species and culture? Our present goals should reflect our human ecology in relation to the rest of the ecosystem and biosphere, I argue.

In thinking about goals today, Wendell Bell noted that life does not stop at our skin. Although we live in an anthropocentric world, the reason to be concerned with other life goes well beyond human-centrism. Humans ought to value the welfare of animals and plants based on their intrinsic worth. As environmental philosopher Baird Callicot says, "Our welfare is shared by the welfare and freedom to some degree with all other life."[46] Everything in the universe, including us, other living entities, and nonliving materials, had a common origin in the exploding star of creation. We share some degree of identity with everything in existence. When we die, the elements that make up our bodies are recycled into other living forms—future plants and animals and other humans.[47] We must remember that we are all part of the same great biotic enterprise on planet Earth.

These goals speak to securing the continued processes of living in a world that is both old and complex. We want to offer our children and future generations a better world full of nonhuman life and dignity for all. Currently, we are wavering on many of our past commitments, as we labor under unprecedented growth and a host of new demands about how the GYE should be used and we have to make decisions. Ongoing and foreseeable challenges in the GYE will test our resolve and capacities. It behooves us to live up to these goals.

Conclusion

This chapter uses an ecological lens, as well as a historical one, to view the GYE and human ecology and culture. This view shows us that the GYE is a huge, open system that is complex and highly dynamic. In Chapter 2 we saw that two of the most significant dynamic system processes are predator–prey relations and large mammal migrations. Other key processes include photosynthesis, succession, weather, and climate. These systems are all changing in response to significant and growing human influence. Today, we understand more of the evolutionary history of this system, the coevolution of many species and communities, and the patterns of disturbance and renewal playing out. Yet this knowledge is little used in policy or education today.

The ecological and historic lenses offer us a path to developing a new relationship with individual animals and populations, their ecologies, biotic communities, and their environments—the GYE. Conflicts arise and persist today among humans because we are not in agreement about the goals for the GYE or how to achieve those goals. There is presently no effective means to coordinate all management policy activities for the common interest. We have an institutional system that is highly fragmented, conceptually limited, and reactive to change and conflicting human interests. We must start thinking more seriously and fundamentally about these dynamics in an effort to escape our current predicament—conventional thinking and shot-term, self-interested behavior. This whole book and the case it makes for a new conservation story might prove the way forward to sound democratic process, sustainability policy, and the common interest outcomes in the GYE. At the least, the GYE offers us a learning opportunity to try to achieve those goals.

Chapter 4

BOUNDARIES AND CONTEXT

The earth was formed whole and continuous in the universe, without lines.
The human mind arose in the universe needing lines, boundaries, distinctions.[1]

Donella Meadows, Global Citizens Columns, 1978

"The fate of Yellowstone National Park's wildlife hinges on lands and events far beyond the boundaries," noted Chuck Preston of the Buffalo Bill Center of the West in Cody, Wyoming.[2] He is just one among a growing chorus of voices making the same point—the GYE is an open system that is constantly in interaction with surrounding areas and as such, a large view of the context is required to effectively attain our conservation goals in the region.[3] The need for a contextual view is valid for all land and seascape systems worldwide, whether parks, reserves, wildlife refuges, or public lands. Context considerations are often underattended to in management and policy.

This chapter looks at GYE's context and allows us to see how the GYE is situated or nested at various scales. In doing so, I target the concept of resources. The GYE is a resource, but what exactly is a resource, how can we grasp GYE resource dimensions, and what are the views we hold about resources? These are more complex questions than one might at first think. Our answer determines what we do to and for nature and wildlife, and ourselves. As we will see, our society has a built-in view of resources that prefigures how we see and exploit nature. A new story of the GYE would change that old view of resource and context to one that truly permits coexistence, sustainability, and adaptability.

Looking at the GYE as a resource, both as a raw (natural) and cultural resource, over varying scales of space, time, and complexity is a subject that ranges from concrete matters (e.g., old faithful) to conceptual and philosophic matters (e.g., what is a resource, who gets to use resources for what purposes, and what are GYE's cultural resources and nonconsumptive values?). Recall that what a resource is changes over time. For example, at one time oil was not a resource, nor was atomic energy. The terms resources, boundaries, and context are notions or linguistic concepts humans made up, so we could use nature in an orderly way. At one time long ago, we did not have those words or ideas. It seems that now is the time for us to come up with new notions and linguistic concepts to guide our use of the GYE.

Resources allow us to create our material culture, and we have come to an unquestioned worldview in our culture about resources and nature. This fact, as we think about GYE's future, brings up matters of scales and interconnection of all kinds of resources, often with indirect connections and time lags.

The established view of resources is being questioned presently because of the climate crises, extinctions, and degradation of ecosystems. By taking a scalar or geographic and philosophic lens to GYE's natural and cultural resources, I hope to highlight the importance of realistic contextual understanding of our situation, come to view resources perhaps differently than we now do, and lay the foundation for a new story of the GYE— our relationship to resources and nature. This chapter focuses on concrete resources and also ranges to basic concepts about our view of nature, our worldview of resources as material objects for our use, and how we learn about and use resources—both raw (natural) and cultural (our mental level, institutions).

Understanding the context of GYEs is crucial for understanding what happens inside its local and regional boundaries and also how it is influenced by forces and factors out to the global context. As the biotic (e.g., elk) and abiotic (e.g., air) move in and out of the ecosystem, as well as within it, they interact at different scales in ways that influence the structure and ecological processes within the GYE. For example, mule deer in the southern ecosystem live in a landscape with over 180 property boundaries that they must successfully navigate each fall and spring along their migration route. These human-imposed boundaries, together with natural obstacles such as rivers, foothills, and sagebrush basins, create a complex landscape for mule deer to traverse.[4] Conserving mule deer is therefore impossible without considering their interactions with this context. Similar contextual matters exist throughout the entire ecosystem and beyond for all species and issues, even the largest issue—the future of the entire GYE.

Resources and Dimensions

Resources, including informational resources, are the primary inputs for all our policy and management, as well as in our daily lives. Resources are essential for us to thrive as a society in healthy environments. There are two kinds of resources: (1) raw (natural) resources and (2) cultural resources. The latter includes knowledge, language, concepts, and skills of individuals and institutions that are developed over centuries and millennia. For those living in the GYE, some resources are nearby (e.g., wildlife, gas fields, forests, rivers) and easy to see. Other resources are geographically distant (e.g., rare elements used in computers and iPhones). Some resources are renewable (e.g., water, grass, forests) with careful use while others are limited and easily exhaustible (e.g., rare earth elements, rare species). Our current dominant worldview advocates for a consumptive, commodified relationship to resources to maintain and improve our material living.

Our survival depends on our understanding and use of resources of all kinds. The importance of a natural resource is determined by society's values at the time and people's expectations about how they can use resources to achieve their demands (e.g., living conditions). The GYE's natural and cultural resources are what attract people to visit or live in the region. For example, Yellowstone National Park's wildlife resources, such as bears, wolves, and bison, are attractions from which businesses profit. Not surprisingly, plant, animal, and mineral resources in the GYE fall across diverse and over overlapping jurisdictions (Figure 4.1). Yellowstone National Park is a cultural resource.

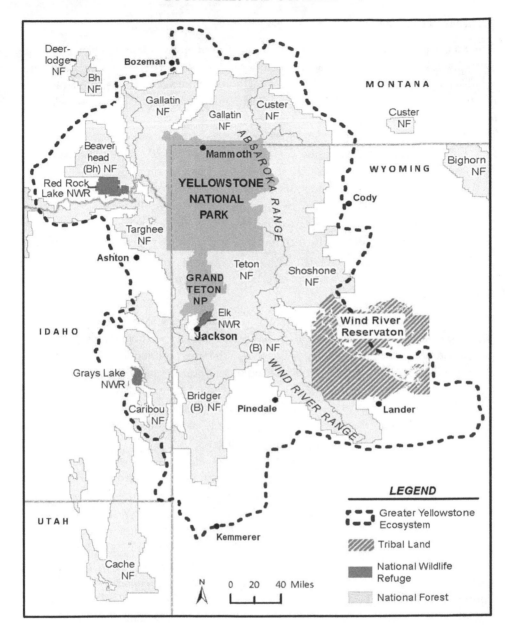

Figure 4.1 Map of the major federal agencies (National Park Service and U.S. Forest Service) in the Greater Yellowstone Ecosystem today. There are other agencies (Bureau of Land Management), state agencies (state lands), the Wind River Indian Reservation, businesses and corporations, and thousands of individuals and families who own or manage pieces of the ecosystem.

Natural resources

The term "natural resources" refers to the biophysical environment in which all human activities take place. In our modern society, "resource" tends to connote an extractive relationship. The chief problem that we face now in the GYE is how to interact with natural resources, including wildlife, sustainably. In most of the GYE, we use wildlife as an extractive resource by hunting.

Let's dig into the concept of resources and see how we think about and use resources. Our view of resources goes to the very foundation of our old story of the GYE, as a resource that we hold today. Seeing resources as concrete things (e.g., grass for cows, oil for society, wildlife for hunting) is a widely shared view today. In contrast, seeing our society's overall cultural formula to use nature is difficult, especially for nonconsumptive, nonmarket uses or values. To see our cultural formula requires that we rely on knowledge of philosophy—concepts and abstract matters—that is no less real, but harder to see and talk about in everyday conversation.

Resource categories

Humans often divide resources into categories such as forests, water, air, wildlife, range, parks, and so on, thus creating physical and cognitive boundaries (e.g., forests vs. grasslands, private vs. public, political vs. economic distinctions). These bounded compartments help us keep things simple and manageable. However, they also constrain our thinking and limit our ability to identify nuanced relationships. We can see problems throughout the GYE in most all management and policy because of this boundary drawing. And the consequence of that is bounded thinking. In turn, it causes problems for wildlife, among the agencies, and with the public.

This "in-the-box thinking" creates a world of "hidden events," where things happened "under the radar" out of sight until they become undeniable (e.g., chronic wasting disease spread and a needed response). In the GYE, we tend to focus on what is before us—immediate, our primary interests—while overlooking or ignoring our effects on other parts of the system, especially larger parts (e.g., climate change) and longtime periods.

Resources and their use vary over time and context. For example, YNP is a natural and cultural resource. Today, as such, it fulfills many of our core values: It is a place to learn about nature (enlightenment); it is a place to exercise and to appreciate (physical and emotional well-being); it is a place to enjoy with family and friends (affection); it is a place that supports numerous hospitality and ecotourism businesses (wealth); and it is a place to share in democratic ideas (power). Some of these values contradict each other and require trade-offs that are navigated through effective management policy process, if we are lucky.

One prominent example of this is that wealth from extracting oil and gas is inconsistent with biodiversity goals for the GYE. It is clear that some of our current practices are at odds with our overriding, most important goals for a healthy environment. Consequently, we need to align our operations (methods), perspectives (understanding), and practices (work) in a comprehensive systems-oriented way within the GYE. We are currently having difficulty doing that. We need to recognize and overcome our "in-the-box" thinking that is currently deeply institutionalized in the agencies and the public's thinking and behavior. We need to move on to more integrated ways of thinking and using resources, all resources.

Cultural resources

Cultural resources are the products and systems of human society, including tools, art, literature, architecture, language, and ideas. YNP is a cultural resource, as are agencies of government (e.g., the social security system, national defense, and the NPS). In some cases, landscapes are cultural resources, particularly in areas where human influence has been prominent and prolonged (e.g., agricultural regions). Our cultural resources are in constant "dialogue" with our natural resources, and our human context is inseparable from our natural context.

Yellowstone was founded as a sharable cultural and natural resource. How should it be shared? And, with whom? When? More and more people, all with special interests, want to use the GYE to their own benefit (e.g., techno-adventure recreationists, businesses, ski resorts). Some of their goals are designed for short-term value accumulation. We call these "exploitative." Others goals are carried out with long-term benefits in mind. We call these "conservation" and "sustainable."

Yellowstone and all of the national parks were designed with many long-term benefits in mind, but today, competing short-term benefits are calling them into question. We are constantly having to make value assessments about acceptable trade-offs for short- versus long-run benefits. Do we allow mushrooming development at the expense of wildlife migrations? What level of backcountry recreation is acceptable before the cost to wildlife and the landscape becomes too high? Such conversations are constantly happening in the GYE, largely in newspapers, promotional publications, in social media, and in agency and officials settings, including the courts. All of these conversations are deliberating the future of the GYE. It is these communication channels that form our views about the way we determine the fate of the GYE. Collectively, these communications and actions are what we call the "policy process."

Making choices about natural resources is a part of the overall human social and decision-making—the policy processes. Already, there are many specialized human social and decision processes focused on the GYE. Although all people value a healthy environment, life, and community, different people want to use GYE's resources differently and are in conflict with one another. Each group develops an understanding of acceptable resource use under the conviction that their preferred uses will best benefit their own special interests and, perhaps, the common interest. There is a massive struggle underway today over how we are going to use resources and who gets to decide. We all participate directly or indirectly in answering these questions and in this struggle.

Values

To understand what we do, it is essential that we look at values. What are values and how are they manifested in our lives? Professor Harold D. Lasswell, one of the most important scientists of the twentieth century, observed that in any given situation, human beings act in the way that they perceive would maximize benefits to them. This is known today as the "maximization postulate."[5] Humans strive to produce or acquire one or more of eight functional values: power, wealth, well-being, enlightenment, skill, rectitude, respect, and affection (no order implied):

- *Power* refers to participation in decision-making. People want to influence the decisions that affect their own lives and communities. Power is exercised in many forms, such as voting, decision-making, or deploying police or military forces for internal or external control.
- *Wealth* refers to the control of resources (natural and cultural). It affects production, distribution, and consumption of goods and services in society.
- *Well-being* includes safety, health, and comfort, both mental and physical.
- *Enlightenment* is the accumulation of knowledge. Information is gathered, processed, and disseminated in many ways. Scientists, journalists, and educators specialize in enlightenment.
- *Skill* refers to the acquisition and exercise of talents of all kinds—professional, vocational, or artistic.
- *Rectitude* is participation in forming and applying standards of responsible or ethical conduct. Rectitude is encouraged by ethical frameworks taught by churches, families, or other institutions.
- *Respect* is recognition, freedom of choice, and equality. Awards, honors, and everyday patterns of deference are practices associated with respect.
- *Affection* is love, intimacy, friendship, loyalty, and positive sentiments among family, friends, and communities.

We as humans need some measure of all eight values to live dignified, healthy lives. That said, each of us, according to our unique circumstances and personal and cultural context, place more importance on some values than others. We use resources for value purposes (e.g., for well-being).

These values are wide-ranging from eating harvested wildlife and using natural gas to heat our homes to the benefits of family vacations in the parks and exploring ethical and aesthetical relations to wildlife and nature. Our record of managing resources wisely to achieve values is decidedly mixed. There have been many successes—the establishment of the two national parks in Greater Yellowstone. There have also been considerable shortfalls, such as the slaughtering of bison historically. The GYE is potentially a shareable natural and cultural resource. Yet there is growing conflict over how to share it and who should decide. Management policy outcomes are determined by the content of the issues at hand, the context, the institutions involved, and both individual and institutional value demands.

To understand the eight functional values, imagine a mixing board of the sort found in recording studios. This mixing board mixes values rather than the sound of musical instruments. It has a sliding lever for each of the values that are calibrated from "absolutely must have" to "not so important." For each of us, individual stories of meaning determine the position of the sliders and, thus, our unique mix of values that we seek at any time. Although the sliders may change positions for different situations and over time, the mix will generally be consistent throughout our adult lives. Again, resources are used for value purposes.

These considerations and uses of resources, beginning with our values, have produced our society, and also our impact on the natural and social world.[6] Early on, in 1864, George Perkins Marsh, an environmental observer, first raised the alarm about our harmful impacts in *Man and Nature*,

> Man has too long forgotten that the earth was given to him for usufruct alone, not for consumption, still less for profligate waste. Nature has provided against the absolute destruction of any of her elementary matter. [...] But she has left it within the power of man irreparably to derange the combinations of inorganic matter and of organic life. [7]

Dimensions

The three dimensions of paramount importance in the GYE are boundaries, context, and patterns. They loosely correspond to three scales—space, time, and complexity. These are not things many people think about, but they are concrete, real, and play out in all that we value, do, and think. The focus on dimensions raises questions about what they mean for science, management, and policy in the GYE? And, also for our understanding of GYE's context? One application of this knowledge is restoration ecology (e.g., reintroducing wolves to YNP).

The three dimensions are concepts about the GYE that have hard concrete consequences in real life and in management policy. How we see and understand these dimensions influences what we value and values in turn influence what dimensions we account for. For example, people may not value a complex, long-term understanding of the ecosystem or their role and responsibility in it. Instead, people may only value the here and now and fulfilling the more immediate value wants.

Boundaries

Boundaries are everywhere in the GYE. Geographic boundaries are the easiest to comprehend. However, other boundaries include ecological, jurisdictional (geopolitical), and social, as well a host of invisible mental boundaries (personality, philosophic, policy—all about our mental level).

YNP predates the 3 states and 25 counties that make up the GYE today. These arbitrary geopolitical lines were superimposed on the landscape with minimal regard for the ecological and social processes at play in the region and cannot contain the entirety of such systems. Why were they formed and whose values or interests do they serve? As a result, the management and administration of these is disjointed and noncohesive. For instance, many contextual matters in the GYE do, in fact, scale up to the global level (e.g., climate change) and historic boundaries do not recognize this fact at present.

Additionally, personal boundaries also influence our management of the region. These include guidelines, rules, or limits that a person creates to set standards for what is permissible for themselves (e.g., where and when to hike or bike in grizzly country). We

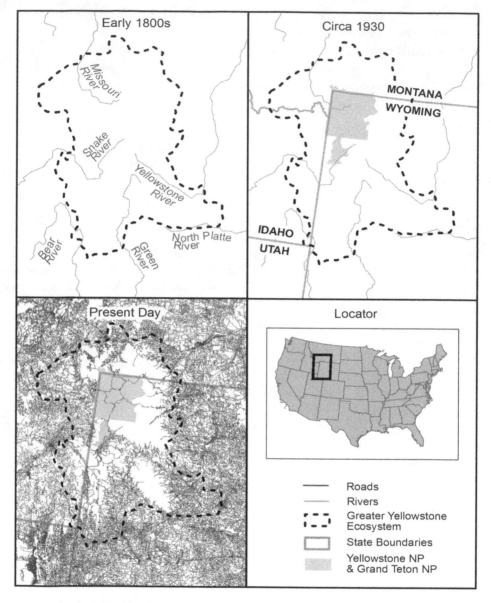

Figure 4.2 A map of some changes in boundaries, context, and patterns and processes in the Greater Yellowstone Ecosystem over last two centuries.

set standards and expectations for how other people ought to behave toward wildlife and ourselves, and how we will respond when someone steps past those limits. Overcoming problematic boundaries of all kinds—geographic, political, personal, and social—is the major challenge we face in the GYE today. This means that we need to pay close attention to what we are doing with geographic, jurisdictional, and functional boundaries in the GYE, as well as cultural, institutional, and cognitive ones (Figure 4.2).

The primary concern with boundaries is that sometime they cause disjointed management and policy. Boundaries impose divisions in an otherwise open, continuous GYE system that is naturally integrated across scales. Boundaries limit our ability to understand critical relationships between places and times, although they do also serve as a useful tool for breaking down the large-scale systems into manageable and more accessible units.

The relationship between spatial scales is of particular importance for understanding landscape ecology. Landscape ecology is the science of studying and improving relationships between ecological processes in the environment and particular ecosystems. This is done within a variety of landscape scales, developmental spatial patterns, and organizational levels of research and policy. The integrity of any piece of land or water is ultimately dependent on the integrity of the landscape surrounding it.

Conservation biologist Reed Noss, a professor at the University of Florida, recommends thinking bigger and more contextually, all the time.[8] His work on landscape contexts, environmental mosaics, species' requirements, and functional thinking is part of what is required to best manage ecosystems successfully. The lessons he offered are useful for the GYE from his extensive works on many ecosystems in North America:

- Many species require areas well outside protected area boundaries. Whenever possible, work with regional planners to assure that key pieces of habitat are functionally connected through habitat corridors or low-intensity intervening land use.
- Ecological history is highly informative. Which habitats and species have prospered and which have declined as the extent and intensity of human activity increased?
- Larger scales are often more meaningful for the conservation of biodiversity. Recognize that management may increase the number of species on one spatial scale while decreasing it on another.
- Do not treat all species as equal. Focus attention on those species most vulnerable to extinction from human actions, those that have declined most, and those that are most ecologically significant (e.g., keystone species).

Context

Context is our entire surrounding environment, including other people and wildlife, at all scales. It includes the interactions of people as they seek to achieve what they value through interacting with one another, society, and its institutions.[9] Context is a never-ending process. For people, the GYE is a value-laden entity as noted previously that people try and organize to solve important individual and collective problems of survival and community (i.e., natural and cultural resource problems).

Context is important for understanding any issue we are concerned about at the moment. A management policy that is appropriate in one place (e.g., a specific time, space, complexity situation) can lead to disastrous results in other contexts. Without contextual sensitivity, effective policies may never be developed in the first place. Context influences the processes that can be used to formulate management policy. Importantly, context changes over time and our systems should be adaptive and responsive to the dynamic world in which they operate.

Ideally and pragmatically, sound natural resource management policy calls for operations, perspectives, and practices that serve the common interest and create a system of practical meaningfulness. Common interests are interests sought and shared by the whole community, are those that provide people the mix of values they need for a healthy, sustainable future.[10] Evidence is mounting that we are missing a high common interest benchmark in the GYE, given how rapidly both the environment and culture are changing, many harmful (e.g., climate change, invasive species, new and old diseases, habitat fragmentation, stress put on wildlife from human behavior, and more).

To make matters more complex, it is essential for each of us to find meaning in the rapidly changing culture. We do this through our individual and cultural stories of self in the world and in the GYE. In some cases, striving to adjust to new changing contexts (e.g., climate, tourists, demands for affordable housing) can lead to anxiety and conflict because they force new meaning, interactions, and stories on us. For example, earlier the GYE region showed a more homogenous society than today. People who reside in the GYE can be clumped into communities of identification and interests. Many overlap. Some communities are made up of citizens, like police and firefighters and allies. Others are the service and maintenance people. Additionally, others serve tourists and recreationists. There are the tourists, visitors, and recreationists themselves. Government employees form another loose community. As well, there are other communities. Some conflict with one another for resources, access, and activities (e.g., wildlife conservation and recreationists in some cases). Finding shared interest and clarifying and securing common interests are becoming more difficult as time goes by. Oftentimes, we do not want to accept the new conditions and adjust our thinking and behavior accordingly.

Patterns and process

Over spatial, temporal, and complexity scales, we do find patterns and processes that are discernible. Pattern learning and recognition are critical, particularly patterns in management policy and society. By analyzing the GYE's problems at different temporal, spatial, and complexity scales, we can identify the linkages between scale orders that act as conditioning factors on the system ecology—natural and human. In turn, we can better understand how different participants in the ecosystem frame problems. Appreciating scales in flexible ways allows us to achieve our primary objective of stimulating further inquiry of the ecosystem's issues and finding shared interests. It also provides insight into the controversy and conflict over these matters in a way that is useful for public policy purposes in the common interest.

Modeling is one popular means to illustrate patterns for scientists. Modeling is an abstract, usually mathematical, representation of a system that is studied to gain understanding of the real system. Models are used to study system characteristics (ecological, economic, other). Using data gathered from the field, ecological relationships—such as the relation of sunlight and water availability to photosynthetic rate or that between predator and prey populations—are derived. These are combined to form ecosystem models. For example, there are models of elk foraging in Jackson Hole and of diseases on the National Elk Refuge.

Models are studied to make projections about the dynamics of the real system. Often, the study of inaccuracies in any model (when compared to real, empirical observations) will lead to the generation of hypotheses about possible ecological relationships that are not yet known or understood. Thus, models enable researchers to simulate large-scale experiments that would be impractical or unethical to perform on a real ecosystem. Ecosystem models have applications in a wide variety of disciplines, such as natural resource management, ecotoxicology, environmental health, agriculture, and wildlife conservation. For the GYE, there are many different kinds of models.[11] Some data-dependent models simulate animal populations, others habitats and connections. Other models look at human growth and its effect on the environment. Still other models look at climate change. We can expect models to be used in the future as we grapple with resources, scales, values, contexts, and management policy.

Overall, the concepts of scale and dimension are of paramount importance to understanding our resource–human interactions and the GYE. The overwhelming volume of data about resources and scales over historic time available to us today provides insight into our organizational and institutional systems, as well as our current value demands. We can use that knowledge to address GYE-level problems in the common interest. Are we currently prepared to address our problems by advancing constitutively toward systems-wide sustainability?

Worldviews about Resources

Our society's dominant worldview (labeled the mechanistic or progressive myth) permeates our entire culture and all its operations. We live in a culture in which this belief is accepted thoughtlessly. Part and parcel to this belief is the power of reasons in the form of positivistic science. These are accepted without question on a scale unprecedented in human history. These determine much more about our lives than we might be aware of. Under this covering worldview, which we take on unconsciously, we assume things exist for us to rearrange, consume, and discard as we see fit. We are the controlling mind in this worldview. Everything beyond our mind, it is assumed, is a material matter for us to do with as we please, or so it is assumed as an assertion, claim, and faith. This worldview perpetuates the false human–nature dualism that ascribes "nature" as something "out there" rather than as something that we are a part of. This false dualism encourages unsustainable resource use and represents a boundary to our thinking about and relating to the natural world.

Resource views

Materiality is an important concept as part of the dominant worldview of our culture as we consider the GYE and its future. The material world represents the things that occur for us to see and interact with. Seeing Yellowstone and the GYE as a material entity provides us a shared, conventional, and accepted way of seeing and understanding the world and our place in it. It gives us a sense of nature and our relation to it. It is a view that can be taken on without question. However, it also traps us, limiting our thinking to convention.

Yet we know now that our quest for individuality, dignity, and transformation continually invites us to see more than just the material (e.g., the conventional, literal, concrete in our lives). Each person works in some way to transcend the material world and to access an immaterial dimension, how even conceived. Perhaps this is why outdoor contact with nature is so important to so many people, whatever form it takes. This work of transcendence is the process of meaning making for all of us. If we stop and think, our stories of meaning about the GYE are the way we begin to move beyond a purely material view of the raw resources that in part make it up.

Material and immaterial

The GYE is more than just the wildlife and mountains that it contains. It is part of our national identity, a symbol for freedom, an idea about self-restraint and appreciation—a cultural story. The ways that humans manipulate the material world are interwoven in the entire social and cultural context and our stories of the GYE. Society and culture dictate why and how we manipulate the material world. It is also through the GIVEN of society and culture for most of us that we find meaning and the immaterial dimensions of life.[12] Materiality lies at the foundation of our current view of the environment and policy systems.

Sociologists often claim that materiality has two major faults. First is the tendency to reduce everything to its material, which is reification. Reification is the complex idea of treating an immaterial entity, such as happiness, fear, evil, and viewing wildlife or the GYE, as a material thing. Treating the immaterial as material is a way of transforming it into something concrete and, therefore, easier for us to understand. For example, a wedding, with its flowers, rings, and dresses, is the reification of a couple's love. In the GYE, we focus on the number of wildlife species, tourists, and business indices that we can count. We incorrectly use the number of visitors or gross income as proxies for the value of the GYE to us as people, given our myth. We have an obvious need to critique this dominant view in society for many reasons.

Critiques

The social sciences have critiqued materiality as a concept and worldview. That critique goes to the heart of our culture and present behavior. The discussion here may seem an academic digression, but I can assure it is not! How we see the GYE is at the center of all that we do.

Materiality physicists tell us the Newtonian world of the material is not reality from a philosophic standpoint, but it is only how we apprehend it.[13] Regardless, the material world is what reality is conventionally to most people and, thus, we live in a biosemiotic (symbolically constructed) world that has become the means to express our conviction that the immaterial also exists and has value. We use the material to make the immaterial (transcendent) easier to grasp and to attain. We construct temples and monuments to symbolize immaterial aspects of the world and our society, and we often claim to experience the immaterial through our interactions with these material things. These built objects idealize culture and its views—freedom, justice, dignity, and more. The things that we call Yellowstone National Park and the GYE are two of these monuments, too.

Many of us experience the national parks as material objects, without question or pause. If we can instead come to realize that the GYE was created out of human experience, history, and culture (semiotics), then we can begin to understand how that process and our experiences can change us. In fact, if we can grasp the GYE as a human phenomenon, then we will be equipped to understand the natural-cultural system we call the GYE.

Our societies and cultures are the product of the endless struggle to understand and comprehend the material world while seeking transcendence and connection with the immaterial. The public debate in the GYE illustrates this fact. The progressive myth tells us that our lives are defined and that the immaterial is attainable through material consumption. If a person owns a fishing rod or the newest pair of hiking boots, then she will be able to experience the grandeur of nature. Today, the central question for many people is this: Are we more than the material things that we have accumulated? Are YNP and the GYE merely interesting collections of material things found nowhere else? A place we set aside, can go, and see those material things (e.g., geysers) and make money off of it? Are YNP and the GYE nature?

History tells us that society's definition of humanity is largely dependent on our relationship to the material world. We define ourselves in relation to other entities, and in doing so delineate boundaries between us and them. In this case to YNP and the GYE, which are material to be sure, but they also evoke or carry a large dimension of immateriality or transcendence. This is seen in the individual stories of meaning in Chapter 1 (see Figure 1.3). So, on the one hand, we see the GYE as a special place full of interesting material objects (e.g., Old Faithful) and on the other hand, we experience YNP and the GYE as immaterial or potentially so.

Despite our efforts to find transcendence, materiality lies at the epicenter of our belief about the true nature of the world, including the GYE. Marxism and capitalism are built on this view of materiality. Both are praxis, strategies, and actions to exploit materiality for the purposes of accumulation and distribution of material things. Our culture created YNP and the GYE, and we treasure it today for multiple reasons, including its materiality. It is true that we created our understanding of the material objects with which we surround ourselves to meet our needs and meaning. However, this cognitive, semiotic, and cultural strategy is limited, though, as it creates a finite matrix for being human, our need for expression, and our potential. This finite view may be the root of the growing meaninglessness that many people are feeling in today's rapidly changing society.

Our sense of materiality, change, and meaninglessness gives us a feeling of growing incoherence. In turn, incoherence feeds more meaninglessness. Thus, we seek the immaterial and transcendence as a way to find meaningfulness of our lives. What we make of the GYE today, and the story of its importance in our society, is a way of dealing simultaneously with materiality and immateriality. If this is true, what will happen to the GYE and all the plants and animals there in the future as our culture evolves?

Accumulation

Today, our stance on materiality allows us to transform the world through consumption and production. We are now in the process of turning the GYE into a cultural shopping mall and a recreational destination. This is having harmful environmental costs to the GYE. This harmful transformation takes the form of threatened and endangered species, overcrowding, and developmental sprawl. Local businesses, and even state governments, are pushing hard for more materiality and accumulation in the region, all driven by the progressive myth. This behavior, and the mindset that comes with it, is a growing cultural and environmental problem that we face in the GYE today, especially if we are concerned about the sustainability of this ecosystem.

Our society also tends to reduce humans to the material. We define ourselves as objects. We also do the same to our social relations (epitomized in the prominence of youthful recreationists since the 1960s—the "fun hogs" and "motor heads").[14] Thinking fundamentally, our material culture is a linguistic device (semiotic concept) that makes up our view of materiality. It also makes us appear authentic—real. Just look at the glossy magazines in Jackson Hole that advertise the region's material nature with a focus on endless opportunities for fun, without including the consequences to plants and animals or other people. This pattern of living is a form of material hedonism. No one person or group is driving this trend. It is buried deep in our society and its culture (and the progressive myth). Yet, paradoxically, outdoor activity can be a gateway to genuine conservation and environmental ethics.

Objects and objectification

This social pattern that we are now experiencing is leading some people to see that things are out of control, destructive, and causing dissonance. Excessive materialism in economics, for example, leads to a loss of humanity and dignity. Some critiques take the form of "back to the earth movements," recycling, slow foods, energy efficiency, and more. It seems clear that materiality and anti-materiality (immateriality) concerns are evident both in our individual and cultural stories of meaningfulness in the GYE. Materiality is a subject central to all that we think and do. It is a notion that we need to engage with seriously, if we are concerned about the future of the GYE.

Process of objectification

Views on materiality typically bring us to a common-sense notion of "things" and "thingness."[15] YNP, GTNP, and the GYE are all "things," as are wildlife and other people. But are they really just things? Are these entities real things, ideological constructions, or symbols of something else (e.g., the immaterial)? I assert that the GYE is all these and more, a mix of material objects like plants and animals and immaterial concepts such as culture and language. Together, these entities help us find meaningfulness for ourselves.

Why do we argue so strongly over grizzly bears or the health of the GYE? A grizzly bear is real, but what does it mean to different people with different notions of materiality and immateriality, and objective and subjective reality? We typically seek meaning by taking a stand on grizzly bears, recreation, and a myriad of other subjects in the public sphere. This is important for us, as Arnold Arluke and Clinton Sanders, philosophers and animal rights advocates, note,

> Although animals have a physical being, once in contact with humans, they are given a cultural identity as people try to make sense of them, understand them, or communicate with them. They are brought into civilization and transformed accordingly as their meaning is socially constructed. To say that animals are socially constructed means that we have to look beyond what is regarded as innate in animals—beyond their physical appearance, observable behavior, and cognitive abilities—in order to understand how humans will think about and interact with them. "Being" an animal in modern societies may be less a matter of biology than it is an issue of human culture and consciousness.[16]

Philosophy of objectification

Considerations of the boundaries and context of our thinking and behavior (and our struggle to find meaning) are rarely appreciated or talked about in today's public dialogue about the GYE. Consider Georg Friedrich Hegel's, the famous philosopher, argument that there is no fundamental division between humanity and materiality.[17] His argument is that everything that we are and do comes from the reflection back on ourselves out of the mirror of reality, as the philosopher Richard Rorty described it.[18] All the images we use are from our cognitive and biosemiotic processes (brain and culture) by which we created materiality in the first place. In turn, we need to ask two questions: (1) What is created by our semiotic (language) process of making materiality? and (2) How has this sequence or relationship captured our conventional, everyday understanding of reality and the GYE today? What are the consequences for us and wildlife?

These questions are a complex matter, both philosophically and practically. Keep in mind that we create cities that are material. Cities, in turn, help create our sense of self and provide meaning through a sense of place and belonging. Cities are cultural objects (resources) that we made out of the materiality of the world. We create cities to represent our views and understandings of the material world. In short, whatever form or medium we choose to use (e.g., science, art, religion, culture, experience), we can only know who we are by looking into the "material mirror." Such matters are central in our look into all things GYE.

Manipulation and instruments

It is abundantly clear that we have objectified and instrumentalized the GYE within our conventional culture of today. We just looked at objectification regarding the GYE, but what is instrumentalization in terms of the GYE?

Instrumentalization

Instrumentalization is the process of making something an instrument or tool in order to achieve one's own value goals (e.g., profit, hedonism, gratification). Neoliberalism, among other things, is the dominant formula in our society that guides or promotes an instrumental use of nature and people. The famous French philosopher Rene Descartes in the 1600s introduced the idea of looking at the world and nature in a mathematical way—mind and body, humans and matter, people and nature, things and uses. With Descartes, this dualistic division affected our perception and accelerated our seeing the world as material, objectified, and then to instrumentalization, as we do today.

Instrumentalization has become part of public citizenship today. The German philosopher Martin Heidegger pointed this out when he said that everyone became "an object which you now can exploit."[19] Instrumentalization is another form of "boundary" drawing between the instrumentalizer and the instrumentalized. One result of these philosophic underpinnings is that we perceive the world around us as separate from us and that makes instrumentalization automatic and seem natural. Nature became a material object, something separate from us and, thus, something we could exploit. Not long after Descartes lived, the Industrial Revolution started; it put his instrumental philosophy into material action. We now have a record of the consequences.

Education

We are all educated and socialized within our culture. As such, we are conditioned to perceive and interact with the world in certain ways—conventionally. As our culture is about materiality, objectification, and instrumentalization, so too do we inherit a dominant cultural view that embodies them. We take our culture as "common sense." And, our culture has developed a system of education to transmit that view, typically without question. It is natural for any society to transmit its major beliefs to the next generation. As we go to school, maybe through to university, we gain the ability to see and understand the world and ourselves in it, consistent with the dominant culture. We become socialized to our culture's norms. We become skilled at reproducing the understanding accumulated by generations before us, who brought us to this place in our history. Philosophers call this the GIVEN.[20] We are a product of our culture's long-term processes—the GIVEN. Consequently, we bring this with us on our visits to the GYE.

Education is socialization and intellectualization of our dominant culture and its myths, beliefs, and formulas.[21] For most people, this education meets the needs of the conventional culture for reproducing itself. It enhances the capacity of the individual to participate in and extend that culture and its worldview. It typically does not encourage the questioning of its philosophical foundations, which includes coming to believe that the way we do things now is the way they should be done in the future. An unquestioning stance is central to our way of living, thus the dominance of conventional, everyday, concrete thinking and living. This kind of education creates the modern person throughout our entire society.

Much of the controversy and conflict in the GYE is whether the mainstream, common-sense conventional version of reality will continue to dominate, as it has in the past, or whether it can be modified or some other view can be adopted to bring us into a different understanding of the world and ourselves in it—coexistence. If we want to begin changing the dominant paradigm, it is important to adapt and transform our education systems. We must open ourselves up to critical reflection and criticism.

Agency (personhood)

Agency is the capacity of any living being to act in any given environment. In the context of our current use of the GYE and its likely future, can individuals escape the influence of the dominant material culture—the GIVEN? As the progressive myth took hold in our society, it called for instrumentalizing people and the GYE. This has huge implications for agency as a defining feature of human individuality (e.g., the "fun hogs?").

Let us consider American transcendentalist philosopher Ralph Waldo Emerson's famous 1842 essay on self-reliance, which is a powerful statement on agency.[22] This essay is about how we see ourselves and our responsibility in the world. His quotes are famous, for example, "Nothing can bring you peace but yourself" and "Be yourself; no base imitator of another, but your best self." Does convention limit our potential agency as individuals? Are we the only living beings with agency? How about animals? How do we express or assert our agency? In the GYE, we must seriously consider these questions and their consequences.

The reality of the world and the GYE is really a "hybridity" of self-constructed knowledge, beliefs, and interactions, including what the positivistic sciences tell us about the presumed real world. This hybridity is the basis of our own sensory experiences and agency. Our dominant worldview (the progressive myth) is the cause of our current unsustainability, everywhere and in the GYE specifically. Signals coming in daily from each dimension discussed above reveal our present unsustainability. People who want to use the GYE in more sustainable ways are running up against our entrenched culture— the GIVEN and its views on materiality, objectification, and instrumentalization. These dominant views are very deeply institutionalized in all of our social and decision-making processes and will be difficult to overcome quickly. Change for each of us will be difficult, but not impossible.

We can change, if we are able to engage more deeply with all the matters introduced here, and find our place in our communities and in nature (our authenticity and agency as stories of striving and transcendence). This personal human project of transformation is about each of us as individuals engaging with nature and each other in fundamentally new ways. Visiting the GYE can create a new consciousness for some people, providing an opportunity to reconsider their values and goals for themselves and the world. However, as noted by Upton Sinclair, the famous novelist, "It is difficult to get a man [or woman] to understand something, when his [or her] salary [and worldview] depends on his not understanding it."[23]

Greater Yellowstone Ecosystem's Context

The GYE is a complex and dynamic adaptive system. As such, its context is also complex and dynamic. The context operates at many different ecological and human scales, some local and others global, such as weather, suburban development, and changing political attitudes (Figure 4.2). All scales and dimensions discussed above interact synergistically, directly or indirectly in the real GYE.

The GYE's human context is especially important, as it determines what happens to the biophysical ecosystem. The human context is growing in diversity of participants and in competing demands on human uses of the ecosystem. The growing volume of people visiting the park and region and their different uses presently are fueling fragmentation and conflict. Consequently, it is important to map and understand this human context analytically and realistically.

The Greater Yellowstone Ecosystem's environment

Overall, the GYE's context is perennially controversial because of the many and growing differences among people's views on how the GYE should be used and who should decide those usages. Take grizzly bear conservation as one example. Some people feel we have enough bears now. Others see that there are too few bears given how most bears die now, at the hands of humans. Virtually every land-use and human behavior in the GYE is the subject of controversy.

We know that numerous contextual factors affect the GYE on scales from local to global. State-level decisions can affect federal conservation efforts across large landscapes for decades, as policy and institutions change. Major political machinations at the national level in the United States also affect the GYE. The region, nation, and world are indeed connected. This is nowhere clearer than in the GYE, especially if one takes a 30–50-year picture of the situation.

Social context

Encompassing a lattice of boundaries, the GYE's stunning mountains, waterfalls, and glacial lakes span parts of Idaho, Montana, and Wyoming. This includes seven federally managed national parks and forests (Figures 4.1 and 4.2). Perhaps, the GYE is the most significant landscape in the American West. Diverse human communities are nestled within the GYE. The human population, as well as many other contextual features, has changed dramatically in recent decades and is expected to change even more and faster than ever.

Over the years, my colleagues and I have documented the GYE's context, albeit incompletely. We have looked at the overall arena and analyzed many specific cases and broader, foundational issues that are at play in the system (e.g., leadership, constitutive policy process, case specific issues). The context presents a mix of interconnected biophysical, social, and decisional dimensions.[24] This whole context is experiencing great change currently—locally, regionally, nationally, and internationally (Figure 4.3).

Local Scale
Individuals – psychology, values, personality,
  adult development, ethics, knowledge, skill
  sociaiity, responsibility
Groups – "politics," interests, solidarity, power
Civic leadership, businesses, local populations
Owner or administrative patterns
Local and county planning
Front and backcountry uses
Relationship and responsibility for wildlife, other

Regional Scale
Regional and ecosystem "health"
History of culture and land uses
Human population and composition
Development patterns
Economy (neoliberalism),other

National Scale
Public and civic organization/order
Constitution, governance, leadership
Congress, Executive, Judiciary
Political parties and associations
Civility, cooperation, competition, other

International Scale
Global environmental-planetary process
Climate change – weather, ocean currents
Western civilization – Enlightenment –
  post-modernism
International relations – cooperation,
  competition, conflict
Science and technology, other

Figure 4.3 Types of contextual forces and factors affecting the ecological and social structure and functioning of the Greater Yellowstone Ecosystem.

What would a map of our current social process look like, if we could take a snapshot of it? Who are the participants. What are their perspectives, value positions, strategies, and preferred outcomes? People can be clustered into groups to uncover their value structures in the GYE. However, the ecosystem has many landscape, social, and political boundaries that cause innumerable divisions and management problems. Among these, our present social boundaries are particularly challenging.

Participant groups

I identified generalized social groups involved in GYE-wide activities and GYE's many local communities and issues. In reality, individuals span several of these groups at the same time. As such, this speaks to how complex people's identities and demands vary and overlap at the same time. All of these groups have individuals who are well meaning, motivated, and care deeply about the GYE and its future. However, their stories of meaning, beliefs, and behavior often conflict with one another. When people act based on their own particular stories of meaning about the GYE and their role in the world, they may conflict with other participants or groups. Generally, levels of conflict seem to be on the rise.

The first group is the "Traditionalists," the "Old West" people who prioritize the local and the mechanistic or progressive myth.[25] They tend to belong to families who have lived in the region for generations. They often rely on traditional job sectors, such as ranching, mining, or other extractive industries. Their view values individual rights and freedom, and is frequently aligned with conservative politics and policies. They perceive themselves to be marginalized by federal power and government. National- and state-level conservative political leaders capitalize from this view. Conflict is played out through diverse issues—local control, gun ownership, religious freedom, and social conservatism.[26]

Second are "Environmentalists," also known as the "New West." These individuals prioritize values of well-being (environmental) and rectitude (their ethical outlook). They tend to advocate liberal views and policies, and the progressive cultural formula, to manage resources. They typically promote protection of the public's natural resources and land base. Frequently, Environmentalists are inspired by scientific positivism as the solution to problems. They see the Traditionalists as an impediment to a better future. Environmentalists sometimes share similar values and perspectives to the "Outdoor Enthusiasts," as described below.[27] This second group also includes animal right groups (or "compassionate Environmentalists").

Third are the "Outdoor Enthusiasts." This group is very diverse ranging from responsible conservationists to little caring individuals for wildlife. This could be said for the other groups too, perhaps. The form of recreation they participate in is equally diverse. Some enthusiasts experience nature solo, others in groups, and some activities are human-powered, others are motorized. There is a lot of crossover across skiing, hiking, backpacking, fishing, hunting, boating, and horse riding, for example. Some enthusiasts use snow machines and other powered means to experience the outdoors. There seems to be less crossover of motorized use with biking, climbing, and paddling.

Outdoor Enthusiasts, or recreationists, prioritize well-being (personal) and outdoor skills (in biking, skiing, rafting). They increasingly come to the GYE from East and West coast population centers. Few become long-term permanent residents. Communities such as Jackson, Wyoming, Bozeman and West Yellowstone, Montana, as well as Cody, Pinedale, and Dubois, Wyoming, and other smaller communities provide the amenities they crave: access to public lands, ski resorts, bike shops, beer halls, and restaurants. These individuals tend to be more liberal and usually favor maintaining federal control of lands with open access for the activities they partake in. They often view traditional land users as exploitative and harmful, and as impediments to a healthy, postmodernized future. In general, Outdoor Enthusiasts know little about the region and seem little concerned about the consequences of their recreational activities on wildlife and other resources. Responsibility and self-management are seldom topics of deep consideration, at least publicly in the broad light of the whole community.

Fourth are the "Conservationists." These are diverse people in value terms, but in general, they value rectitude, knowledge, power, and skill. This diverse group overlaps with all the other groups.[28] The unifying element present in this group is a desire to protect the GYE, as it exists today. As such, this group is similar to the outlook of Environmentalists but is different in that it is based more on ecological and historic knowledge. Some people share a blend of the two views. Values driving these interests differ across subcamps. For example, a participant in the "Old West" camp may hope to conserve land in order to protect ranching livelihoods. A participant in the "New West" camp may hope to conserve land for recreation or business opportunities. Thus, both participants are, in a sense, conservationists, yet they vary in their approach to what and how to conserve.

Fifth are the "Agencies." Federal and state agencies share control of the GYE as a public resource. Collectively, they provide the superstructure for allocating authority and control in the region. Agencies show varying bureaucratic perspectives and a fixed decision process (i.e., policy preferences, standard operating procedures). These agencies, though not necessarily their employees, tend to prioritize power and conformity over flexibility, openness, and pragmatics. Each agency has a history that comes into play when nominally promoting "collaboration" in the public interest. The U.S. Forest Service is the oldest. The first forest reservation was established in the GYE in the mid-1890s. First, the land east of Yellowstone National Park came under federal protection. Then, shortly thereafter, Shoshone National Forest was set up. The forest service prides itself with its professionalism and multiple-use management. The National Park Service was established in 1916, although Yellowstone National Park was established in 1872. Grand Teton National Park was established in 1929. The Park Service with its large protection mission grew out of the U.S. Calvary management in frontier days. Other federal and state agencies have different histories, organizational cultures, and missions. Their interaction has consequences.

In short, the agencies are often in competition with one another for public legitimacy. True cooperation is largely limited to technical, ordinary issues in response to specific problems (e.g., feeding elk on the National Elk Refuge). The Greater Yellowstone Coordinating Committee (GYCC) is an illustration of the above, showing numerous limitations to ecosystem-wide policy needs. Its public slogan is "Protecting the integrity of the Greater Yellowstone Area landscapes."[29] At present, there is much room for GYCC to engage itself with the complex dynamics of the GYE, its science, management, and policy, in an integrated, farsighted way.[30]

Sixth are the "Elites" (elected officials, judges, experts, business leaders, and super wealthy). While few in number, elected officials and the super wealthy hold the most influence. These individuals value power, respect, and wealth, seemingly above other values. Despite sentiments about the democracy of public lands, they control most of the decision-making processes in the GYE. Frequently, they cater to special-interest groups, even when they are in conflict with the common interest. Powerful local and regional interests, both private and public, look to the Elites to protect traditional power structures and social hierarchy.[31]

Seventh are the millions of diverse annual "Visitors." They come from all 50 states and many countries worldwide. The GYE is a widely known draw for millions of visitors annually. These people enjoy sightseeing, new experiences, and landscapes and comment on their adventures, photographically and in social media. Visitor numbers are growing, and local and regional communities are building more accommodations and recreational opportunities for them. Although this is the largest group in number, they hold very little influence on specific decision-making in the region.

Eighth are the "Animals," although most people do not include them in the set of actors in the GYE. There has always been disagreement about the way we should consider animals. As such, there is ambivalence over how we treat animals in the GYE. What is the legal, political, and social standing of animals. What should it be? These are highly controversial matters. Do we objectify animals and use them for fun and profit? A growing number of people are committed to an ideal of "humanness" for animals (e.g., animal rights groups, compassionate Environmentalists). How we think about animals, as well as ourselves in relation to nonhuman animals, is changing as society changes. This change offers hope that our inconsistent and poor treatment of animals may also be resolved.[32]

Further, some people see that animals and plants have an existence value, a right and claim to exist into the future. Some people even see that clean air, pure water, and healthy soil should have standing in our management and policy deliberations. To me, this is really only a matter of how much we want to take animals, plants, and ecological processes and health into consideration, as we use the GYE. Some of our uses presently directly conflict with the long-term health of the GYE. These conflicts need to be adjudicated through sound democracy, reliable scientific knowledge, and a practical ethical stance.

And ninth, there are "Minority Communities." These exist as such due to their size, history, and present political standing. These are First Nations People—the American Indian community discussed earlier. The largest resident native American population in the GYE is on the Wind River Indian Reservation in central Wyoming. Further west, the Blackfoot Reservation is in eastern Idaho. Other native peoples have standing in the GYE.

The Latino community is another community that works in the service industry, construction, and other professions in the GYE. Some Latinos are residents, others transient. Each summer, migrant workers from eastern Europe flow in to work in grocery stores and other businesses. As well, various minorities work in agriculture and diverse positions throughout the GYE for varying lengths of time. The African American community exists as individuals and small groups in the larger cities of the GYE and beyond in the region (e.g., Billings, Montana). There are also religious communities present. Some established themselves a century ago, such as Mormons. Other diverse formal and informal religions are present—diverse Christian communities, Jews, Buddhists, and others.

Overall, there is no comprehensive contextual accounting or map of the regional-to-local social context that identifies participants, their perspectives, arenas of interaction, strategies, and outcomes. Instead of an accurate and reliable contextual map for all to share, many of us have a complex mix of impressions about other people and groups. This gives us a scant and incomplete picture though of the humans in the region, visitor or resident, and what their connection to the GYE actually is. We are left with learning the context from various reports in the media, from agencies, and through scientists, public officials, interest groups, and others in fragmented ways. Much of the available intelligence about the context is limited to biophysical matters. Officials could gather systematic data on the human community, subcommunities, and their use of GYE's resources for all to use. Such data would be especially useful in management policy.

Regional context

The GYE is part of the tristate region composed of Montana, Wyoming, and Idaho, and is also part of the American West, which encompasses both a geographic space and a piece of the American imagination (Figure 4.3). This region faces a diversity of environmental issues.[33] Today, in addition to the traditional environmental issues of interest, there are growing and divisive conflicts arising among groups who seek to use resources and the landscape: "Outdoor Enthusiasts" (a very diverse group seen as part of the "New West") and Traditionalists (long-term residents, local peoples, and those personally knowledgeable of previous decades, the "Old West"). Many of the newcomers are techno-adventure recreationists, interested in biking, hiking, skiing, and climbing. These land uses can and do conflict with established uses such as agriculture. Agencies, leaders, and the public are struggling to address the rapid growth, growing competition, and magnifying conflict in the region (Table 4.1).

The dominant myth about the American West as frontier fuels individualism. The benchmark of this traditional Western view goes back to the mid-nineteenth to late twentieth centuries. At that time, the culture was dominated by values of an asset-holding class of rural ranchers and pioneers.[34] This view continues today and is portrayed as a highly romanticized image, popularized in literature and film of the American West. It is also heavily promoted by tourist bureaus in Jackson, other gateway towns in the region, and across the three states.[35] For example, state representatives influence federal agency policy through the president's cabinet and Congress to support local businesses and states' rights agendas.

The myth of the Old West dominates throughout rural Wyoming, Montana, and Idaho. People who hold this view presuppose that their values are grounded in tradition. Yet, newcomers challenge that view by their lack of acknowledgment and respect for this community. Throughout the GYE, a massive cultural shift is underway presently, reflected in the changing economies and demographics, and in the reallocation of power from rural to urban areas. The change causes anxiety as these shifts are translated into political and cultural conflicts.

Most of the GYE is contained in Wyoming. Wyoming is a largely agricultural state, with approximately eleven thousand ranches and farms averaging a total of 11,000 acres, but it is also notable for its incredible vistas and topography. The Continental Divide runs northwest to southeast through the state, creating immense mountain ranges, most notably the Grand Tetons. Here, it has been said, America is "high, naked, and exposed."[36] The geology and paleontology of Wyoming are unparalleled. It has poor soils for agriculture and a tough climate, but teems with world-class wildlife.[37] Importantly, Wyoming was

Table 4.1 Time line—major phases (in bold) in the human history of the GYE since Euro-American discovery and colonization.

1800–50	1851–1900	1901–50	1951–2000	2001–50
Exploration and Exploitation	**Colonization and Carving Up the Ecosystem**	**Natural Resource Exploitation**	**Modern Period (Mixed Views)**	**Postmodern Period**
Region used by 25 native tribes	Indian reservations established	1914–18 World War I	1939–45 World War II	Major cultural and environmental change (e.g., climate warming) underway
1804 on Euro-American exploration begins	1860–65 Civil War	Continuation of traditional resource exploitation	Great changes occurred at all scales 1959 Grizzly bear studies begin *1973 Endangered Species Act 1976 National Forest Management Act*	Expectation is continued growth, demands for greater access to public lands, recreation, business
1820–40s Fur trappers and traders	Massive slaughter of wildlife in west	Rural economies and lifestyles dominate	Population growth	Large population growth expected
1840s on—missionaries, surveyors, explorers, military	*1872 Yellowstone National Park* established	*1916 National Park Service*	Influx of businesses, adventure tourists, and more	All scales of context in flux (contours unclear)
	1890s Idaho, Wyoming, Montana statehood	192930s Great Depression	Origin of greater ecosystem concept	Domination of capitalism (neoliberalism)
	1893 Frontier declared closed		Agency responses	Commodification of the ecosystem
	Railroads—transcontinental		Environmental groups—formation	Constitutive and ordinary decision processes in flux
	Population growth, immigration, cities, counties			
	Massive industrialization nationally			

Selected federal legislation and policy, some of which occurred prior to recognition of the GYE concept, are in italics.

one of the last states to be settled and therefore it still maintains deep roots to notions of the frontier and has preserved vast expanses of open land. It has appeal to visitors and recreationists.

Wyoming is the least populated U.S. state, and the culture places high price on frontier ideologies like individualism.[38] The Wyoming Humanities Council's guide states that people who are new to the state will find that "both long-term residents and those who choose to live here cherish a deep attachment to the state. Wyomingites take pride in the 'Wyoming way' and strive to preserve the qualities that make our state special."[39] These traditional and "Old West" ways of life are changing quickly, however, as rivers are dammed, forests are cleared, agriculture is becoming industrialized, and urban and suburban development spreads across the landscape. These changes are cause for concern for many long-term residents of Wyoming and have sparked increasing conflict in the region.[40]

Local context

Although I could highlight any one of the counties or local communities in the GYE, I feature Teton County, Wyoming, and Jackson because they are the geographic center of the GYE. They show all of the many dynamics and conflicts evident elsewhere throughout the GYE. As well, Teton County is 97 percent public lands and contains a national park, wildlife refuge, forests, scenic rivers, and more, all close together. Finally, I feature the county and city because there is so much written about them compared to other counties and cities.

Teton County, Wyoming, with its county seat in Jackson, is a major center of the state's tourism. Jackson and surrounding developed areas make up one of the largest urbanized areas in the GYE.[41] Other growing urban areas are Bozeman and West Yellowstone, Montana, along with Idaho Falls, in Wyoming, Pinedale, Lander, Riverton, and Cody. Some smaller communities such as Big Piney, Kemmerer, and Meeteetse benefit from being in the GYE. Thus, it is crucial to understand the trends, conditions, and context of Teton County and Jackson if we want to see the big picture of the overall ecosystem.

Most residents of Teton Country embody the "New West" thought system. As such, they hold the people, amenities, and pleasurable activities as high priorities. The environment, scenery, and wildlife are considered part of the amenities, so while the environment is talked about frequently, it is not attended to or protected to the extent required to keep it healthy in the long term. Consequently, environmental quality and wildlife populations are eroding. The county's strategic plan states that the primary goal is to secure "a healthy community, environment and economy for this and future generations."[42] Given the current conditions, however, this goal is impossible to achieve because the county tends to emphasize the economy at the expense of the environment.

The dominant worldview in Teton County is similar to that of the broader regional and tristate area—a belief and commitment to economic growth using neoliberalism.[43] The county actively promotes tourism because of the easy access to its natural amenities. This strategy has led to sustained, but unsustainable, population growth.[44] From 1990 to 2000, Jackson grew 63 percent and from 2000 to 2010, it grew nearly another 20 percent.

Meanwhile, nontraditional immigrant populations, such as Latinos, have increased over 1,000 percent in that period.[45] However, while the region has traditionally relied on rural manual labor jobs in agriculture or mining sectors, the new population has instead congregated in urban areas and relied on the recreation and tourism sectors.[46]

The tension involved in the urbanization of the GYE and the many newcomers to the region is now a major political and social issue. Bozeman, Montana, is the epicenter for much of the tension, so to speak. Hannah Jaicks, a social scientist, noted that beyond Jackson and Bozeman, "the GYE is also home to several rural communities that are geographically remote. These communities still exist on the fringes of urbanized American society," and this is where the tension is magnified.[47] Dr. Jaicks noted that the implications of land ownership in the region are profound. For residents, the relationships between people are integrally tied to asymmetrical owner relationships. The government is the dominant landowner, owning over 50 percent of the GYE land base. The tension is over how lands should be used, which user groups should have the say, and, more broadly, who should be involved in decision-making about human uses. Every policy decision made by federal authorities significantly affects the lives of locals, many of whom harbor distrust or resentment toward our governmental institutions.

The GYE economy is currently dominated by economic expansion, population growth, and infrastructure changes, and these trends seem to be continuing. Over the last few decades, improved technology and travel infrastructure, most notably the regional airport in Jackson and Bozeman, created convenient access to the GYE.[48] The enhanced access to lands, combined with increased public amenities and outdoor recreation opportunities, has bolstered visitation to the region. The area now receives well over 4 million visitors annually; and from 1990 to 2010, the resident population in the GYE grew over 54 percent.[49] However, the new residents and visitors often do not reflect traditional demographics of the region. Many newcomers were either wealthy and are called the "amenity migrants" who purchase second homes or ranches and also the laborers they require to staff the service industry or "New West" liberals pushing Environmentalist agendas and outdoor recreation.[50] These ally themselves with existing communities and interests.

National context

Understanding the national context is vital to understanding the present situation and for shaping GYE's future. There is great social and political change underway in the United States, and the consequences of that change are unclear in the intermediate or longer term, yet it affects the region and the GYE.[51]

National system

One evident national trend is a rise in conservatism, which is also labeled populism and nationalism in the national discourse. At the time of this writing, the dominant political and economic trends continue to promote "American exceptionalism," calling for a return to the days of hegemonic dominance. There are two dominant elements of the national materialistic and progressive myth shaping the current political discourse everywhere, including the GYE. Overall, what we are seeing, I believe, is a major attack on our constitutive institutions and democratic processes that have existed for over two centuries. Our shared society and culture may be coming apart, unglued around formerly shared identity, goals, and civility. If correct, these trends and conditions portend a shift in the relative economic dominance of the U.S. economy globally, with consequences for all Americans in all aspects of their lives, including what happens to the GYE.

Globally there is much change affecting America. Our national culture and economy need to adjust to reflect this reality. For example, one study found that of 18 of 20 critical renewable resources worldwide, including maize, rice, wheat, soy, and animal products, all peaked in 2006.[52] However, global demand for food and water is expected to rise by 35 percent and 40 percent, respectively, by 2030.[53] These contextual matters are of major significance to the United States.

Cultural ideology

Ideologically speaking, the GYE is situated within three of the most conservative states.[54] Presidential elections, useful for gauging changes in the overall political orientation of the U.S. population, tell a story of increasing conservatism, a reverting back to the old mechanistic or progressive worldview. From analyzing differences between the 2008 and 2012 presidential elections, for example, we observe that much of rural America swung right, while only the urban centers held more liberal views, which often place higher price on environmental consciousness and social justice.[55] During this period, we have seen the rise of the Tea Party Republicans, Alt-right, and white supremacists (KKK, neo-Nazi, and more), fueled by anxieties over the many forces and factors of change beyond America's sole historic control. Ongoing social factors and elections may or not change these trends and conditions.

As the old gives way to the new political order, whatever that may be, entrenched interests coalesce around shared anxieties, conservatism, and other ideologies, which then push on our governing institutions (Congress, courts, media, markets, universities). This influence promotes a class of political leaders whose identities are forged by their representation of those who feel deprived by change—social and environmental. Certain interests have a strong voice in public policy, including in the three states of the GYE nationally. What is new, tough, and disturbing is that the common interest, which was once considered a functional principle of our democracy, is harder to ascertain (e.g., the wilderness study case at the beginning of this chapter shows this). We maybe fragmenting into many special interest camps. Time will tell.

Currently

These contextual forces and factors are playing out in the GYE in real time. At the national level, neoliberalism dominates the worldview of the elites (the progressive myth). The free market produces winners and losers in this contest. This creates problems as many believe that the market ensures that everyone gets what they deserve.

The United States is becoming more fragmented and ideological.[56] Some observers claim that there are now seven major regions: the Pacific Coast, Inland West, the Great Plains, Gulf Coast, the Great Lakes, the Great North East, and the Southeast Manufacturing Belt. Each region has large cities and its own unique history and social and political outlooks. If this trend continues, it will be increasingly difficult to manage and govern the United States together as a single cooperative entity. In general, the western states are asserting their states' rights views. Wyoming, for example, asserts its authority and control in most public lands and wildlife. It lays claim to decision-making and this proves problematic when it comes to cooperation and joint management policy.[57] To conserve the GYE and its wildlife, we must work cooperatively across scales, political positions, and jurisdictions.

In sum, the neoliberal story dominating our society (the mechanistic or progressive myth and formula) is part of wider global system that is producing significant environmental and social problems, including those in the GYE (e.g., climate change, biodiversity loss, injustice). When emphasis is placed on individuals and competition, we fail to consider broader consequences and to act in the interest of the common good. Recognition and emphasis of problems is fueling a second myth (the problematic or environmentally aware myth, which I discuss later), which counters or competes with the dominant worldview. This problematic myth seems to be on the rise in some localities, perhaps the GYE. The dominant view is so powerful that it can override science as a source of authority, as it often does, for example, through climate deniers.[58] We are living in a time of "later neoliberalism," according to some economists, which has produced unprecedented problems, stresses, and anxiety on our cultural system and for many individuals.

Global context

Finally, there is the largest context, and the key element in this context. We are all interdependent in most spheres of human activity on the planet, whether we know it or not. We live in a single planetary ecosystem—the biosphere—comprised of a broad range of social and ecological subsystems. The GYE is merely one of these subsystems, within the global context.

Global system

Several global trends are arising from globalization, including the reorganization of the economic and power system. These trends are exerting pressures on the planet and on the GYE in various ways.[59] First is more and more people will demand a greater share of raw and cultural resources on a finite planet. The ensuing population stress will place a heavy burden on all natural and cultural resources.[60]

Second is a factor that eclipses population growth and resource problems. It is the rise in material and energy consumption levels worldwide. Climate change will greatly affect trends.[61] Projected changes will affect all societies and their ability to produce food, secure cities from rising seawater, and withstand increasingly extreme and unpredictable weather events. This, in turn, might destabilize global life support systems.

And third, for now, a growing global middle class along with the super wealthy will seek out increasingly rare opportunities for recreation and connection to high-quality natural environments, such as the GYE. We can expect aggressive promotion of the GYE as a tourist and recreation destination, thus supporting economic growth and the commercialization of nature.

Responsibility

The emerging changes in the GYE, along with our growing knowledge of our environmental impacts, are forcing us to grapple with a new understanding of reality.[62] This requires all of us to rethink matters and raise questions about our responsibilities. Given regional, national, and global statistics, what are our new responsibilities that we need to assume?[63] This emerging data is forcing us to consider new approaches to progress, prosperity, and responsibility. To successfully do so requires new forms of social organization and collective action (see Figure 1.3).

Social and individual responsibility is a notion essential to advancing the public good in the GYE. We must reconsider the social role of leaders, scientists, officials, and citizens. We need to start asking what kinds of knowledge our educational systems should emphasize. How do we prepare people to understand and live in contemporary society and to actively participate in its positive transformation? And, what ethics and values should we transmit in the educational process? Performing a deep analysis of the complex constellation of societal needs at every level will allow us to rethink the current role of knowledge, power, and other functional values in society.[64]

The GYE faces complex contextual challenges ranging from the immediate local concerns to the circumscribing global dynamic. Our present unsustainable economic growth continues to fragment authority and control, magnify competitiveness, and encourage reactionary behavior. These, in turn, erode realistic problem orientation and democratic decision-making processes at the expense of human dignity and environmental adaptation. We all know that humans favor simplification, typically ignoring or overlooking the different scales of our being and interacting. We do so at our own peril, especially so under current trends and conditions.

Conclusion

The GYE is a shareable natural and cultural resource. Using a geographic, philosophic, and sociological lens as I do in this chapter gives us a scaled picture and brings dimensionality (boundaries, context, and patterns) to our understanding of the GYE. Lines and boundaries of all kinds—mental, geopolitical, administrative—have subdivided the GYE today. This partitioning has served to fragment the GYE. Consequently, this partitioning prevents the whole system from being seen as an interconnected, functioning, coherent system that is open to the rest of the world. In reality, there are many kinds of boundaries, besides the obvious geographic ones. They all limit our present understanding, knowledge, and cooperative conservation work needed for the future. The most important of these boundaries are the mental, conceptual, and philosophic boundaries in our mind and culture—our bounded rationality.

Practically, what happens in the GYE is constrained by our thinking and actions. In turn, our thinking and actions are constrained by society's dominant worldviews, philosophic views, and our behavior (i.e., the mechanistic or material progress paradigm and its supporting stories that make up culture). Our dominant worldview presently sees the GYE as a unique assemblage of material objects that can be instrumentalized and manipulated for profit (money) and fun (recreation). The most important boundary problem we face is not material, it is in our minds, our thoughts, and our conventional actions.

Equally important to us in thinking about the GYE's future is appreciating the context of the larger ecosystem. The GYE does not exist separate from its context. Typically, we fail to pay adequate attention to context when making management policy decisions. We need to reimagine ourselves not as entities separate from nature but rather as actors and participants in the natural world with responsibility for the environment and wildlife.

Part 2

PEOPLE'S STORIES

People, including our personal subjectivities and values and our shared culture and institutions, play a paramount role in all of the conflict and success in the GYE. The people level and our differing mental outlooks affect all that is Yellowstone, all the time. This overall subject is the most neglected dimension of Yellowstone's management policy. Currently, people are influencing GYE's present and future, guided by the stories they use to inform themselves. It is these stories about people and nature that are endless in variety and content. Our stories are about what we understand of the biophysical world we experience in the GYE. At the same time, our stories reflect who we see ourselves to be. People are at the very center of GYE's big story and what we make of it.

Consequently, we need to zero in on people, their psychology and meaning making, and simultaneously the larger societal context and policy processes that we are all part of. We need to do this, if we are to understand the GYE as conceived and conveyed through our individual and cultural stories and behavior, which are rooted in historical contexts. I want to zoom in on our individual and social selves, without losing perspective on the big, long-term perspective. How can knowledge and insights about our biology, psychology, and sociology help us make sense of current affairs and of the immediate challenges we face. Can philosophy help?

To get a realistic grip on people and their individual and collective actions requires that I bring in the humanities as well as social and integrative sciences in this part. Consequently, I draw on various disciplines that I have found helpful to understand the people dimension. My agenda here is somewhat global. I look at some of the major forces and factors that shape us as individuals and our societies all over the world. These are likely to influence the future of the GYE and the whole planet.

The three chapters in Part 2 look at what people do and why. We know that reality is composed of many threads. In this part, I try to cover different aspects of ourselves as people and society. I do not claim to be exhaustive. I intend to offer a selection of subjects. I do offer examples and cases throughout this section, many from the GYE, to illustrate what we do and why. However, your own experiences offer rich case material for you to mull over and I challenge you to reflect on your own situation and social interactions for additional case materials. Please do so as I present knowledge and insights about people from anthropological, psychological, and sociological traditions in the next three chapters. I do not conclude with simple perspectives or answers. My aim here is to stimulate further thinking and help readers participate in some of the most important conversations of our time.

To address the questions I raise, I examine what professionals in various fields have found about why individuals, groups, and society behave as they do. These fields seek to measure and explain the regularities of people, groups, and society (e.g., anthropology, psychology, sociology, history, philosophy, policy sciences). These fields of knowledge have built up centuries of scholarship and insight by observing human behavior and diverse cultures, and they are invaluable if we want to understand our own behavior and that of other people and groups. The knowledge and insights in the next three chapters may appear abstract, but I can assure you they have hard, real consequences in the GYE when we act on them. I invite you to approach these chapters with an open mind.

In presenting this material and relating it to the GYE, I raise the basic, foundational, and all-important questions that I introduced in Chapter 1 (see Figures 1.2 and 1.3). These questions are more conceptual or abstract than everyday conversation typically allow, but they are vital for us to ask and answer. We are in conflict with one another over our different experiences, stories, and answers to these foundational questions. For example, whose version or story of GYE's reality counts—the rancher, environmentalist, business person, or recreationists (ontology)? How do we decide (epistemology)? How do we determine whose are right (axiology)? Struggling with and comparing answers to these questions keeps us grounded (pragmatic).

The fields I draw from in Part 2 are in contrast to the natural (biophysical) sciences that I explored in Part 1, but if we are truly committed to conserving the health of the GYE, we need to understand both of these thought traditions, and begin to integrate lessons from both to put knowledge to good use in the GYE, as I argue throughout this book. This overview in Part 2 is directly relevant to GYE's future and conservation policy.

Chapter 5

CONTROVERSY AND SOCIETY

In conservation we have always assumed a dialogue between ourselves
and everyone else; a civilized adversary proceeding in which reason,
logic, and the meticulous argument, liberally laced with horrible precedent,
would persuade just men and women to our position.
We have invested enormously in that assumption.
Unfortunately for reasons and logic, for ourselves, and for wildlife,
it has not worked. One would like to know why.[1]

One evening in the summer of 2014, I was driving in Grand Teton National Park. In the fading light, I spotted a medium-sized grizzly bear eating in the sagebrush just 50 feet from the road and about 150 feet from a bike path. Fortunately, no bikers were on the path that evening. The next morning when I passed the same spot, the bear was gone, but several bikers were speeding up from Cottonwood Creek, only a short distance from where the bear had been just 12 hours earlier. What if those bikers had zipped past an unsuspecting grizzly? Would it have triggered a defensive response? Bikers have tragically been killed by grizzlies elsewhere.[2] Given the potential for and danger of such a situation, why is there a recreational bike path in grizzly bear country at all?

Debate about recreation in national parks is just one example of the long-standing controversy about what values take precedence in deciding how we use resources and how we relate to nature. Society is conflicted over what parks and public lands are for, what our own behavior ought to be, and especially about what our responsibilities are to wildlife and nature. Are parks and other public lands just places for human recreation or are they something more? What do our present attitudes and behaviors—our mental level—suggest about the importance of the GYE's wildlife and future to us? Let's examine the sociology of our situation as it offers insight into our society, its culture, and institutions, as well as our individual behavior and "mental configurations."

To gain insight into these subjects and the cases of interest to you, this chapter focuses on the most dominant species in the GYE—humans and our mental level. First, I look at controversy and conflict in society using a case study approach. I delve into some of the core beliefs (myths) in our culture, which are grounded in our doctrine (basic beliefs, largely invisible to people) that we use to justify our practices (the formula for our living) in those terms. I then discuss the many symbols we use to communicate our doctrine (beliefs) and formula (the how to live) for our lives, of which few people seem explicitly aware. Finally, I focus on our institutions, our civil society (family, associations, government, schools), and how these constitute and mediate our relationship to each other and nature. In doing so, I hope to shed light on a missing piece of the conservation question in the GYE currently, which should be about the sociological, cultural, or institutional dimensions in discussions about wildlife, appropriate uses of the system, and other management efforts.

Society and Individuals

Human social behavior (i.e., sociology) is at the center of the GYE's story, whether that story is personal or cultural.[3] And, it will always be so. Each and every case tells us this is true. Today, we contest with one another over what YNP and GYE mean, how we use them, and how we manage them. We often simplify these discussions down into conflicts about material resources like wildlife, water, or recreation space without considering how our deeply entrenched personal and collective values influence our goals and hopes for the GYE. We typically get lost among the details in diverse cases enmeshed in social dynamics, powerful symbols, and our personal self-interests. This makes it hard to fully understand what we are talking about and what is going on in each case, much less overall. As a result, it is harder for us to identify and work toward our shared and common interests.

Sociology and other social sciences (e.g., anthropology and psychology), as well as the humanities, help us understand the behavior of individuals and societies. Society is a form of collective consciousness. It is a kind of self-awareness built up over the centuries and it is still evolving. That collective consciousness is our culture, which is shaped by those who have gone before us trying to adjust to a changing biophysical and political world and which we continue to constitute through our beliefs and actions today—our mental level. Society and its institutions make up the context in which we live out our lives and struggle to find meaning. We as a society know a lot about our world. Individually, though, we know significantly less. This is evident in all cases about the GYE.

Society—who's version of reality counts?

Given the breadth of challenges that we currently face in the GYE, it is remarkable to me how little is understood about the sociological dynamics, context, or our society's beliefs, cultural formulas, symbols, or institutions, and how rarely these are considered in management and policy. Instead, we have focused our efforts on technology and biophysical levels and science as solutions to the multifaceted and broad-ranging problems in the GYE. We get caught up in the conflict and controversy that emerges from these social processes and search for technological solutions without giving much thought to the human root causes. Much of the sociology in the GYE is seen as "politics," as partisan and special interest battles over particular, bounded issues. Some of my comments that follow are generalized and may be viewed as too critical about these sociological subjects, but examples are in cases like bison management, the recent wilderness study area public collaborative case in Teton County, Wyoming, and the Snow King development decision in Jackson, Wyoming, among hundreds of others.[4] In order to resolve the conflicts that

I explore in this chapter and move toward a more sustainable future for the GYE and the people in it, we must come to see these individual conflicts within the big picture of biophysical and social realities.

Society and reality

We use society, culture, and beliefs as ways to organize our individual lives, to make meaning for ourselves, and to cooperate and contest with one another. As we live our lives at the individual level, we operate primarily in an immediate and affective way. We seldom stop and think about society, culture, and our beliefs or how they figure into our thinking, interests, and behavior.

Nevertheless, society influences our development, feelings, and actions in manifold, and often hidden, ways.[5] We belong to many subcultures. Different subcultures emphasize different values, including different definitions of the "good" life and of success. These subcultural differences play out in the GYE across the many diverse individuals and groups in play and thus make it hard to ascertain shared and common interest. Above this, though, our society shows stable patterns of operation and beliefs through its institutions. The most conspicuous institutions in the GYE are those of government and law and the many civil organizations that form a complex interconnected network. We as individuals comprise these institutions and confer authority to them in particular scenarios through our participation in them.

Our ability to change the world or the outcome of any specific case (e.g., elk, bison, bear conservation) rests on our understanding of broad social dynamics, and our own personal dynamics—our mental level. To come to a deep understanding of these, we must cross the divide between disciplines, between theory and practice, and between thinking and doing as best we can. We are interested in realistically understanding our society, its culture and institutions, and our presence in them. Recall the practical heuristic in Chapter 1 about *people, meaning (values), society (institutions), and resources (natural, human)*. Applying this heuristic in actual cases enables us to extend our knowledge of self and society well beyond the boundaries of convention, ordinary everyday thinking, and subjective individualism.[6] It allows us to functionally integrate.

Although we talk about YNP and the GYE as though they are just a concentration of things that are "out there" in the world (e.g., geysers, rivers, wildlife), they are actually artifacts of the way we make meaning through our subjective and intersubjective reality sharing (our culture) as we grapple with the material world. This is what we do as people engaged with other people about GYE's things, places, and processes. This does not deny the physical existence of the objects.

Sociology and reality

Sociology is the academic study of social behavior, including its origins, history, development, organization, and institutions. As a social science, it uses various methods of empirical investigation and critical analysis to develop a body of knowledge about groups and society. It studies social order, social disorder, and social change. Sociologists provide insights on society from decades and centuries of observations. They discern repetitious patterns in our behavior, and they offer grounded theories to explain what is observed. This information gives us knowledge, insight, and theories for managing challenges in our group living. We can all benefit immensely from possessing this kind of knowledge. Sociology can aid the management policy dynamics in the GYE on all of its many controversial issues.

However, there is very little active sociological work in the GYE today, although it is directly applicable to the challenges we face. The sociological window into the happenings in the GYE is especially helpful because it requires us to focus on groups (and subgroups) and interrogate how and why they interact and with what consequences. It requires all of us to dig below and behind the conventional facades of our living and conflict and get to the root of the foundational questions structuring our lives. However, it is seldom used by officials, pressure groups, or individuals at present. This is a real detriment to the work being done in the region. To use sociological knowledge, we must come to grips with our own complex psychology and sociology, which many people and organizations are loath to do.

Sociology tells us that what we think and do is merely a reflection of the groups that we belong to. The GYE is full of concrete examples illustrating this.[7] Given the critical role of group and social dynamics in the GYE, a sociological understanding of individual and group behavior would help us gain moral and political sensitivity to real situations. It would help us see prejudices, biases, and their consequences. It would reveal the real-time experiences of solidarity and community commitment. Finally, it would make clearer how to change undesirable conditions for the better. While this advice is rarely the silver bullet people hope for, it is the only way forward.

Individuals

All group behavior is a product of individual acts. The question before us is how best to understand ourselves. However, individual acts may be strongly influenced by the overriding behavior of dominant or persuasive individuals in a group, or by the majority of the group (e.g., group think). As individuals, our interactions typically have symbolic importance, and involve deep identification and personality investments. These acts involve the ego (self), and all are *interpersonal acts*. The acts can be passive or intense, substantive or cathartic. If we stop and think, we can see this is true.

As individuals and as group members, we all have a *perspective*. Our perspective stems from our *identity*. It sets up expectations that may be met or not, may be realistic or not, may be practical once acted upon or not. You can see this in cases in the GYE. Our expectations motivate us to make *demands*—expressions of what we want to see happen. Perspectives and demands are value statements that convey to other members of society what we value and who we see ourselves to be. Demands have both symbolic and substantive dimensions. We try to join our individual demands with others to increase their strength.

Further, we are sentimental beings, which influences our demands and expectations by influencing our group alignments (and everything else in our lives). Sentiments are about our felt emotional relations. Because most of our interactions involve sentiment, one way or the other, many people assume their interests are shared or not.

Overall, humans are extremely complex, both as individuals, as groups, and as a communal species. This complexity is evident in any GYE environmental issue, so when we consider the challenges we presently face, it is critical to consider the role of "self" and our interpersonal relationships in the demands that we and others are making. Conventional thinking focuses on the other, the "out there," rather than focusing on ourselves as the real issue. When this happens, we limit our ability to understand the conflicts we face, and this is contributing to much of the current unresolved controversy. In this chapter we will look at a few cases in the GYE, including conflicts over wolves, bears, recreation, and analyze the sociological dimensions at play.

Levels of society

We know that there are levels of reality, well below the surface culture that we see played out in everyday life. We must explore the value motives related to social life, individual life, and the institutional structures that cause people to behave as they do, functionally, in the deep culture, if we are to understand what is going on in interpersonal interactions and the controversy we see in society (Figure 5.1). There are many factors at play, some visible, but many not. For example, many of us use and respond to the news, social media, and cultural symbols (e.g., flags). We also draw on and use concepts, notions, attitudes, and word symbols in our communications. All of these social cues affect us as sentimental beings, whether implicitly or explicitly, and therefore affect our values, demands, and decision-making. Most cases in the GYE ignore our deep culture at play.

Our current understanding of society in everyday culture is a result of a particular kind of consciousness—conventional. Fortunately, high-order systematic thinking allows us to move beyond surficial, conventional views of culture to begin exploring how deep culture influences material culture, and how this plays out in the conflicts in the GYE. Our lives are full of social interactions, which are part of a multilayered web of meanings, identities, expectations, demands, beliefs, actions, and sentiments. GYE cases show this fact in varying ways.

It is the role of sociologists and other people to look beyond the conventional culture to examine these interactions from a functional perspective, one that allows us to understand more deeply how values and cultural prescriptions influence human behavior. This functional view permits us to understand the roots of the problems we face, and thus enable us to move beyond technical, misguided ordinary solutions toward multifaceted solutions that address problems at the systemic and constitutive levels. Overall, we urgently need to move to a more functional understanding of our actions and society in the GYE and the consequences of our current paradigm.

Individuals, groups, consciousness

What is the relationship between individuals and groups according to sociologists? And, what form of conscious awareness does our culture give us about these? What do GYE cases say about this relationship?

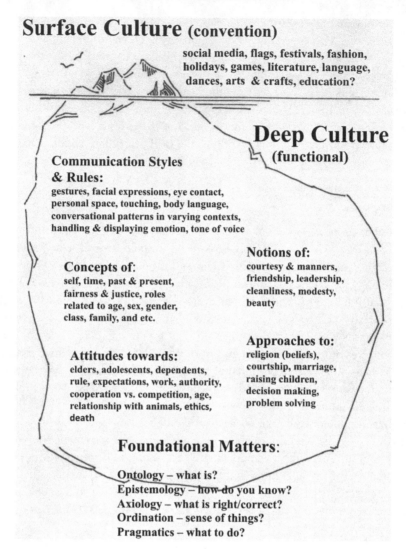

Surface Culture (convention)

social media, flags, festivals, fashion,
holidays, games, literature, language,
dances, arts & crafts, education?

Deep Culture
(functional)

**Communication Styles
& Rules:**
gestures, facial expressions, eye contact,
personal space, touching, body language,
conversational patterns in varying contexts,
handling & displaying emotion, tone of voice

Concepts of:
self, time, past & present,
fairness & justice, roles
related to age, sex, gender,
class, family, and etc.

Notions of:
courtesy & manners,
friendship, leadership,
cleanliness, modesty,
beauty

Approaches to:
religion (beliefs),
courtship, marriage,
raising children,
decision making,
problem solving

Attitudes towards:
elders, adolescents, dependents,
rule, expectations, work, authority,
cooperation vs. competition, age,
relationship with animals, ethics,
death

Foundational Matters:

Ontology – what is?
Epistemology – how do you know?
Axiology – what is right/correct?
Ordination – sense of things?
Pragmatics – what to do?

Figure 5.1. The cultural iceberg showing two realms of culture—the conventional, surface culture and the functional, deeper culture. It is vital to understand the deep culture because we live our daily lives inside conventional culture.

Models of individuals

Sociologists describe three models of the individual in society (i.e., prisoner, drama, puppet models) that help us think about ourselves in relation to the many issues about the GYE.[8] Specifically, these models shed light on how and why we interact with other people the way that we do. As you read the following section, I invite you to explore which model best fits your thinking around a particular issue. And, as you read daily newspapers, watch TV, or scroll through social media, you can learn to see which combination of these three models people are implicitly using to see themselves and other people as individuals in society. Let's look at these three models briefly.

The first model is the *prisoner model*. It suggests that we are trapped in the world inside our heads and inside our culture, even while we think we are dealing with the one full reality "out there." Are we really prisoners of our culture and personalized thinking? The fact is, we are only able to make sense of things by using the concepts that we communicate through—words and gestures. The concepts or symbols are our subjective "windows" conditioned by culture. Newspapers are full of "he said, she said" words and gestures (communicating competing views of the world, as seen through our different "windows"). Yet we know that the things that we see as "out there" in the GYE (e.g., bison, geysers, rivers) only appear to have an outward tangibility because we perceive them as apart from us in our seeing and meaning making. We reflect this presumed tangibility through symbols ranging from public monuments to T-shirts, photographs, words, and gestures. Stickers to save grizzly bear #399 and her family are one example.

There is no escape from this condition, which I call the "in-house" problem, as we live inside our heads, subjectively, with all the images we use to find meaning. To be sure, society gives us a (sub)culture, which is a value perspective and a cognitive outlook that we mostly share with one another. Usually we take on our dominant (sub)culture automatically and unconsciously. Despite many of us being imbued with the same (sub) culture, we do not all have the same beliefs and experiences or make the same meaning of them. In fact, quite the contrary. This is why we are all unique individuals and "see" things differently. However, even in light of all of our variations in experience and thought, we may still be—as this model suggests—prisoners of our society and, as such, unable to escape or change. In a sense, this is true, but it is a prison we have made for ourselves, a product of our thinking, context, and a long history of social process.[9] Perhaps this is why there is so much conflict in the GYE.

The second sociological model of individuals in society—the *drama model*—speaks to our emotions and mobilizes us to action. The GYE story is full of drama and emotion for many people. If you doubt that, follow any incident or issue for some weeks in the news and look at the emotive words used to heighten drama for participants. In the GYE, conflicts over grizzly and wolf conservation, migration corridors, business, and recreation are all dramatic issues for many people. That drama is simultaneously an individual, group, social, and political drama. It manifests in personal and group conflicts, celebrations, hopes, and tragedies. Most drama is really about social and political forms of interaction—good versus bad, right versus wrong, justified versus unjustified. The assessments and claims we make about forms of interaction speak to the personal nature of the issues for some people.

Drama can be exhilarating, exciting, and energizing. It can bring awareness to an issue and increase conversation. Perhaps emotion and drama are necessary for our own meaning making and sense of being. But, dramatic stories often evoke a highly emotional, egocentric response that can prevent cooperation and effective problem solving. In the GYE, much of the drama and conflict that we experience results from competition among the people involved. That drama is evident in virtually any GYE case you care to look at, including grizzly bear and wolf conservation, for example. The conflict in the GYE is really over whose stories will dominate and become reality. This is, naturally, a highly emotive issue.

Drama and conflict involve deep emotional involvement, typically in uncertain situations. The conflict over wolves, for example, is a competition between two dominant stories—the dominant Old West frontier myth and the emerging conservation narrative. Ranchers do not want wolves in the GYE because they threaten cattle and other livestock, while environmentalists support wolf populations because they serve a critical role in the ecosystem. These two stories about the value of wolves collide in the GYE. At the surface, this is simply a conflict over the number of wolves that we allow in the system, but underlying these desires is a fundamental question about whose stories of the American West will persist and become "truth." Is the GYE a place for frontier ideologies and sprawling Old West ranches? Or is it a haven of pristine wildness, abundant wildlife, and mixed landscapes among a sea of ever-growing development and recreation? Are these two images necessarily in conflict? Is there room for integration? This dialectic is a drama, and there are many other cases (big and small), playing out in the GYE, all the time. Often, the participants operate using conventional, everyday thought, which employs popular, superficial images and follows a prescribed reactive strategy for addressing conflict. However, if we are ever going to make meaningful changes in the GYE, we must move beyond the prescribed script and rewrite it to be more productive and more fulfilling for all involved.

As we learn more about human conflict in the GYE, we are increasingly able to sidestep or avoid social dramas in ways that allow us to genuinely address our differences. Writing op-eds or attending community meetings are some ways to move beyond the conventional drama and begin to move into a rational and cooperative problem-solving sphere. We can also avoid drama in management policy by being pragmatic, yet principled. Social creations such as the National Environmental Policy Act (NEPA), citizen science, and community problem solving are all tools that we have created to help us address complex issues directly instead of only addressing the superficial side effects of conflict. To do so, we must become systematically "problem oriented," as I will discuss later in this chapter.

This brings us to the third model of individuals in society—the *puppet model*. The puppet model asks how much awareness, freedom, and scope individuals really have. Are we all just puppets wherein the strings are pulled by society and culture? Or are we independent agents capable of recognizing the restraints and conditions of society and of making our own choices and opinions within them? To answer this question, we must focus first on individuals and group membership, because it is our individual and collective actions that determine what happens in the GYE and elsewhere in any particular case.

To further attend to these models and questions, I offer the basics of human experience. We all know that as humans, we have "subjectivity," a mind that continually runs in our heads. Each of us has sole access to our own thoughts. We are, more or less, aware of our own thinking, feelings, and desires. We are able to convey some small portion of these thoughts and feelings to others through words, art, and other symbols, but these are only representations of the actual world within our heads—our self-stories. Largely then, other people watching us have to infer our subjectivities, with some indication from our words and actions. Most people are unaware of how they occur to others, who also have their own subjectivity. This can get confusing as we interact over cases in the GYE.

Consequently, one is never certain about what is true, how they know, or if it is right, all of which are subjective matters (see Figure 1.3). This is one of the most difficult parts of group problem solving. To begin understanding the complex interaction between self and society, it is helpful to pay attention to how we think differently when alone or in a social group. What model do you see to be true for yourself, and what are the pros and cons of each model as it pertains to ongoing public interactions that concern you in the GYE or any particular case? Would the people who hold the same position as you on dramatic issues in the GYE agree? What about those who oppose you?

As we participate in groups in social processes we have the chance to see those roles in our own interactions. Watching ourselves and others in particular cases is a way to find out who we are and how much of our thinking and behavior is programmed into us by our group membership, society, and culture. Are we really the free individuals we often see ourselves to be?

For me, I range across the three models, depending on the case at hand and the opportunity available for me at the time to actualize my individuality and contribution. In some cases, I am forced to play the part other people demand (puppet). It other cases, I am powerless (prisoner). And, in some cases, I engage with energy and a sense of agency (drama). In reality, all three models are likely at play all the time in most people and cases in the GYE. Some people may adopt one role while someone else adopts another, even within the same conflict or arena.

So, how should we view our society, its culture, and its institutions overall and in the GYE? Our society is clearly a social drama that includes conventional traps, at both the level of the individual and the group. Conventional traps include fixed roles and rules that structure our behavior and allow us to operate blindly without explicit attention to the foundational questions (Figure 1.3). These serve to confuse and distract us from reality by prescribing certain ways of being or acting in the world. Fortunately, if we can recognize the traps of convention in advance of our actions, we might be able to avoid them. If successful, we will be better able to understand our own behavior and that of other people and to use interactive social processes most likely to help us integrate our differing perspectives and values in the common interest.

Groups

We are, without doubt, a communal species. We belong to all kinds of groups— families, communities, countries, religious groups, and more. Groups are formed when like-minded people with similar perspectives and values on a particular topic band together. Group membership provides mutual support, requires loyalty, and melds sentiments. Groups may be closed or permeable (as members circulate in and out).

Groups have both internal and external relationships that either accommodate or conflict with other groups. All social groupings have their own culture, which includes the norms that constitute the social order for the group. There are many hundreds of groups active in the GYE presently, from government agencies to NGOs to businesses to recreationists.

All of us have opinions that are formed both from our cultural contexts, our personal experiences, and the groups we belong to. When we as a society reach consensus of opinion, our views are taken as incontrovertible and authoritative—the "truth." Ideally, public opinion supports our general, shared interests, but quite often, individual, group, and organizational opinions reflect special interests rather than those of the collective whole. What then is true? What GYE groups offer what kind of opinions?

Some groups have *principle interests* (e.g., those that are rooted in ethics and explicit values) and others have *expedient interests* (e.g., those that a group or organization uses as an instrument to achieve its goals regardless of means). In reality, most groups have both. Organizations can also be highly coordinated with formal organization or be more flexible. Understanding group dynamics in the GYE is essential for high-order leadership to do its job successfully. The role of leadership should be to clarify goals, mobilize sentiments and loyalties, and engage groups and organizations in a cooperative enterprise to conserve the greater ecosystem using a flexible co-learning strategy.

Society as consciousness

Society is a tool that humans invented and built over the course of our evolution. We use society to facilitate social living and to enable altruism and cooperation. The creation of society is an amazing feat, given that we are very different individuals with our own self-interests that often compete with one another (i.e., self-preservation, tribalism).[10] All people want certain value outcomes that benefit them and their groups (e.g., respect, influence).

Society serves as a social contract, in which our culture prescribes norms of behavior and patterns of interaction that we call institutions. We abide by these norms, which are enforced by sanctions, a form of cultural punishment. We do not have to think about these matters. We are simply born into them and assimilate them as we develop. Our culture is GIVEN to us. However, awareness of the ways that society influences or limits our thinking through convention is critical when considering the state of the GYE and its future. The content of our culture, the GIVEN, greatly influences how and even if we can develop as individuals. It affects how we participate in any particular case in the GYE.

Society—Convention and Function

We are individuals with different perspectives and value demands. Depending on whether these perspectives and demands are shared with other individuals or groups, we either compete or cooperate. We cannot understand the cooperation or competition in our society, or in the GYE, without digressing a bit into controversy and politics. This is where stories and meaning making collide. We must look behind the facades of our everyday understanding of society into how we create collective meaning.

Influence, power, practices

Living in our society requires influence, power, and individual, group, organizational, and institutional practices. The extent to which a person has influence in their social sphere and beyond depends on the potential of that person to affect other people, groups, or institutions. Typically, influence is a function of how effective they are at distributing their values. *Values* are the goal events, things, or processes that we strive toward. Our practices show how we use values (Table 5.1), which are central to how and why we behave like we do. For this discussion, it is important to extend our understanding of the eight values introduced in the previous chapters and delineate between welfare and deference values. We can see these in all GYE cases, if we know what to look for.

Welfare values are those that are necessary conditions for the maintenance of physical and mental activity of a person. Values such as well-being, wealth, skill, and enlightenment are considered welfare values. *Deference values* are those that take into consideration the acts of other people and the effects of those people on yourself. They are inherently relational and cannot be attained without social interaction. Power, respect, affection, and rectitude are all deference values (Table 5.1).

Welfare and deference values always influence human behavior, although not always at the same intensity or in the same ways. No doubt different individuals and groups prioritize welfare or deference values differently. Fundamentally, our stories are about our values—they convey information about our demands, hopes, and identities. The societal importance of stories depends on the people involved, the situation, and the narrator.

Sociology and facades

In the GYE, sociology can be used to elucidate more about the ongoing conditions—the social context. Whereas convention limits our ability to thoroughly understand the actual challenges we face, examining the level of functional culture can be incredibly useful. It is the job of the sociologist to look beyond the rigid façade of convention to correctly identify and explain what is really going on.

In the GYE, many people are trying to do this in their own way often without explicit knowledge and insights from the humanities, social sciences, and integrative sciences. All the wrangling over grizzly bear conservation is but one example. Different people with different value holdings and demands are contributing to the public dialogue about how people affect the bears' health and future. There are thousands of letters expressing those views, seeking influence, and advocating for various outcomes. There are many hundreds of newspaper articles on the bear issue. On the surface, these letters and articles appear to be about bears, but in reality, they are about what people value and demand and, as such, reflect their identity, stories, and meaning making.

Table 5.1 The eight functional values that all people seek, desire, or require for the good life. Welfare and deference value interactions are always in play at all times.

Values (Human Dignity)	Definition	Groups/ Institutions	Raw and Cultural Resources
Welfare—Values Well-being Wealth Skill Enlightenment	Health, recreation Goods and services Talent, expression Knowing, research, communication	Mental, physical health Production, income, savings Training, occupations Education, discussion	Resources of all kinds are used by people and groups to get the values indulgences that they seek. Everyone is involved: Government Pressure groups Elected officials Businesses Individuals
Deference— Values Power Respect Rectitude Affection	Influence, decisions Standing, deference Ethics, moral, rightness Warm, friendly relations	Government, political parties Social class, honors Practices, standards Family, friends, groups	Resources of all kinds are used by people and groups to get the values indulgences that they seek. Everyone is involved: Government Pressure groups Elected officials Businesses Individuals

Modified from Harold D. Lasswell and Abraham Kaplan, *Power and Society: A Framework for Political Inquiry* (New Haven, CT: Yale University Press, 1950)

The whole bear controversy is less about bears themselves and more about people demanding a certain relationship among humans and wildlife. Those who want to see grizzly bears delisted are asking for a utilitarian relationship, where wildlife can be hunted for sport. Those who are asking for grizzly bear conservation are fundamentally expressing a desire for human–wildlife coexistence under a different formula. The bear saga is a social or policy process over value differences and whose values will dominate (see Table 5.1). There are many participant groups in the bear case: local and federal government agencies, sport hunters, conservationists, environmentalists, local residents, and more. Each group makes its own value demands, and the relationship among these demands determines whether certain groups conflict or cooperate.

When addressing the complex issues in the GYE, we are equipped with many possible *frames of reference* or lenses (i.e., psychology, philosophy, policy, conventional vs. functional). It is the role of sociology and the other social sciences to expose the real story behind the everyday conventional accepted views in the GYE today, regardless of the case. The integrative sciences (the policy sciences) could help in this regard too, if used to guide problem solving. Ideally, we will use these lenses to raise our humanness, humanity, and dignity in healthy environments, both in the GYE and well beyond in upcoming cases.

Controversy and conflict

Controversy and conflict are semiotic symbols (really linguistic symbol complexes) that depict struggle between opposing stories, values, and interests. Often, highly emotional thinking drives controversies and conflict. When considering the challenges and controversies in the GYE today, it is clear that the environment and the GYE are important to all of us, and we all have some emotional stake in the outcomes. However, there is a multiplicity of contentious voices. Many of us are drawn into controversy because it fits the prescribed narrative of a noble struggle over injustice, in which the weak are pitted against the strong. This conventional story line frames controversies such as wildlife conservation, migrations, or oil and gas development as a competition between two opposing groups.

A clash of beliefs (paradigms, worldviews, myths) explains the destructive conflict about elk and bison management in Jackson Hole, for example. We already have a rich case study of this issue. A government contracted facilitator, Dr. Emerson and colleagues from MIT, who are the authors of an unpublished situational analysis about this case, offered data from 130 interviews that speak directly to the basic dynamic that has led to unproductive conflict, a lack of public trust in government, uncooperative interagency relations, and an inability to address ecological problems.[11] The study's authors did not know about functional value dynamics as used in this book, but nevertheless their study is useful for examining value dynamics, at least in retrospect. The core problems in this case are differing myths (beliefs, formulas, and stories) about legitimate uses of the power held by different government agencies and an inability to reconcile them. No law, or science, can fix these kinds of issues simply or cleanly. Nor can any conventional effort.

Researchers who study natural resource management policy, especially those interested in integrated (via interdisciplinarity) problem solving, examine society, culture, and group dynamics using the policy analytic approach I recommend in this book, which includes looking at the functional value dynamics. For example, Dr. Christina Cromley studied bison management in the GYE for four years.[12] She documented the different myths at play, how they collide with one another, and how the myths affect people's interactions in bison management. In particular, she looked at the scientific belief or myth (i.e., about positivism, objective reality, and valid knowledge) and how wildlife biologists, veterinarians, and range managers all draw on this myth to justify what they do. That belief now dominates our society in the form of positivistic (biophysical) science. The myth that all rational claims can be proven by science and that we can use reason to arrive at an objective truth is accepted without question by most people and technical experts. Despite widespread belief in positivism, the bison conflict has not been resolved. Each of the participant groups in the bison case is drawing on their disciplines, grounding beliefs (i.e., myths), and group loyalties to arrive at very different conclusions about bison management. However, Cromley found that most of these people were unaware of their own cultural myth (story) and how it influenced their thinking, interaction, and demands.

She also studied grizzly bear conservation, in which there was profound disagreement over when it is appropriate to move "problem" bears, what the management zoning system meant, and what the relationship is of bears to livestock and other wildlife. She questioned whether the scientific myth that dominates the worldviews of many wildlife managers is actually the most effective outlook for addressing complex social and political problems. She asked, "How can biological science resolve or even understand the many problems caused by a myth clash in the power or decision-making process?"[13] She concluded that it cannot, but recognized that officials and land experts resorted to standard, conventional operating assumptions and programs, nevertheless.

Fortunately, Cromley suggested a way to break through the impasse to find an integrated solution in the common interest. She recommended addressing the normative (value) issues at play. This includes clarifying expectations and demands about the conditions under which moving or killing bears is acceptable and about related matters like who gets to participate in making those decisions. Cromley emphasized that creating more open and inclusive decision-making processes in which participants can listen to and address one another's expectations and demands is key. My own work over the decades across many different issues and conflicts in the GYE has over and over again demonstrated that this approach can create more open and inclusive decision-making processes. However, we continue using convention producing ever more controversy and conflict in their wake. As a result, bison management and so many other conflicts remain unresolved today despite the fact that we now have a comprehensive and practical understanding of the dynamics underlying them and a better way to approach and hopefully resolve them.[14]

A way out?

Is there a way out of the destructive pattern of drama, controversy, and conflict in the GYE? We may not be able to change society, culture, and its institutions, but we do have several options.[15] One way is to personally withdraw from the process. After all, detachment is a form of resistance. Another option is to work within the system for common interest outcomes in pragmatic ways. Humans are good at circumventing limitations of the system, undermining it, and creatively introducing innovations. This strategy has proven potential to bring about constructive change in novel ways and through new alliances. Another way forward is through advocacy. At the least, sound advocacy makes the job of those who seek to maintain the status quo more difficult. It often forces the agencies and officials to do a better job than if left to themselves and their conventional approaches. The pressure of collective demand can be a powerful force for shaping the behavior of our representatives and public workers. If any of these strategies are successful, a transformation in consciousness and actions might occur. Such a transformation would lead to better conservation management and policy. Likely, all approaches are needed.

Culture and Symbols

I use the concept of cultural myth (worldviews, paradigms, belief systems, systems of thought) not to describe a false belief system but to identify a foundational social dimension in our lives. *Myth*, as used here, is a mechanism for creating meaning out of our experiences with other people and with nature.[16] Each culture and subculture is organized around a basic myth or set of beliefs. These myths are unifying and constructive because they give us the superstructure for our lives in which we live in our ordinary, everyday living. They unify groups with diverse experiences by unifying ideas of how the world is, how it ought to be, and who we are within it. However, myths can also limit our capacity to think critically and realistically about the world, and at times can get in our way, so it is important that we are aware of the influence that these cultural myths have on our everyday thoughts and actions.

Society and culture

To understand our society and culture in the GYE, we need to examine these basic belief systems (paradigms, frameworks, worldviews, philosophies) and how they figure into our sociology and more. Beliefs are codified in our communities through cultural norms and rituals. Because myths are the mental and social tools that people use to make meaning of their experiences, understanding myths is essential to understanding why people behave as they do in the GYE. While we are not always conscious of these, they have far-reaching consequences on land and wildlife conservation management policy. A patchwork of belief systems has influenced conservation outcomes in the GYE and will continue to fuel the increasing conflicts and controversies. Just as people cannot be separated from the land, they cannot be separated from their myths.[17] Creating coexistence and a sustainable future for the GYE depends on changing our current unsustainable beliefs and the damaging practices that ensue. In short, convention and its foundational beliefs need to be adapted to our changing circumstances.

Beliefs as myths

Our beliefs are at the heart of our meaning making through myth. Various individuals and groups clearly illustrate this in the GYE, as do we all. We, as individuals and as a society, create myths as the principal way to coordinate community actions. Myths are what tell people what is moral or justified, what is right and wrong, what is truth, reality, and what is fake. It gives us grounding in a world of flux, as landscapes, plants, animals, and human communities undergo constant and rapid disruptive change. Myths function as cultural glue that binds people into workable communities over time. Cultural reality, which consists of shared ideologies, stories, codes of behavior, and symbolic identities, is created as myths become real to the people and the groups who invent them.[18] Some people are tightly socialized to society's dominant myth whereas others are less so. One example of a shared foundational myth is democracy.

Myths are the basis for all cultures and are shared within many types of communities at many scales. Myths operate in entire nations, in groups with common interests, in neighborhoods, and even in professional communities.[19] Members of one culture or subculture tend to see their myth as the right, moral, and true way to see the world. They tend to see outsiders as foreign, different, and untrustworthy.[20] This way of understanding community is as a relic from an earlier evolutionary time when tribes of people were in direct competition with one another for resources. However, in today's connected world, the problems that we face are increasingly globalized, complex, and require greater cooperation to solve. In the GYE, this means integration across all authorities and jurisdictions (Figure 1.3). In order to meet the challenges of this globalized, complex, and dynamic world, we need to learn to see the ways that competing myths play out.

One good example in the GYE of social myths and their consequences is in Dr. Hannah Jaicks's study of how different people understand their connection to YNP and the predators within it.[21] In Jaicks's multiple-year study, *The Conflicts of Coexistence: Rethinking Humans' Placements and Connections with Predators in the Greater Yellowstone Ecosystem*, she spoke with people from many different backgrounds, livelihoods, and perspectives. She dug deep into what is behind the conflicts between humans and large carnivores in the GYE. She spoke with hunters, ranchers, citizens, and biologists, and went into the field on both hunting trips and scientific research trips. Jaicks examined trends of development and conflict that are expected to continue and intensify, leading to continued divisive social interactions among people.

Jaicks's work differentiated the full range of values in the human–carnivore relationship. Her work revealed the limitations of the current scientific and policy paradigms (positivism) to mediate that relationship. Importantly, her work pointed out that our relationship with carnivores is a symbolic expression of our larger concern about our place in the world and our relationship to nature, however understood (Figure 1.2). She concluded that both the physical and symbolic concerns that some people hold remain unaddressed by the conventional management policy system currently in place. Consequently, this partially failing system has polarized people's attitudes and the public policy debates regarding wildlife management, which take place amidst changing local, regional, national, and global contexts.

In the end, Jaicks identified several relatively successful people and projects that do, in fact, constructively manage the human–carnivore relationship. These projects are successful, in part, because they broaden participation through community-based initiatives, apply practice-based conservation, adaptive governance (not adaptive management), and develop support for these while also managing opposition to them. These are concrete steps that can be taken throughout the GYE to make real progress on the challenges we face, rather than remain stuck in the swirling eddies of recycled convention and conflict.

All myths about society and nature can be understood to consist of three interrelated parts—first is *doctrine*, or ideology, that is largely taken as a matter of faith and is essentially invisible to the people who hold it (Table 5.2). Second is *formula* about how the doctrine is to be applied in everyday life. And third is symbols that stand in for the doctrine or formula (e.g., the American flag, grizzly bears, or something else). The three dimensions come to us in our culture as a package, although different people adhere to difference doctrines, formula, and symbols in varying degrees and ways.

Our doctrine about how we live, treat each other, and govern community life in America (agreed-upon goals based on what we believe to be true in the world) is expressed in part in the Declaration of Independence and in the preamble to the US Constitution,

Table 5.2 The heuristic about people, meaning, society, and the environment (our myths) used throughout this book.

People	Meaning	Society	Environment
With perspectives	Individual and social group affiliations	Myths/worldviews/ paradigms	Nature/ environment/ ecosystems
And identities	Symbols (words, deeds)	Doctrine (basic beliefs)	Big carnivores, for example
And expectations	Personal stories of meaning	Formula (how to live the doctrine)	Other symbols thereof
And demands	Cultural stories of meaning	Symbols (representations)	Behaviors, actions
And value outlook	Sociology of knowledge (claims/counterclaims)	Institutions as system of thought and embodying the cultural myth	
Seeking value outcomes	Respect, affection, power, wealth, rectitude, skill, enlightenment, well-being		
Coping with reality of everyday	Conventionally, functionally, or otherwise	Through innovations and new institutions	Evolution (adaptation, extinction)

while the Constitution itself functions as the formula for American democracy (how to achieve our agreed-upon goals). In turn, the doctrine and formula are reinforced in daily life by symbols, such as the American flag, stories of folk heroes, the pledge of allegiance, and uniforms. Our lives are rich with symbols, though we often overlook their social and political functions.

Myths can be made visible by looking for the symbols used in any situation (e.g., bears as referent for nature in YNP). To find symbols, we have to analyze ordinary objects and the words and behavior of people to look for signs of hidden cultural interest. Professors Harold Lasswell, Abraham Kaplan, and Jack Solomon, a policy scientist, philosopher, and semiotic scientist, respectively, offer principles for analyzing symbols in politics. They are:

1. Always question the "common-sense" view of things because common sense is really "communal sense." That is, it is the habitual, ordinary, conventional opinions and perspectives of a community.
2. The "common-sense" viewpoint is usually motivated by a cultural (mythic, doctrinaire) self-interest that manipulates consciousness for ideological reasons. This is the form that myths of power take.
3. Cultures (and individuals) tend to conceal their ideologies to themselves and others behind the curtain of "nature," defining what they do as "natural" or "common sense" and condemning other people's cultural practices as "unnatural" or bad.

4.   In evaluating any myth or system of cultural practices, one must consider the values
       (of self-interest) behind it. These interests can be described in terms of eight basic
       values all human beings want: power, wealth, enlightenment, skill, affection, respect,
       well-being (mental and physical), and rectitude.[22]

Myths, our beliefs, and the formulas and symbols we use to represent them in our
everyday lives all serve a political and social function. While it does not always play
out as such, in America, our governing myth about power is that all people should
share power. Our formula should be enacted to implement this myth. It calls for
everyone to have a vote in our democracy (ideally). The formula also includes that
within our governmental institutions, power is dispersed across multiple levels (local,
state, federal) and branches (legislative, judicial, and executive). Our dominant power
myth also tells us that decision-making should be open, reliable, comprehensive, fair,
and accessible to all who are interested. Finally, it tells us that power should be used
to serve all people, not just the few most powerful or privileged in the private sector
or government.

Of course, this is only one of the myths in America today, and conflicts arise when
people do not have equal access to the power structures that we ideally share. Additionally,
some people do not act in the common interest and instead advocate for their own desires
disproportionally through wealth and power. In the GYE, the public controversy in all
cases is over the myth, formula, and symbols, which are the standards for how the region
is to be used. In this light, the major contentious issues in the GYE are just staging
grounds for the ongoing battle to establish the dominant myth (doctrine, formulas, and
symbols). The wolf then is more of symbol to be fought over than an actual biological
problem to be managed.

Our shared myth or basic beliefs in this country also tell us that power monopolized
by the few is wrong. Many of us feel this to be true when we experience discomfort over
the growing wealth and power inequality in this country. Other myths tell us how the
other basic value categories (knowledge, respect, loyalties, ethics, etc.) are to be produced
and enjoyed within a community. Differences among people on these matters are at the
heart of the controversy in the GYE and across our nation today. When these many
values are not shared, achieved, and enjoyed widely in a community in the ways called
for by the myth, people get upset and react. In America today, the benefits of a decent
wage, adequate education, and health care are not shared, achieved, or enjoyed fairly
by many people. Consequently, controversy and conflict ensue, and decision-making
processes become politicized, as we can see easily in the GYE across many issues.

Symbols

Attending to symbol analysis in the GYE is essential to understand myths (doctrines and
formulas). As we perceive the nonhuman world through our individual senses and cultural
belief systems, we create different notions of nature, wildness, and wilderness. These
views are subjective, but no less real than a geyser or a bison. Reality is a combination of
objective and subjective elements. Yet there are many belief systems in the United States

and throughout the world, and each offers a different conception of reality and human agency. Many controversies and conflicts among individuals and groups—from local to international—are nominally about objective elements (e.g., GYE's grizzly bears, wolves, bison), when in actuality they are about differences in people's subjectivities.

We can learn to see our own myths or beliefs, including the formulas and symbols that we use, in any dialogue in the GYE. One example comes from Dr. Justin Farrell in his book *Battle for Yellowstone: Morality and the Sacred Roots of Environmental Conflict*.[23] Sociological and anthropological researchers such as Farrell have studied how myths are structured, as well as how they function in different (sub)cultures. "Yellowstone is a symbol for so much more than just what's happening inside those invisible boundaries," says Farrell, an associate professor of sociology at the Yale School of Forestry & Environmental Studies. If you doubt the powerful symbol of Yellowstone, just go to any souvenir shop in any of the gateway communities to see the T-shirts and other items stamped with the Yellowstone image on sale.

Another apt example of symbol use was when the bison was chosen to be our first ever national mammal in 2016, taking its place next to the majestic white-headed bald eagle as a symbol of American ideals.[24] The act of choosing the bison as a symbol for America made explicit the relationship between the actual animal and the abstractions of American frontier history we make when we see or hear it. However, this decision makes little difference in how bison management is carried out across borders and is ultimately just a symbolic gesture. Yellowstone superintendent Dan Wenk noted, "The iconic [GYE] bison deserves our best efforts to assure its place on the American landscape."[25] It is clear that we have a way to go to achieve that effort.

Carnivores as symbols

Myth is directly relevant to the carnivore controversy in the GYE. Most of the debate is carried out symbolically, notes Steve Primm, who works with People and Carnivores, a successful environmental project of NRCC based in Wyoming and Montana that works to find practical coexistence beyond mere symbolism.[26] For example, Primm's colleague Matt Barnes, a range specialist, works in Wyoming's Upper Green River. He is working with ranchers to minimize grizzly bear–cattle conflicts.[27] Scientists, environmentalists, "wise-use" advocates, conservationists, and even recreationists all interpret nature and carnivores as symbols, according to their own mythologies. In turn, these mythologies mirror each person's value outlook and interests.

In a 2015 special article on wolves, *Jackson Hole Magazine* asked, "Why are they still controversial?"[28] The article noted that even 20 years after their successful reintroduction into the GYE, wolves are still a lightning rod for social debate. A similar article on grizzly bears that was published a year before asked the same question. These articles focused on the raging controversy after decades of conservation.[29] Importantly, the controversy is not about wolves and bears themselves, although they are ecologically important animals.

How do our myths about large carnivores play out in today's world? There are two extremes: save them all or kill them all. Each side justifies its views differently, both grounded in their respective myths. In our collective efforts to improve management policy of grizzlies, wolves, and mountain lions, for example, we try to continually consider objective elements such as population numbers, ranges, mortality factors, hunter success, and conflicts. This analysis is limited, however, because it fails to address the complex human dimensions of conservation in these cases. Moving toward a more holistic, integrated model will require managers to dig deeper and look at our arguments, sociology, and myths in a comprehensive, pragmatic way.

Yet, there is much resistance to incorporating social science into decisions about the GYE management policy because of disciplinary hegemony, but if we wish to address these issues in a pragmatically meaningful way, it is a necessary step.[30] Disciplinary hegemony is about the present near-total control of decision-making by a few powerful disciplines, such as conventional wildlife biology in the service of state agencies, for example. But wildlife biology cannot tell us about people's underlying feelings and beliefs about their relationship to animals, to the land, and to nature at large and, as such, is not sufficient for goal clarification, conflict resolution, or coexistence. One way to begin to understand these issues and questions and begin moving beyond disciplinary dominance is to examine popular stories (symbols) about these animals.

Grizzly bears, for example, are a key referent species in the region and we have many stories about them. At one time, the grizzly symbolized unnamed nature itself, and to some people it still does. But other views of the bear came about as they were decimated and nature itself seemed to recede before the weight of Manifest Destiny.[31] Views of grizzlies range over those of Native Americans, early explorers and mountain men, emigrants, ranchers and sportsmen, to modern conservationists.

Society and Institutions

Our institutions hold society together. In one sense, institutions are invisible. However, at the same time, they are everywhere, all the time. Institutions, broadly, are the "well-established and structured patterns of behavior or of relationships that are accepted as a fundamental part of our culture."[32] The institution for health, for example, is the system of doctors, nurses, hospitals, medical knowledge, and skills that serve us when we need it. Other institutions produce the values we enjoy (Table 5.2). Institutions as used here are not physical places but simply a web of individuals and organizations that structure our society and allow it to function (e.g., family, associations, government). Some institutions are problematic, clearly, but many serve us well. A stable society has institutions that show stable patterns of operation and beliefs that benefit its people.

Institutions

What is the institution for conservation in the GYE? Can you list the organizations, groups, and key individuals that make up that institution? We must come to see and understand our institutions, if the GYE is to have a healthy future. But the institutional dimension of our society is too often overlooked in conservation efforts in favor of our attention on animals, the environment, or other people.

Our lives are determined largely by how our values, organizations, and institutional components fit together and work for us through culture. The public policy dialogue today in the GYE, and throughout the country, evidences they are not working as needed. Otherwise, we would have less harmful conflict and more integrated outcomes. How are we to understand our institutions?

People and technology

Most people give their consent to the existing society and its culture through their participation, often unconsciously and conventionally. Cultures are held together by an overall shared political myth. In any culture, there are counter-norms, sentiment, and violations of the mainstream cultural expectations for behavior (e.g., crimes). The social order of our society is in those practices of custom, morality, religion, and so on (i.e., our institutions) that determine what happens to a person when they depart from any of the norms. These basic institutional dynamics, and inherent differences among people, undergird all that we see in the controversy over the meaning of the GYE, its current management policy, and future directions.

Today, technology, which dominates our attention and makes life easier for all of us, is simply the tool that we use to manipulate both raw and cultural materials for our benefit. For example, we use technology to manage our forests, food, and parks. The GYE is possible as a tourist destination only because of technology. Visitor accommodations, roads, communications, and wildlife techniques such as radio telemetry allow the place to function as it does.

But technology is not without issues. A major source of people's anxiety in places such as the GYE is about the rapid and persuasive encroachment of technology in all aspects of their lives. Simply put, technology is rapidly and profoundly changing the way people live and come to their own stories of meaning. This dynamic is clearly operating in the Greater Yellowstone Ecosystem, the tristate region, and the nation at an alarming pace.

The upgrade of the Moose-Wilson road corridor in GTNP is a prime case of technological controversy. In this case, there was a proposal to enlarge the road and add a bike path, which would increase public access to the most critical wildlife habitat in the park. This popular wildlife viewing road is occasionally closed because of grizzly bears foraging nearby but the corridor was considered for upgrading, as it is a major thoroughfare for tourists headed to the massive Teton Village ski and recreation resort.[33] The "upgrade" management focus was driven in part by modern technological improvements increasing the ease and cost-efficiency with which we can construct roads, and in part by people's changing attitudes toward the relative value self-interest (e.g., biking) versus common interests (conservation in the park), their relationship to nature and wildlife, and their priorities and responsibilities. These are all part of this complex policy dialogue and is no different than most and other policy processes. These policy dialogues play out in formal institutions, including businesses, environmental nongovernmental organizations (ENGOs), and governmental organizations.

There are significant environmental issues with the proposed road upgrade. When I think back to the grizzly bear surveys that I did in GTNP, Jackson Hole, and Bridger-Teton National Forest in the late 1970s, I remember finding no bears south of Teton Wilderness, except an occasional individual on Spread Creek. Now, there are about 60 bears using the park, some using the Moose-Wilson corridor.[34] This represents real progress in grizzly bear conservation, and the proposed road developments could threaten to reverse the advancements that have been made over the past 50 years. What is to be done?

Despite the ecological risks, talk moved forward to enlarge the road and make a bike path, driven largely by the many political interests that are involved. Political and economic interests, working though the Chamber of Commerce and the governor, were mobilized to ensure that the NPS will enlarge the road, tourist volumes, and add a bike path—all of which are in their favor by increasing potential for revenue (wealth). Still, many remained opposed to the proposed upgrades because of the significant carnivore and other wildlife conservation problems. Eventually, a "compromise" was made on the road that included elements wanted by each of the different participants in this highly politicized decision.[35] However, a formal NPS assessment concluded that none of the "four alternatives (in the Environmental Impact Statement) in their entirety would minimize human safety hazards from bears or the impacts to bears or their habitat."[36] This road controversy is only one of many contested issues in the GYE. How the corridor upgrade plays out over the long haul will be telling of who we are as a people, what we value, and how we see our responsibility to wildlife and nature. This controversy shows how many subissues of the GYE's "problems" all have the same contributing factors— our sociology and myths—even as the details of each case differ.

Power mythology

How does the myth of power actually work in our culture and institutions when carnivore issues are publicly deliberated and decided upon? Our current myth derives from the human-centric and exploitative view of nature that was brought over by the Europeans and shared by other cultures. This myth emphasizes a relationship to nature predicated on grazing, logging, hunting, and other extractive technologies for human benefit, which is inherently about the power of humans over nature (see Table 5.1).[37]

Directly and indirectly, many of these beliefs and accompanying practices came to affect grizzly bears, mountain lions, wolves, and virtually all other species and ecosystems in destructive ways over the last few centuries.[38] In North Americans under the practices of the dominant culture (with its frontier mentality), much of the continent's biology has been irrevocably altered. In the West, the megafauna has been greatly reduced or killed off entirely. At its peak, the North American continent had more wildlife biomass than Africa. The American bison, for example, once ranged across the American West with numbers as high as 65 million. By the early 1900s, however, through a federally backed extermination effort aimed at asserting control over indigenous nations, they had been hunted until only 24 remained in the GYE.[39] Since then, bison numbers are on the rise as herds have been reintroduced and domesticated, though their numbers remain a mere fraction of what they once were. This is a story that has played out across the GYE time and again as folks work to restore populations of many species under the 1973 Endangered Species Act, for example.

New mythologies are at work in wildlife conservation in recent times, as evidenced by restoration efforts and enhanced species protection. No longer does a purely extractive belief hold sway. Our new beliefs and formulas about how we ought to use and relate to wildlife have come about as we come to grips with ecological limits and our own unintended destructive impacts. With dramatic shifts underway, we are now moving toward a more favorable public attitude of wildlife, and especially carnivores, one that favors human–wildlife coexistence. This is not true everywhere though, and in some places, there are backlashes to coexistence. This trend is only a few decades old, but it is evidence of changing myths and institutions. If the trends continue, there is hope of coming to a more realistic and flexible relationship with the land, carnivores, and other life forms. Such an outcome—a new doctrine, formula, and symbols—is starkly needed today.

Reality

As long as myths and institutions accurately reflect biophysical and social reality and adequately explain people's experience, then they are retained over long periods of time. This results in stable societies. But when beliefs and practices no longer serve the community as a whole, people usually change or adapt their myths and behaviors to the changed circumstances. This seems to be what we are experiencing in the United States and GYE presently, as evidenced by political unrest, environmental movement, and rise of the conservation institution. Although ever changing, myths are rooted in real people and their real experiences, in real places, and because of real events. Today, there are many people clamoring to adjust the overall American myth and formula and its institutions about how we relate to nature and each other and our responsibilities.

Knowing how to identify and discriminate societal myths that influence people's thinking and behavior can be enormously helpful to those who are interested in changing public decisions about coexistence. It is important to know how to interpret myths and institutions for practical purposes in political discourse. Also, it is important to know how to participate constructively in any decision processes in ways that serve common interests.

Large carnivores will be conserved and managed sustainably only when participants create, acknowledge, and respect an appropriate belief system and institutional arrangement that supports these animals' conservation. Such a system must motivate people to constructive coexistence. Knowledgeable and skilled leaders can help communities achieve genuinely efficient and effective management of carnivore species. We need to make sure that people's conservation identities are really helping wildlife and nature, and not just taken on so that people can just feel good about their self-proclaimed conservation identity.

Analyzing institutions

Many institutions overlap to determine what happens in the GYE. It is my sense that most people focus on other individuals or groups, and overlook the importance of themselves, their organizations, and the institutions they comprise. Institutions all serve one of the eight foundational values referenced above, or a combination thereof. It is easy to see

Table 5.3 A framework for analyzing social institutions and developing skill in communicating, creating, and supporting constructive change.

Task 1	Task 2	Task 3	Task 4	Task 5
Trends leading to creation of institution being analyzed	Specific values affected by institution (*List indices*)	Appraisal of specific practices of institution (democratic or antidemocratic) (*List indices to support conclusion*)	Probable future effects of institution if practices are not modified (*List indices to support conclusions*)	Alternative suggestions for institutional modifications designed to widen value distribution (*List indices to support suggestions*)

these values in operation in the GYE, if you know what to look for. When examining our own institutions, we should ask ourselves whether they are contributing to value sharing or to value deprivation. The core function of institutions is to mediate the exchange of values. Ultimately, institutions determine whether a society can strive for freedom and well-being, or whether it is subject to human control and exploitation of both cultural and natural resources. A problem-oriented framework for analyzing institutions is in Table 5.3. Use it to look at a GYE case of interest to you.

People's perspectives

Institutions come out of the perspectives of people. Although our major institutions were created many decades ago by the people of our society, they still have influence over us today. And we continue to shape them as we go into the future. As such, one of our most important responsibilities as individuals in our society is to participate in the decisions that define and shape our institutions. Social institutions do not just happen—they require careful individual participation and ongoing attention.

In moving on, recall that *people* seek *meaning (values)* through *society (institutions)* using and affecting *resources (natural and human)*. In this section, we are focusing on institutions. From the local to the global level, we must take direct actions to bring about course correction in our institutions, if problems are detected. This means that we must develop the mental and verbal tools necessary to understand the personal and institutional makeup and behavior of our society. The problem-oriented, contextual approach described and used in this book is recommended to do just that (Table 5.3). Part 3 goes into all this in practical detail. The GYE is a prime case to learn about our institutions. In doing so, we can gain analytic maturity, knowledge, and skills through our inquiry into institutions. This requires engagement with other people and the salient issues in question.

And, for us to meet our individual and group responsibility as citizens, we need to look at our own predispositions and perspectives in the context of our institutions. Our individual perspectives determine how we respond to other people and to the problems that we face. We need to behave responsibly in interpersonal relationships in ways that contribute to the wide distribution of values (e.g., respect). And, we need to know how to analyze and appraise institutions and adapt them as necessary (Table 5.3, Part 3). Our outlook very much determines how we participate within institutions, either reinforcing or challenging them.

As incoming signals take form about how well we are living with each other and nature in the GYE, we need to adjust our perspectives, decisions, and modify our institutions as we go. Humans are gaining more and more power and control over our world. At the same time, we are more closely connected to each other. Our individual and cultural stories, indeed our whole society and its institutions, need to keep up with changing realities. Our knowledge of the biological and social realities of our lives can be used to stay ahead of the learning curve. We need to be on the leading edge of that curve. This requires mature, knowledgeable, and skilled people.

Institutions in context

With the rise of some social movements, we are developing new theories and new ways of understanding our role in the world. As we construct new theories, we must consider the context of the present moment: the history of the human struggle and how culture and its institutions have changed through time. We must guard against overly simplistic epistemological and ideological views, however popular, as we move into an unknown and uncertain future. To move events and processes toward a more mature citizenship that takes its responsibility seriously, we need to develop a way to analyze institutions, bring about constructive change, and make them work for us in the context of nature (Table 5.3).[40]

In our quest to achieve some semblance of coexistence and sustainability, we should strive to meet our present needs without sacrificing the same opportunities for future generations. The societal flux or transition presently underway is affecting everything, including our stories and institutions with unknown outcomes. Our success or failure for coexistence and sustainability depends on our ability to think in terms of institutions and systems, whereby we can come to integrate environmental, social, and economic considerations. This transition could be leading us toward a new high ground. As we move deeper into this century, with ever clearer signals coming in about the future, it is obvious that we need a new cultural story for ourselves, one based on reality and truth, yet sensitive to our universal need for meaningful and transcendent stories of affirmation. The GYE is one place where that struggle is taking place.

Conclusion

This chapter looked at the GYE and its future through a sociological lens. Using the sociological lens forces us to see our sociology, culture, and institutions more clearly than we might otherwise. Sociology tells us a lot about ourselves as individuals in society and culture. It helps us understand how and why we make meaning of the GYE and behave as we do. Importantly, this lens opens up new avenues for each of us to contribute to a healthy future for the GYE by becoming more aware of ourselves, our groups, society, culture, and institutions and how best we can interact with one another most constructively. The sociological lens is especially needed to address GYE's challenges coming in now and those emerging in the signals coming in from the future (e.g., climate change).

Yellowstone is controversial. What we are seeing in the mushrooming controversy over bears, wolves, and bison, for example, is our society's deep, unresolved conflict about our relationship to wildlife and nature. Recall that referents (bears, wolves, bison) are the stand-ins for the big issues (i.e., relationship of humans to wildlife and nature, our culture, and its institutions), which we grapple with all the time. Arguing over bears, wolves, and bison is nothing more than a struggle over the present and future meaning and use of the GYE.

Using sociology as a way to understand people's behavior helps us see the underlying social, cultural, and institutional issues at play and at stake. Big-scale change is underway in the GYE, though it may not be clearly visible to everyone. Much of this big change is sociological. It is clear that we need to better integrate environmental knowledge with our society and its culture to come to a better relationship with wildlife and nature. In many ways, this is an individual challenge. It is also an institutional challenge. In short, we need to take ourselves through a smooth transformation to a new conservation paradigm that moves us closer to a healthier relationship with wildlife and nature. This means that we must change if we are to secure coexistence and sustainability in the GYE.

Chapter 6

PEOPLE AND STORIES OF MEANING

The problems we are worried about are caused
not just by other people but by ourselves.[1]

As we experience and learn more about the Greater Yellowstone Ecosystem (GYE), we are easily captivated by its beauty and importance. When visiting, travelers and residents alike are sure to see herds of bison and elk. They can go from one roadside vista to the next breathtaking view in minutes. There are limitless opportunities to experience the park and its animals either by car or on hiking trails. Visitors may decide to eat ice cream at Tower Falls or relax on the front porch at Roosevelt Lodge. Those looking for adventure can catch a stunning view of the Teton Mountains from the top of Mount Washburn. Many even get to see grizzly bears and wolves. While all these opportunities are "out there," experiencing them is an internal, personal process—a psychological and perhaps transcendent one.

People do not arrive at YNP as blank slates. Each person is influenced by their own background, personality, and knowledge base that affect what they choose to see, how they feel, and how they make meaning of it all. This is true for both residents of GYE and short-term visitors who visit the parks or other parts of the ecosystem. We each value things differently. We are at different stages of experience, existential maturity, and adult development. We vary in our reflectivity, insights, and judgment. We show differing degrees of emotional intelligence or sensitivity to our surroundings. All of these dimensions influence how we interpret our GYE experience.

We are the center of our own stories of meaning. After our Yellowstone experiences, we may tell our friends about our trip and show them pictures. This is one way of creating and sharing our meaning of the GYE. As I discussed in the last chapter, the many controversies so evident now in the GYE are really amplifications of our colliding stories of meaning. In the end, this boils down to our individual psychology and social living. It is likely that the future of GYE's wildlife will hinge less on wildlife itself and more on our own psychology, sociology, and philosophy. Consequently, in this chapter we will range widely and generally over psychology, philosophy, and history, gaining perhaps new knowledge and perspective on people and their behavior in GYE. I hope the applications of this knowledge and the new perspective it offers are useful for GYE through my descriptions and examples.

Eternal Questions and History

When we come to Yellowstone and the GYE, we are attuned to both eternal questions and the answers to those question that are already built into our personality and culture (Figures 1.2 and 1.3). There is an endless intergenerational discussion that goes back at least 2,500 years in Western civilization about the meaning of life, reality, and much more. We today are the inheritors of that discussion. These very questions about values and responsibility are the ones that informed the establishment of Yellowstone National Park many years ago, and which we continue today through debates for how we ought to manage natural resources.

Foundational questions

Throughout this book, I address the five foundational or philosophic questions introduced in Chapter 1 (Figure 1.3):

1.   What is the world, as it occurs to us? (ontology, reality, and truth)
2.   How do we know? (epistemology, knowing)
3.   What is right? (axiology, rightness, correctness)
4.   What is our sense of things? (ordination, concurrence)
5.   What is practical for us to do? (pragmatics, what, how, and why)

At present, the answers to all these questions are built into our culture and its institutions. For example, the conventional understanding is that there is an objective truth that we can come to understand through rigorous reasoning (scientism) and that a good life is defined by success in a free and open market (neoliberalism). Because these assumptions are imbedded within our culture, we seldom do the work of engaging them for ourselves. But our current institutions fail to precure the common interest in many cases and are instead generating many of the problems that we face in the GYE. To truly address these challenges, we must examine these foundational questions ourselves, as individuals and as a society, and come to understand the consequences of our answers.

Progressive history

There are two major paradigms of thought in our society and culture: the progressive, which is dominant today, and the problematic, which is becoming more popular. The progressive worldview is a narrative of our journey as continuous forward progress through time and into the present and future. This myth is typically about "man" as an active agent in the world who is able to arise out of the limitations of time and nature by grasping his own dignity, agency, and future. This view imagines human beings as entities separate from the rest of the world around us. It promotes progress above all else, especially material progress. Today, given the scientific and information age we live in, we see ourselves at the apex of this progress (and separate from nature). We have engineered our way through many of the problems that we have faced historically, including disease, limited agricultural productivity, and more. Our technological advances have reinforced our view of progress as control over the material world around us, without much interrogation into the ill effects. Many people today are totally immersed, blindly and conventionally, in this dominant cultural narrative called neoliberalism.

The progressive worldview is troublesome because it rests on a philosophic foundation that separates us from nature. This view promotes human advancement, rather than stable coexistence, as the point of civilization. We have only recently begun to see the harmful, unintended consequences of this progressive myth, such as overexploitation, unfairness, and environmental destruction. At present, not everyone is willing to examine the limitations of our dominant paradigm, remaining deeply buried inside and committed to the conventional progressive narrative. These people are often politically conservative. Many people, however, are beginning to critique the progressive worldview and draw attention to the damages it has wrought. These people are beginning to forge a new outlook. These two worldviews are colliding in GYE at present.

Problematic history

The second major paradigm—the problematic myth—is gaining prominence now. This view of human evolution, history, and consciousness acknowledges the harmful consequences of the progressive myth. It argues that we overlook important matters when we operate inside a neoliberal paradigm. Take the example of the marginalization of diverse voices, such as Native American communities in the GYE, due to lack of economic power. The focus of the problematic myth is on critiquing the shortfalls of progressivism and advocating for a new outlook that is more inclusive and that promotes human, nonhuman, and environmental dignity. However, the focus of the problematic worldview is often limited to bigger, distant concerns (e.g., climate change, ocean pollution, extinctions), though it is beginning to be applied to analyze local challenges. Overall, the problematic paradigm focuses on the detrimental side effects of the progressive myth. The problematic worldview recognizes the current impoverishment of human life and of the environment—biodiversity, ecosystems, and wildlife. It emphasizes growing fragmentation and incoherence in society and the environment. This second outlook— the problematic view—is gaining ground in GYE.

Each of these two paradigms has implications for how we view the world and how we assess the problems at hand. In order to understand the ways that these paradigms are at play in the GYE, let's explore how different people interact. This explanation gives us a much broader perspective on GYE than is otherwise available.

Individual People—Psychology

As a species, we have been shaped by evolution. It has honed our senses and intellect, enabling us for daily living, self-awareness, and abstract thought.[2] These qualities are the tools with which we process our experiences and social contexts, as well as shape our personal stories about who we are and what we value. As a species in society, we codify our stories of meaning and continuously refine them by experience. And when we convey them to others either in-person, on social media, or otherwise, we reinforce social norms and expectations about appropriate behavior, human–nature relationships, and communal goals. Ideally, these encourage the development of responsible, reflective, mature, and moral individuals. We all have these evolutionary and social tools to varying degrees, and we can use them to confront the limitations of convention in our storytelling when appropriate. This will enable us to take informed and pragmatic action to address problems that we face in GYE.[3] Unfortunately, this informed, pragmatic action is not always manifested today, despite all the work, activism, and public opportunities that exist to aid conservation in the GYE region.

If we want to ensure a healthy future for the GYE, or anywhere, we cannot go on pretending that the social and environmental challenges we face are "out there." Much, if not all, of the challenge ahead is inside us, created by how we think, what we see, and what we do. The challenge is about how we as individuals understand and interact with other people and nature. It is especially difficult when people have different stories than our own. How do we find cooperative ways to work toward a common interest in GYE nature conservation?

The ecosystem's visitors and residents

What do we know about the visitors to the GYE? Visitation to the region has been increasing, exceeding 4.25 million people in 2016. Visitation broke records in 2020. This trend is expected to continue. The majority of visitors to YNP come between mid-June and late August. Increasing visitation challenges park management whose goal is to expose people to nature and to provide them a wholesome, authentic experience, without sacrificing wildlife and the ecosystem in the process. Currently, the massive influx of visitors (more than thirty thousand per day during peak season) overburdens the infrastructure (e.g., roads, sanitation, rangers). Visitors sometimes violate park rules by harming or harassing wildlife, walking off boardwalks in geothermal areas, littering, speeding, or poor sanitation. As the number of visitors continues to increase, the problems of facility overcrowding, traffic jams, degradation of the environment, and disruption of wildlife behavior will only get worse.

Visitors want to be submerged in a natural experience. They want a memorable trip filled with stories about nature, particularly charismatic wildlife. One survey found that visitors were willing to pay significantly more to enter YNP if they were guaranteed to see a bear.[4] Another survey of park visitors showed that their main interest was in having a "sensory experience," followed closely by "shared experiences" and "photography opportunities."[5] The 2016 Visitor Use Survey showed that 96 percent of people reported "viewing natural scenery" as their most important reason for visiting the park, while 83 percent reported "seeing wildlife" as most important.[6] These reports suggest that the GYE and YNP are important because of their aesthetic and "referent" (moral) value. The beautiful scenery and wildlife remind visitors of the American frontier ideology and fulfills a deeply rooted desire for a connection with the natural world, a phenomenon dubbed "biophilia" by the ecologist Edward O. Wilson.[7] However, if part of the value of nature in the GYE is its novelty, this novelty also presents a problem—many visitors are not educated on how to behave around hot springs, geysers, and wildlife (e.g., bison and bears) and may accidentally harm wildlife or themselves.

In addition to the millions of visitors that come to the region each year, the GYE has around three hundred thousand permanent residents, who span 21 counties in three states. The permanent residents in the GYE are a mix of Old West traditionalists like ranchers and New West recreationalists. These two groups often clash over values and demands. Additionally, the population in the GYE is growing rapidly.[8] As more people move into the GYE, they place increasing pressure on the natural resources in the region. Increased population density increases the number of human–wildlife interactions and burdens the environment through additional demand for resources and recreation space. Together, the visitors and residents are rapidly changing the ecology, economy, and character of the GYE, often for the worse. Efforts to curb these changes have been largely unsuccessful.

Multiple perspectives

Is the GYE's future really dependent on what humans do? If so, how do we "fix" problems to achieve our intended conservation goals? The answer depends on who you ask. Ask an ecological scientist, and he or she may tell you that the future of the GYE can be ensured by focusing on the biodiversity (bears, elk, invasive species), landscape connectivity (corridors), and abiotic conditions (soils, fires, climate). Ask a recreationist, and he or she may tell you that the best way to protect the GYE from industry and development is to build an economy around outdoor sporting opportunities, such as biking, boating, and skiing, and expand these opportunities even into wilderness areas and national parks. Ask a conservationist, and he or she may tell you that to fix problems, we should take down interfering fences for migrating animals, build under- and overpasses on roads to facilitate wildlife movements, and put up road-crossing signs to preserve wildlife migrations and slow drivers down. Ask a businessperson, and you may hear that we must increase commerce and profits to bolster the economy of the region. And, so it goes with every participant. Each person has a different story about the proper way to relate to the GYE, a view that is informed by their own personal context, lens, interests, and perspectives, including their degree of self-awareness and their personal ethics.

Personal and adult development

Our individual maturity, psychology, and stage of personal development also determine how we experience the GYE. Personal development occurs throughout childhood and adulthood, marked by progressive psychological changes. Psychologists recognize five stages of adult development, and the differences in these stages influence how adults interpret and understand themselves and the world around them.[9] Although there are five stages, most adults do not progress past the third stage—the socialized mind—and this has ramifications for how we deal with conflict in the GYE. Adult development, personality evolution, existential maturity, reflective judgment, emotional intelligence, and moral development are subjects, or at least should be, all relevant to people concerned about GYE today and in the future (Table 6.1).

Originally a field focused on infants and children, developmental psychology has recently grown to include adult development and the aging process. Many researchers have investigated adult development and its relationship to depth of understanding, problem-solving skills, empowerment, and sense of agency. This research is valuable for understanding the state of the GYE today because we are challenged with solving its complex problems through cooperation with people who have different tools in their emotional and intellectual tool boxes. These differences often prove problematic in itself.

A leader in this field of research is Robert Kegan, a professor of Adult Learning and Professional Development at Harvard. Kegan's 1994 book *In Over Our Heads* looks at how people feel when new "orders of consciousness" emerge (or fail to emerge) from their experiences in the domains of parenting (families), partnering (couples), working (companies), healing (psychotherapies), and learning (schools). Kegan repeatedly points to the suffering that can result when people are presented with challenging tasks and expectations without the necessary individual and social support to master the needed skills to address them successfully.[10] This perspective is especially helpful in understanding people in GYE.

Whatever the virtues of higher orders of consciousness, no one should expect to master them before they are ready or when they lack the lived experience and necessary support from others (Table 6.1). Thus, the task before us is not only to respectfully interact with people who operate in different developmental stages but also to encourage development and transformation in us all. We can all help ourselves and each other move along the development track through the many ways suggested in this book.

People active in the public debate on appropriate present and future management policies in GYE are at different stages of adult development. Consequently, they can fail to communicate effectively with one another both because participants are not aware of the differences in emotional and intellectual skill and because few situations exist to address these differences in constructive ways that enhance effective communication. In order to manage problems in the GYE, we must address this basic challenge and promote development and transformation within individuals, where possible.

Existential psychology

Existential psychology is another major subject of importance to us in GYE, if we want to understand people. How we make meaning, how we cope, and how we ground ourselves all affect our personality, motivation, and experiences in life, including our visits to the GYE. We can either try to understand this existential dynamic and manage it constructively or we can let it manage us destructively.[11] Our anxieties are aggravated when we are confronted with situations that challenge our sense of thoughtful confidence, responsibility, and connectedness. Sometimes, our anxieties are managed poorly, so we discharge them onto other people and the issues we face. This is of limited use and can

Table 6.1 Three positions of adult development describing key features of each and noting that individuals seem to "plateau" at one position in life and not move to the next higher level.

Development Position	Description	Self	Expression
Socialized Mind (Majority of people)	We are shaped by the definitions and expectations of our personal environment and those around us.	Our self coheres by its alignment with, and loyalty to, that with which it identifies, typically social peers.	Expresses itself in our relationships with people, with "schools of thought" (our ideas and beliefs) or both.
Self-Authorizing Mind (Minority of people)	We are able to step back enough from the social setting to come to an internal "seat of judgment" or site of personal authority that evaluates and makes choices about external expectations.	Our self coheres by its alignment with its own belief systems/ and ideology.	Follows a personal code; has ability to self-direct, take stands, set limits, and create and regulate its boundaries on behalf of its own voice.
Self-Transforming Mind (Rare in people, perhaps 1%)	We can step back from and reflect on the limits of our own beliefs and ideology or personal authority. We can see that any one system or self-organization is in some way partial or incomplete. We are friendlier toward contradiction and opposite.	We seek to hold on to multiple systems of thought rather than projecting all but one onto the other systems. Our selves coheres by its ability not to confuse our internal consistency with wholeness or completeness.	We behave by aligning with dialectic rather than adhere to any pole.

Adapted after Robert Kegan and Lisa L. Lahey, *Immunity to Change: How to Overcome It and Unlock Potential in Yourself and Your Organization* (Boston, MA: Harvard Business Press, 2009),17.

promote conflict. Instead, we should seek to understand the source of our anxieties and develop healthy skills to cope with them. The field of existentialism helps us to do this.

Existentialism targets the very foundation of how all people deal with the certainties of life: death, isolation, responsibility, and meaning making. These four sources of anxiety are the bedrock of our existential psychodynamics. Existentially mature humans are able to deal with their anxiety about these inevitabilities in relatively healthy ways, but not everyone has the skills to do so. Just because someone is an adult does not mean they are existentially mature. Different people cope with these four anxieties in different ways. As I have discussed, we as humans make meaning through our personal and cultural stories. Thus, when we visit Yellowstone or the GYE, we are engaging with a cultural story and furthering our narrative about who we are and what we value. It is an existential story and, in some cases, a transcendent experience. Perhaps that is why the GYE is so important to us individually and as a culture—it gives us a story of awe, grandeur, and larger meaning, well beyond ourselves.

Existentialism, in contrast to clinical psychology, views anxiety as a constructive emotion. It is little more than an inevitable part of the human experience. As independent agents, we are challenged with the task of leading authentic and personally meaningful lives; however, we recognize that of the infinite possible iterations of a life, we can choose only one. Existential anxiety arises as we grapple with this reality and with the fear that surrounds the question—what if I chose the wrong one? When considering the role of existentialist thought in how we interact with the GYE, it is important to remember that existentialism does not support any of the following statements:[12]

1.   The good life is one of wealth, pleasure, or honor.
2.   Social approval and social structures override the individual.
3.   Accept what is and that is enough in life.
4.   Science can and will make everything better.
5.   People are good by nature, ruined by society and external forces.

These five statements are part of the accepted conventional story line, but existentialism stands in marked contrast to that convention.

This discussion of psychology overall, and of existentialism in particular, shows us that a healthy future and individual freedom for each of us exists if only we are sufficiently self-aware. Unfortunately, we are too often constrained by convention. We blindly accept that we are free, that we see reality as it truly is, and that we are able to think with clarity and independence. But these assertions are simply not true. Based on what we examined above, we know that each of us is limited in how we think and in what we have experienced. In turn, these limit our freedom of thought and action. It is possible to be more self-aware and, at the same time, to come to a more sustainable existence with other life forms. We should always be aware that we are limited by our acceptance of the prevailing paradigm (e.g., convention) and our particular outlook and history. To attain GYE's sustainability, if that is the goal, we must work together to promote individual growth in all of us—developmentally and existentially.

Fast and slow thinking

Another feature of our cognitive system that we can see at play in GYE is that it contains two systems of thought—fast and slow, or System 1 and System 2 thinking, respectively. These two thinking systems were described in Nobel Prize–winning psychologist Daniel Kahneman's 2011 book *Thinking, Fast and Slow*.[13] The premise is simple–there are two routes to persuasion, based on two basic modes of thinking. System 1 operates automatically, intuitively, involuntary, and effortlessly, such as when we drive, read an angry facial expression, or recall our age. System 2 requires slowing down, deliberating, solving problems, reasoning, computing, and concentrating, such as when we calculate a math problem, choose where to invest money, or fill out a complicated form. People in GYE vary in their use of the two systems, varying perhaps case by case.

These two systems often conflict with one another in GYE. System 1 operates on heuristics or mental rules that may not be accurate. System 2 requires effort to evaluate those heuristics and, therefore, is prone to error. Kahneman describes how to "recognize situations in which mistakes are likely and try harder to avoid significant mistakes when stakes are high."[14] In order to address the growing environmental problems, we must rely on and improve our System 2 thinking capacities. There are many opportunities to be more thoughtful, deliberate, and rational in GYE's social, management, and policy issues. These dimensions present themselves daily in the news (as in op-eds), public hearings (in speaking out), and working groups on various issues (actively joining a group).

Motivational psychology

People in GYE are differentially motivated. Why do people act against seemingly good energy plans, as did the National Wildlife Federation in 2011?[15] Why does a person seek directorship of the Jackson Hole Wildlife Foundation?[16] Why would a person take on a challenging climb in the Teton Mountains?[17] These are examples of behaviors that reflect a person's personality and motivation.

In thinking about people in GYE, consider the three following views on motivational psychology, which are rooted in how and why we as humans seek to achieve the eight core values described earlier. First, Alfred Adler talks about "the basic human motive."[18] He says it includes an "aggression drive," a "will to power," and a "striving for superiority." This striving for superiority is functionally equivalent to Abraham Maslow's "self-actualization," by which people desire fulfillment in ways unique to themselves.[19] Kegan, Yalom, and Kahneman each support Maslow's view in their own ways.

Second, in 1961, Carl Rogers recognized only one basic human motive, which he called the "mainspring of life" or "the actualizing tendency."[20] He defined this as "the urge […] to expand, extend, become autonomous, develop, mature—the tendency to express and activate all the capacities of the organism."[21] He described self-actualizing persons as fully functional, possessing an increased openness to experience, an increased tendency to live fully in each moment, and an increased trust in themselves. Motivated by their own intrinsic interest in selected activities, such people lead lives filled with meaning, challenge, and excitement.

Third, what really matters to a person is not what objectively exists but what he or she perceives to be true. In other words, what a person thinks is their motivation. According to Herbert Vaihinger, people possess a "fictional finalism," an imagined goal or ideal self that functions to unify a personality over a lifetime.[22] It represents a specific, ultimate ambition or source of personal meaning. Our imagined goal is related to who we perceive ourselves to be and what we value.

So why are these ways of gaining insight into human behavior so important for us to account for in our examination of GYE? Again, the answer is that people, with their varying interests and personalities, are absolutely central to what happens in GYE. Possessing sound knowledge of people's psychological dimensions can help us work with each other constructively. This knowledge can help us learn about our own feeling, thinking, and behavior to become more functional and existentially mature. Overall, because psychology influences the goals and expectations of individuals, we should all make a special effort to pay attention to it. This means that we need to focus on ourselves and our relationships with other people as we work to ensure a healthy environmental and social future in GYE. We need to make sound inferences about the motives, values, and behaviors of others, especially those involved in social processes and the GYE's future.

Biology and Folk Stories

Let's face the facts ... but what are the facts? We are a biological entity with an interesting and variable psychology and a long evolutionary and cultural history that has led us to where we are now. A long time line has shaped our brain and social mind, which allowed for the invention of culture, symbols, and semiotics. All of us are conventional to some degree. It is evolutionarily beneficial to have a unified sense of the world and norms for social behavior. We play from our biology and draw on our "folk knowledge," our everyday knowing and understanding of things, in many instances. As individuals, we and our accepted folk stories are thrown together in questions over Greater Yellowstone's future.

Social psychology

The people in the GYE—both permanent residents and visitors—are part of communities and social networks. And many of the people who come to experience GYE do so with other individuals. Social behavior is healthy and necessary to our living. However, it can also lead to "groupthink" and other distortions.[23] Social psychology explores how people's behaviors, thoughts, and feelings change because of other people. As social beings, our thought and behavior are constantly shaped by all the people around us. Consider how your behavior changes depending on who you are with. What parts of your personality do you emphasize? Consider, for example, the different personas that you embody with your family, your partner, at work, or with friends. Even when you are alone, you are influenced by other people.

The GYE operates as a cultural and institutional entity because of our collective social psychology. When you come to GYE, you immerse yourself in a social institution established by people who lived over a century ago. You engage, either directly or indirectly, with our shared values and expectations. By thinking of the GYE as a collective social experience, we may be better equipped to make decisions in the common interest. Let's look at some basics of human existence shared by all of us.

Human biology and the "struggle for existence"

As humans, we are influenced by the interplay of many things, including genetics, evolution, physiology, epidemiology, ecology, nutrition, and sociocultural factors. The influence of these has been written into our DNA by evolution, and although we have developed through society, culture, and civilization, we often still follow this evolutionary guidance because, ultimately, it is our biology that permits us to live as we do and enables us to think and experience the world around us. We are our bodies, comprised of physicality, sensory inputs, a nervous system, and other systems that, together, allow us to operate in and adapt to a changing environment.

Our evolutionary history is a fascinating story that unfolds in detail as we discover more paleontological and genetic evidence. Most interesting, perhaps, is the evolution of the brain. In the four million years since the emergence of our most recent predecessor, *Australopithecus*, the average ratio of brain volume to body size (the encephalization quotient) of humans has almost tripled. And along the way, our brain has been significantly reorganized, with much of this change occurring in only the past 100,000 years. The architecture of our brain now makes it possible for us to do what we do and be who we are. It permits our complex social interactions and capacity for problem solving. It allows us to designate YNP and conceive of the GYE given our society.

One way to see how biological and cultural evolution work in our daily lives is to consider social competition. Ongoing competition today centers on our individual and social attempts to achieve access to and control over salient resources (e.g. energy, markets, information). Local contexts function as control mechanisms and are dependent on conventional understanding and social norms (folk knowledge). The kinds of social cooperation that dominates our society are still evolving, as we can see from the changing governance strategies throughout history.

Our mind evolved into its present form in order to better anticipate the future. This, in turn, has led to consciousness and the ability to take oneself into account in the struggle for survival. Thinking about the future can help us adapt to likely, foreseeable changes over our lifetime. Thus, we scheme, work, gain education and skills, and plan for old age. In short, environmental and social variables have driven the evolutionary selection that produced the human brain and mind of today, a mind that is capable of critical self-reflection and intentional adaptation that will help promote a new, sustainable paradigm, hopefully. We see evolution in action in our social and policy struggles in GYE, and need to leverage our intellectual potentials to move toward human–nature coexistence.

Social struggle

Over evolutionary time, it is our social challenges that have most shaped our minds. As noted by Dr. Robert Green, a neuroscientist at the Harvard Medical School, "We're beginning to see a striking aspect of the brain [...] that brains are wired for social interactions."[24] Human beings, along with other great primates and some emotionally advanced species, such as the African elephant, have even evolved a specific kind of neuron, the von Economo cell, that is involved with social interaction.[25] Human beings are hardwired to operate in social groups and to manage social competition. Today, we compete to control access to resources we need as individuals and as a society, even at the level of the nation-state. Much of our present struggle with one another concerns basic human needs, but these basic needs are masked by the demands of our conventional daily lives.

Folk stories

Human evolution and our culture have given each of us an implicit understanding of the world that anthropologists refer to as "folk stories."[26] Folk stories consist of the basic assumptions, words, and concepts that allow us to go about our lives in more or less cooperative ways without thinking very much. We learn folk knowledge as we are socialized and educated by our culture and its subcultures. It is given to us. The GYE is full of examples of folk knowledge at play, such as those that tell some of us that wolves will kill off all of the elk and that we can't have both wolves and livestock.

We organize folk stories into three domains: social, biological, and physical. The social domain directs our attention toward other people and interpersonal relationships. It allows us to divide our social world into categories, such as family, friends, and unfriendly social groups. We have a strong capacity to form "in-groups" based on shared beliefs, values, and other identities. Doing so affirms our psychological outlook and our social affiliations. This group formation capacity allows us to organize and cooperate at large levels (e.g., communities, nations) to better compete for control of ecological and cultural resources. However, we also have a strong predilection to vilify and intensify competition against out-groups, which promotes conflict and tension. One effect of learning to understand the workings of people and social groups is to relate to them in new ways, and provide context that diminishes the tendency for out-group aggression.

The second domain is folk biology. Folk biology is about the conventional, often widely accepted stories about how our bodies work, who or what other animals are, and about how nature works. In contrast to many folk stories, our bodies are biological, just as is wildlife and many features of our living (e.g., pets, livestock), and systems in nature. This domain unconsciously supports our ability to understand other species and allows us to categorize and discern variables of biological significance, such as birth, growth, and death. Although folk biology is common, it often is wrong and misleads people into thinking they understand humans, wildlife, and nature. In turn, folk biology aids people's beliefs about the use of ecological resources and the value of nonhuman organisms. An example, again, is views about wolves, their behavior and relation to elk and other wildlife, and, finally, their role in ecosystems and nature.

The third domain is folk physics, which helps us to understand the physical world by mapping space, time, and forces (e.g., gravity). This allows us to better navigate these entities with accurate mental representations in terms of our work, ecologies, and landscapes. While other species share knowledge of their physical world to find food, shelter, and mates, human folk physics also allows us to make tools, which has allowed us to transform the world to an unparalleled degree. It allows us to broadly share a view of the world, how it works, and how resources are understood and exploited. Folk physics allows us to share a view of reality, without thinking much about it.

Overall, these three folk domains allow us to function successfully in the conventional world, even without a formal, conscious, or explicit knowledge of their content. So far, these folk domains have served us well, in a long-term evolutionary sense. We can easily see folk knowledge at play in GYE's social and policy dynamics in the many stories in currency and in competition.

Integrated people

So far, we have explored the psychology of people and the profound ways in which they differ with respect to personal and moral development. Yet we share many features, including the culture of convention in which we live.[27] We now can see, perhaps, why individuals and organized political groups think and behave as they do in relation to YNP and the GYE or any other issue. Particularly, how they convey their beliefs and how they handle conflicting viewpoints. Understanding this will inform how we see the ongoing conflicts that we face in the region.

Communication

As a social species, we have an incredible capacity to invent and use concepts, words, and other symbols to communicate. Language is the principal medium through which we share our inner psychological selves. We can tell others how we feel, what we think, and how we interpret the things we see. This capacity has given rise to complex social organization, culture, and institutions. Our capacity for communication permits us to cooperate with billions of individuals across the globe, people who are often unlike ourselves.[28] It is also a tool that we can use to manage these differences and ourselves in GYE constructively.

Human acquisition and elaboration of symbols and language has evolved over the past 100,000 years or so, and continues to evolve today. Over many thousands of years, our language has become much more finely descriptive, discriminating, and nuanced (e.g., we now talk about intangible concepts, subatomic particles, and galaxies far away, none of which are visible to us in everyday conventional life as "folk physics"). But the structure of our language affects the ways that we think about and perceive the world.

We understand reality today in a dyadic relationship, including only subject and object. Thus, we conventionally accept that signs are reality, overlooking that fact that our signs are in fact only a representation of an object, a fabrication. In Nietzsche's *On Truth and Lie in an Amoral Sense*, he argues that language (words and symbols) is merely a representation of the objective world we experience, and as such, it is not the Truth: "The 'thing in itself' (for that is what pure truth, without consequences, would be) is quite incomprehensible to the creators of language and not at all worth aiming for. One designates only the relations of things to man, and to express them one calls on the 'boldest metaphors.'"[29] Our failure to recognize this leads us to underestimate how language and communication influence our meaning making.

So, what does all this psychology and philosophy mean for the GYE and its future? Professor Noel Castree, an environmental philosopher, says that when we talk about nature, we are really talking about ourselves.[30] This fact is likely unconscious, perhaps incomprehensible, for most people, who seem to overlook the fact that we humans made up the semiotic category of nature (Figure 1.2). The propensity for categorization and externalization is a tool that we use to define the boundaries of ourselves as humans, for it is only by determining what we *are not* that we are able to begin to define what we *are*. But the distinction between humans and nature is a false dichotomy, and "nature" does not exist a priori. We as humans have defined and constructed "nature" within our own minds. This misconception leads to misdiagnosing the problems and sources of the problems of the GYE and the world. Already, it has led to rendering perceived problems as solely technical, for experts to deal with rather than all of us to address.

Personal growth

Our ability to address problems depends in large part on the mental models that we create. These are related to general intelligence. Good model making requires suspending the unconscious and conventional folk ways of thought in exchange for conscious, explicit representation of the available information. What are those cognitive systems and processes that allow one to escape convention and become consciously aware of internally and externally generated information?

When we are proficient at practical modeling, self-awareness, and problem solving, we are at our best as humans. The policy sciences, discussed later in this chapter and used throughout this book, offer a meta-framework to significantly upgrade our cooperative problem-solving capacities in GYE. In effect, these tools are a systematic way to mentally and empirically generate models of the present and of likely future conditions. From these models, one can seek alternate responses and rehearse behavioral strategies to address challenges.[31]

Confronting convention

The challenges that we now face in GYE require a new way of thinking, such as that which Walter Lippmann, a great American social commentator, discussed in his seminal 1922 book *Public Opinion*.[32] Broadly, we need a better understanding of what we as conscious agents bring to the GYE—our experiences, minds, and problem-solving strategies. Many people before and since Lippmann have identified the need to align our internal world with external reality, although we often fall short of this goal. This oversight or "selective inattention" remains a barrier to success and integration.[33] When people debate the GYE's future, they seldom, if ever, talk about how people know about themselves and the biophysical world, and the proper relationship to wildlife and nature

(Figure 1.3). However, human actions are based on our answers to these questions, so they are vitally important to what happens in the region.

Through culture, we have constructed a human world for ourselves, one based on material (physical) and moral terms (ethics). We now know that our human world is out of sync with the ecological world, to detrimental ends. We have constructed a false semiotic duality between humans and nature, which has failed to promote a sustainable relationship with the environment and all other life, and has instead advanced an exploitative view of natural and living resources. Many of the challenges that we see today in GYE are rooted in this exploitative narrative, and as population growth continues unchecked, we are using and affecting resources to an unparalleled degree. We are now, without doubt, the most ecologically dominant species on the planet. But some of the promised or presumed solutions to social and environmental problems are not working.

Hence, our human world needs some adjustments. Fortunately, we as a species differ from all others in our extraordinary ability to be self-aware and self-reflective, to use language symbols, and to modify resources from our ecological surroundings. We use these resources and semiotic symbols to organize ourselves and make meaning, and they can serve as valuable tools as we seek to change the current problematic view of ourselves and nature toward a more sustainable paradigm. Are we developed enough to take on this task in GYE?

To view ourselves as primal constitutors of meaning is to challenge everyday reality as we ordinarily and conventionally experience it.[34] We constitute the world in such a way that it appears independent of our frameworks of thought. We typically see the world and GYE as external, real, and permanent. We believe there is an objective Truth that we can identify and understand. Yet, we know that the world of our understanding is subjective. It is influenced by our own personalities, motivations, experiences, cognition, and development. We have many truths. Given this, how do we act in the GYE?

When considering the many problems that we face in the GYE, it is important to maintain pragmatic hope. This hope is rooted in our knowledge that it is possible for us to grow in self-awareness, cognition, and communication skills. We are capable of developing our intellectual and emotional tools to modern and postmodern positions that permit us to understand and address complexity in a better fashion than we are now doing. We are capable of advancing and gaining greater executive control over our intelligence, experiences, and problem solving. We can overcome our own resistance to change. We can learn to see our own self-blocked psychology and learning that constrains our personal growth and keeps us from engaging productive and effective actions in search of the common interest in the GYE.

"Fake" Facts and Knowledge

Let's now look at society's view of knowledge, language, and various representations of reality and truth that compete for our attention today in GYE. This includes the whole notion of fake "facts." Is global warming a hoax? Are Yellowstone wolves killing all the elk? Do we have enough grizzly bears? What are we to make of the many different answers to these and other questions? As well, what are we to make of the claims of "fake" facts? How is knowledge mobilized in answering these questions by one person or another, one ideology or another? Each answer implies claims to reliable knowledge, claims that are social phenomena. The issue here is in part about what the claimant believes to be real and true. In the current politicized social climate, who is to be believed? These questions take us directly to what professional sociologists call the "sociology of knowledge." A sociological perspective falls somewhere in the middle of two extreme positions that knowledge is either entirely "objective" or entirely "subjective."

Alternative facts

Consider all the conflicting claims that are advanced as true knowledge in the GYE. There are many. These claims are put forth in newspapers, opinion pieces, public conversation, interest group newsletters, scientific publications, business pronouncements, official sources (e.g., National Park Service), environmental communications (e.g., Greater Yellowstone Coalition, Jackson Hole Alliance), the press (e.g., *Jackson Hole News & Guide*, *WYOFile*, *Mountain Journal*), and on social media. Depending on the issue, knowledge claims are typically asserted as reality and as Truth. But if we took the immense time to look at all the claims about grizzly bear conservation, for example, we would find that many claims are contradictory, and others are patently false.[35] How can we work through the opposing claims, so we can make up our own mind and make sound public decisions?

It seems clear to me that there is a reality "out there," the biophysical world that we experience and filter through the matrix of our senses and thought systems. For example, there are grizzly bears out there, and there is a finite number of them. However, even objective reality, such as gravity, is sometimes in dispute, as are the number of grizzly bears, though not as often as social facts. In addition to objective facts, there is also subjective reality, but how do we tell the difference? How does indisputable reality and facts interact with the subjective construction of reality and truth?

Spheres of reality

Is there an objective reality in GYE that we can learn about if we use appropriate methods (e.g., traditional positivistic science in certain controlled situations), or is all that we think of reality and knowledge just a product of our "mind"—senses and concepts? The latter view is about subjective reality, which is generated by social agreement. This kind of agreement about reality is a different kind of truth from that of gravity. One draws directly from the biophysical world and relies on institutions of science and reason to make conclusions about what is "real," while the other relies on the felt experiences of people and their understandings of them. In reality, these are not entirely dichotomous. Even in objective observation, we are unable to remove ourselves as the observer, and even in social contexts, we are not free from the realities of the external world that constrain us practically and evolutionarily.

Throughout our lives, and throughout historic time, our views of "reality," "knowledge," and "facts" have changed in important ways. Once, reality and knowledge said that the world was flat and that humans are the center of the universe. We now know both are untrue. How do we explain that change in facts? Are we moving closer to an understanding of "reality" throughout history, or are we just making up contemporary stories of meaning to suit our needs, given the context and the growth of knowledge, without being aware that that is what we are doing?

We construct reality socially (intersubjectively), although this fact is underappreciated in everyday, conventional life. All human "knowledge" is produced, processed, transmitted, and maintained in the minds of people. Consequently, the sociology of knowledge is also a biosemiotics issue. We know that knowledge is produced and used collectively through our language and social institutions to achieve larger cooperative, cultural ends. Is reality, or what is accepted as reality in everyday conventional life, the "consensus" that emerges from the interpersonal and public competition over claims to knowledge? If so, whose reality counts most? What are the consequences?

Let us consider the wolf case as an example. We attempt to manage wild populations of wolves based on population numbers and our understanding of the role that wolves play in the ecosystem. We believe that if we can understand the sum of the connections between a wolf and the ecosystem, then we will be able to determine a discrete number of wolves that is sufficient to total the ecological function that we desire. But do we have enough reliable knowledge to know how many wolves there are in the GYE? And, is the present knowledge adequate for setting ideal management goals?[36] Is function for an ecosystem a reasonable basis on which to determine this question? Monitoring of wolf numbers is ending in Idaho and Montana, and likely Wyoming, soon. All three states are part of the GYE and are responsible for the same wolf population. At present, we haven't adequately answered the questions that I raised above, and now managers are faced with managing populations with even less knowledge of their status and behavior than we have now. Are we willing to risk the ecological damage that we may be unknowingly causing if we are wrong, or if our knowledge base is incomplete? This is a prime example of the dominant paradigm at play in management of GYE. Positivistic science is relied on to justify management objectives, and people with power within the institutions who advocate for policies that do not realistically address the challenges of the wolf case, leaving management worse off.

Experience and reality

There are different spheres of reality at play in society and in people's understanding of the GYE. We take our own experiences, and some mix of other's spheres, as the true reality in our daily living. In the end, we have many diverse spheres of reality that exist in society at any one time. These differences become more obvious when people's worldviews are made explicit through conflict over particular issues or when incidents happen, but realistically, these differences are present all the time. We get these different ways of knowing and understanding as we grow up and participate in our communities. The important point is that these differences among people have hard consequences for how we develop our towns, communities, and regions. Often, we conclude that people whose spheres of reality differ from our own are wrong, misguided, or malevolent.

However, this view promotes dissonance and conflict, rather than cooperation and solutions in GYE today, and is therefore unhelpful.

All of us, as members of society, find meaningful conduct in our lives through our subjectivities and stories of meaning. We build and use knowledge to do so, typically without explicit awareness. In our world, where stories and meaning originate in our thoughts, mind, and actions, they come together to constitute our reality. We maintain them as real through our individual, group, and societal lives. That is, after all, the purpose of our culture and its institutions—to make a shared worldview. In a real sense, we find meaning through a complex mix of objective and subjective knowledge and through our shared intersubjective understandings, interactions, and stories. What we hear in the multiple perspectives being dumped into the public dialogue of the GYE and its future are different stories, based on selective use of objective knowledge and subjective spheres of reality. Needless to say, public dialogue is "messy."

The mix of elements in our knowing is the basis for the many diverse and often conflicting individual and groups claims about reality that we hear in the GYE. Regardless of the subject, this raises hard questions about the interplay of objective reality, subjective reality, and intersubjective reality in the GYE. We must work to come to an understanding of these spheres, but acknowledge that ambiguity and differences will always be with us.

Doubts and sanctions

Given that we do not have a single epistemology or understanding of reality shared by everyone in our society, we often manage the contradictions and conflicts by dividing our lives into containers. By subdividing our lives, we are able to address different problems with whatever way of knowing is most applicable at that time without truly having to face the inherent contradictions and conflicts. However, as we seek to advance in our adult development, we should learn to recognize these contradictions as equally valid, and not necessarily opposed, functions. GYE is full of opportunities to do so.

We erect psychological boundaries and heavily ego-defended borders to our reality. This makes it difficult to engage in productive conversation with other people who understand things differently. For example, how do you convince someone that climate change is real when their reality tells them that is not happening? How do you convince someone of the reality of ecosystem problems when their current sense of self and story of meaning depends on not believing that reality? Some portion of these issues arises from conflict over the value of evidence-based knowing and respectful interaction, and it is the eternal task of education and public policy to advance these as values in our society.

One of the biggest challenges facing us in the GYE today is our inability to communicate effectively across our differences. We more often talk past each other than with each other. It used to be that most of our daily interactions were face-to-face, but today, technology has intruded into this domain. The rapid increase in social media use has changed the social landscape. It is increasingly becoming remote, distant, and impersonal. Our widespread use of technology and social media also conditions us to engage only with worldviews similar to our own. Algorithms carefully curate the content that we see, and as a result, we come to view the world as more single-minded than it truly is. We come to expect sameness and are increasingly unpracticed in engaging with differences in people's opinion constructively in respectful conversation.

Grounding One's Self

We come into the world with big brains, a long cultural history, and a lot of built-in evolutionary predispositions. Thinking and reasoning are extremely useful to the future of GYE, especially since the threats to the region are more philosophic and psychological than they are physical, material, or environmental. Philosophy and psychology may sound like abstract things when thinking about GYE, but philosophy is "what everyone does when they're not busy dealing with their everyday business and get a chance simply to wonder what life and the universe are all about."[37] The purpose of philosophy is to come to "correct" principles of judging, choosing, and acting through a process of critical self-examination, enhanced self-awareness, and understanding of self in social life. Given all the controversy over the GYE today, it seems we could all benefit from finding "sound" principles for judging management policy.

Ideas that shaped our thinking

Our ideas that we hold came from somewhere and arose for certain reasons. In this section, I briefly offer some history so that we can best understand why we think, talk, and do what we do. Specifically, I draw on the history of Western philosophy as the foundation of thought for most of the political and social structures in the United States (and therefore in the GYE). Although different societies have their own history and culture, these Western ideas dominate much of the global dialogue today—east and west, north and south, developed and less developed. Our current mix of modern and postmodern philosophy (and political organization) has emerged from three major eras in Western cultural history—classical, medieval, and modern. I will explore each of these explicitly in the section below. I draw heavily on the philosophic historian Richard Tarnas in the following account.

Classic Era: The Greco-Roman worldview

I start with the Greeks, who brought forth an extraordinarily rich culture over 2,500 years ago, and are credited with the origins of what we now call Western civilization that influences all societies most everywhere today. Socrates, Plato, Aristotle, and many others made Greek culture and its philosophy famous. They sought to clarify universal truths, such as mathematics and geometry, out of the disorder of life that they observed. These foundations also included notions of the Good, the Beautiful, the Just, and various other moral and aesthetic features. We are still living the cognitive contribution of classic Greek thought throughout the world today.

The Greek thought tradition was born from a dialectic of naturalism, skepticism, and secular humanism. It is a system of thought wedded to reason, empiricism, and mathematics. With their philosophy, the Greeks provided explicit answers to the five fundamental philosophic questions in Figure 1.3. Their answers and the culture they built around those answers set the template for all that has happened since in Western civilization and much of the world. Our current paradigm is rooted in the philosophy established by these Greek philosophers, and our culture is permeated by its effects. American institutions including democratic government and our sense of what is right and wrong were influenced by this system of thought.

With the rise of the Roman Empire, a new system of thought came to dominate Western civilization. In many ways, we continued our way of knowing and defining our relationship to nature as was established by the Greeks. Like the Greeks, the Romans believed that the universe embodied and was governed by a comprehensive regulating intelligence. Importantly, the Romans believed that this same intelligence was reflected in the human mind. As such, humans were capable of knowing the universal order of nature and natural laws. However, the Romans seemed to be more pragmatic than the Greeks, applying their Greek legacy to law, politics, administration, and military operations. This pragmatic approach is reflected in the American political and legal system today.

Medieval Era: The Christian worldview

Partway through the Roman Empire, distant regions in the mid-east, specifically in Galilee and Judaea, gave rise to a religious movement. Jesus, and later the Apostle Paul, brought forth a heretical sect that challenged the Roman worldview. This movement diffused throughout the Roman Empire, taking form and constituting itself into a church, centralized administration, and overseer of doctrine and practices that today we call Christianity. This movement brought a system of thought that gave new personal meaning and stories about life's purposes. In the early fourth century, the Roman emperor Constantine adopted Christianity, and consequently, Christianity became the sanctioned state religion.

Philosophically and practically, there are several key differences between the Greco-Roman and Christian worldviews. First, the Christians established a monotheistic hierarchy in the universe. Christianity emphasized dualism, the separation of human spiritual matters from nature. Early Christianity undercut the idea that people could find objective Truth about the natural world on their own. This subordinated rational and empirical inquiry, in favor of emotional, affective, moral, and spiritual concerns. Lastly, early Christianity renounced human capacity for independent, intellectual activity, relegating these inquiries to the church and the Scripture. In essence, the Pope, as a voice of God, was the only source of final Truth. This Christian worldview dominated the West for centuries and offered an alternative epistemology to the one arising from the Greco-Roman legacy. While the philosophical and technical (e.g. architecture, military) accomplishments of the Greco-Roman Empire were lost in light of the growing Christian worldview, these philosophies and ideas would come to the fore again in the intervening centuries, as they now form the basis for our understanding of the world and ourselves in it.

The Modern Era: Reformation and Scientific Revolution

In the first half of the Medieval Era, there were a few people invested in scholarly work, but the diligent efforts of some, such as the Reverend Bede, to relearn the Greek and Roman languages and to understand their vocabulary and grammar resulted in a resurgence of their ideas and philosophy. Around 1000 CE, Europe attained a degree of stability and security that brought forth renewed cultural activity—populations increased, agricultural advances were made, and trade and interchange grew, including contact with Islamic and Byzantine cultures.[38] Arising along trade routes, new cities were centers for the wealthy, literate upper classes.

Universities were founded as a growing desire for learning emerged. All these new social formations, from guilds to universities, brought forth change. The West underwent a complex mix of what we now call the Renaissance and the Reformation, and was entering the Scientific Revolution at that time.[39] The Scientific Revolution led to a rebellion against the old Christian order because it laid out the two epistemological foundation blocks for the modern mind—empiricism and rationalism.

Science now reigned as the source of authority for reality and Truth, a condition that has persisted for centuries. However, the authoritative power of science is once again beginning to break down as we are now living in a mixed modern and postmodern world. Today, the dominant modern (progressive) view and the postmodern (problematic) view are in competition for our attention. Each of these worldviews have different answers to the basic and foundational questions (Figure 1.3). Thus, the two create a dissonance that makes identifying the common interest particularly difficult today. This competition is stark in the crucible that is the GYE.[40]

Knowing and thinking

How do we know what we think we know about nature, ourselves, and Greater Yellowstone? Let us look at our system of culture and our framework of mind to address this question. In this section, I explore how two prominent American philosophers address these "given" matters, which are both deep philosophic issues that have consequences for the life and future of GYE. Like psychology and history, philosophy is at base the reasons we think and do what we do. Let's dive in.

Science and economy

Modern thinking has brought forth the concept of science or scientism, a strong belief in the universal value of the scientific method to understand the world and ourselves. It is an approach that tends to crowd out all other ways of experience and all other epistemologies. Scientism assumes that unlike other systems of knowledge (epistemologies), it is value-free and objective, capable of understanding and describing the world as it truly is. Of course, we know that we are not capable of identifying a truly objective reality, because we cannot dissociate our thought from our subjective experience. Again, consider the German philosopher Nietzsche,

> What, then, is truth? Is it a mobile army of metaphors, metonyms, and anthropomorphisms—in short, a sum of human relations that have been enhanced, transposed, and embellished poetically and rhetorically, and which after long use seem firm, canonical, and obligatory to a people: truths are illusions about which one has forgotten that this is what they are; metaphors which are worn out and without sensuous power; coins which have lost their pictures and now matter only as metal, no longer as coins.[41]

In philosophical terms, scientism is an extreme form of logical positivism. Under this view, all problems come from a point source and are able to be solved if we just know enough. This view is widespread in the GYE, where virtually all issues are rendered technical and relegated to experts and technicians to address through research and management. This approach has value. However, it ignores the myriad ways that the world of our experience exists not as an independent entity but instead in continuous dialogue with our individual experience and thought and with the societies in which we operate. Scientism does not promote a deep examination of the foundational questions in Figure 1.3 and thus is not capable of affecting the deep meaningful changes that we hope to see in the GYE. This is one key aspect of our current dilemma. It seems to me, the task before us is to learn to integrate across these worldviews.

Of course, not all scientists hold scientism rigidly. As the famous scientist Carl Popper said, "It is all guesswork, doxa rather than episteme. […] Science has no authority. It is not the magical product of the given, the data, the observations. It is not a gospel of truth. It is you and I who make science, as well as we can. It is you and I who are responsible for it."[42] So long as we maintain this realistic view of science, we will be better equipped to address the challenges in the GYE. It is true that science plays a key role in helping us find the common interest, as it is essential to our understanding of ourselves and the GYE. However, while necessary, science is not sufficient for these things, so the challenge before us is meaningful integration. We must be open to critiquing and examining the role and value of science in understanding and decision-making.

In addition to these considerations, our views of nature and ourselves manifest themselves in this economic system. Under strict scientism, nature exists only as a source of resources we can use for material advantage. In recent centuries, the rules for using nature and natural resources have been given from the outlook that today we call neoliberal capitalism—a philosophy and practice of the wealth market economy that is under increasing scrutiny for its exploitation of both people and the environment to a detrimental end.

This version of economism is starkly present in the growing economy of Jackson Hole, for example, where consumer goods and recreational experiences are marketed to tourists and adventurers as a way to participate in the beauty of the GYE. Economism takes nature as valuable insofar as it serves human needs. It is assumed that nature exists for our purposes. Therefore, nature as a concept is meaningless until we confer meaning onto it through use in the free market, through extraction or recreation—for example, by commodification. Other epistemologies (the mind's way of knowing) and sensory experience are heavily discounted by this view of nature.

This combination of scientism and capitalist economy converges to produce our present system of thinking and social order. These traditional views advance an exploitative, rather than sustainable view of nature and they are fundamentally behind our current environmental problems.[43] In order to suitably address the problems they create, we must examine the institutions and ways of understanding that influence how we interact with nature in the GYE and elsewhere. It seems that we need a broad, contextualized view of science and different means of managing goods and services.

The GIVEN and the mind

Each of us is born into a culture that is given to us as we grow up and become thoroughly enculturated. This GIVEN has been studied by many scholars. Chief among them is Wilfred Sellar, whose 1975 book *Empiricism and the Philosophy of the Mind*, discusses the "GIVEN," or that which is presumed to be true by our culture. The GIVEN in our society today is the conventional view that there is something in us that gives a true picture of reality, a kind of ontological arrogance. We assume that there is an eternal truth, that we understand "rightness" and "wrongness," and that the knowledge we have is certain. Most of us simply live out the GIVEN without questioning its origin or current value, or considering alternatives.

The GIVEN presupposes no learning, no grounding, and no firming of associations (as we often see in political and religious ideologies today). This presumed knowledge provides the foundation for our folk stories of the world, ourselves, and nature. This conventional view dominates almost all the dialogue about the GYE today. We are blocked by our conventionally accepted view of the world and because of this, we cannot move beyond it to a new more sustainable worldview, though we need to, and quickly.

In his book, Sellars rejected all the central arguments that had been put forward to support the existence of the GIVEN. He suggested that we can learn about the world and what is rational, and use a self-correcting approach without the assumptions of a GIVEN. Absolute certainty of how things are is not possible. His work undercut positivism and has tremendous implications for our living and management policy. Sellars also discussed the ambiguity in sense-based versus data-based (empirical) theories. Sensory data is immediate and accessible through common experience. Data theories are derived from inference, which may not be accessible through common experience. These are two additional ways of knowing that tell us different things about ourselves and the world. The term "GIVEN" refers merely to what is observed—sensory input—and assumes that it mirrors reality and Truth, although this is not always the case.

If sensory input is all there is, then there would not be disagreement over "data." As I have discussed, even our sensory experiences must be interpreted by our brains, and are therefore subjective insofar as our thought systems are limited and unique, influenced and informed by our personal experience. If direct and immediate observation were the only window into reality, then the question of what is knowable would be put to rest. But we know this is not true. Direct and immediate sensory input is limited. It does not tell us about ecosystems or greater ecosystems, for example. It does not tell us about emergent properties or dynamic system. These must be elucidated through an amassment of sensory input, examined by reason to infer the truth from the pieces.

Another great American philosopher who has implications for our work in GYE is Richard Rorty, who wrote *Philosophy and the Mirror of Nature*. Rorty wanted to bring philosophy into the mainstream for layperson citizens. Philosophy is clearly foundational to culture. As Rorty noted, culture is merely the collection of claims to knowledge, and philosophy is the tool to arbitrate those claims. In doing so, we benefit from a very thoughtful and deeply considered view of the foundations of our civilization, our notion of people as individuals, and our role and responsibilities in nature and the world. Rorty argued we need to join these philosophical conversations if we are to arrive at a more sustainable and mutually beneficial paradigm. At present, most of us do not know how to do so.

Philosophy, it is said, is where we find justifications for our activities and experiences in places, such as the GYE, nature, and our human-built environments. It is claimed that philosophy is the only place to discover the significance of our lives. And, as the GYE is a space that reminds us that we are a part of systems greater than ourselves, that we are agents that affect all others, both humans and animals, it is also a place for engaging with philosophy in this sense. Therefore, the GYE invites us to question whether our lives exist for us alone or if we have a greater responsibility or purpose toward others and the places we inhabit.

This book began with stories from a mix of personal, experiential, literary, poetic, artistic, recreational, economic, political, environmental, science-based, citizens, officials, and government perspectives and claims. Who do we see as the privileged authorities to make "truth" claims about knowledge and the GYE? I maintain that all the controversies that we currently see concerning the GYE, YNP, and beyond are merely the competition over whose stories of knowing and meaning will prevail. It is difficult to reconcile these claims effectively, practically.

What do we need to know to better understand the GYE and to ensure it has a healthy future? Philosophic knowledge is vital, to be sure, as is knowledge of psychology, sociology, and biosemiotics (language, communication). We also need science. However, we need contextualized science within a broader framework of knowing because I believe that science is always needed, yet never complete. The conventional scientific way of knowing should be respected and employed, although not as the sole answer to questions or problems. The new dialogue that I am calling for in the GYE seeks to establish a much larger integrative framework for examining basic philosophic issues and their practical implications for human living in a rapidly changing world, informed by scientific evidence. This enlarged dialogue is essential for the GYE and its future.

We need to get clear on the philosophy of knowledge and knowing.[44] As long as knowledge is conceived as an accurate representation (the mind as a mirror of nature), then such accuracy requires a theory of privileged representations. That theory must account for the relationship between science, which is presumed to be intrinsically accurate, and Ideas or subjectivisms, which are about contention, conversation, and social practice. We can understand knowledge only when we understand its social justification, and thus have no need to view it as an accurate representation. Perhaps there is no line between philosophy and science. In fact, many of the greatest scientists were great philosophers, and there are fields of thought dedicated to bridging scientific and philosophic ways of knowing, especially in physics and in philosophy (metaphysics).

When we think about engaging in philosophic discourse as a means to move forward toward sustainability in the GYE, confrontation should be replaced with conversation and meaningful dialogue. The new integrative dialogue should be about wholeness, unity, and coherence and should be undertaken with a realistically problem-oriented approach.[45] I believe we can rise to the high imagination needed for the kind of public dialogue and problem solving for which I am calling. Perhaps I am overly optimistic, but when tied together, optimism and pragmatic hope can help.

So, what to do?

I have laid out how psychology, sociology, and philosophy are vitally important in finding a system of knowing and a pragmatic working strategy for us in GYE. The implications of these knowledge areas are only now just entering the discourse about GYE's future. We have a very long way to go, if we are to catch up to the leading edge of this knowledge and its integrative standards for problem solving. Presently there are many forces and factors hindering the use of sociological and philosophical knowledge in the GYE that were explicated earlier, most notably the entrenchment of people and institutions in given, conventional thinking and practices. I believe that the social science and humanities knowledge areas offer tools that will allow us to look beyond technical problems, a biophysical emphasis, and the surface of our culture to learn to see the actual constitutive problems that we face in our deep culture (see Figure 5.1).

We were all brought up in social structures and are generally loyal to them, conforming our thinking to their conventional norms. As a consequence, the reality of everyday life appears already "objectified" and justified as "just the way it is" to most of us, as we grow up. We assume that these objects and concepts of reality existed before we apprehended them and, in fact, that they always existed. The world of the close-up and personal, the here and now that typically dominates our perceptions, thinking, and meaning, is the central motif of the formula for the good, ordinary living, including our prescription for a relationship to nature and other life forms. But as humans, we understand that each of our individual realities is an ordered reality, and that our patterns of thought rearrange over time due to shifts in our individual experiences and knowing. We can learn to think beyond the given conventional structure to arrive at a new way of knowing and being in the GYE. We should work toward a future of sustainable coexistence, but we will not get there without action and critical thought. Which future do we want? Are we willing to do the work to move forward?

Conclusion

We live now in an age where in we have a much clearer picture of how our actions have wrought unintended harmful consequences to the environment and to ourselves.

These problems arise from historic actions and our individual and collective thinking. The conventional views of positivism and neoliberalism are beginning to break down in light of the damage they have wrought, and what we are seeing now in America, on a national scale and in regions like Greater Yellowstone, is the struggle to realize new stories (webs of meaning) about nature and people, and their interrelationship with a focus on justice and genuine environmental sustainability. We can use these stories, old and new, to understand and express our relationship with nature and with each other in more justified and sustainable ways. Our stories are a direct consequence of our psychology, sociology, philosophy, and history, whether we know it or not. This is why I dug into these subjects in this chapter, despite the fact that they are seldom talked about in GYE policy dynamics. We think and express ourselves through language, and as such, it is a mirror of who we are (or think we are) and of what nature is and what our relationship to it ought to be. Language mediates interpersonal relationships and is our window into each other's psychology, philosophy, and experience. This is our reality, individual and collective.

Psychology, sociology, philosophy, and history are about gaining deep insights into humanness by grounding ourselves existentially and getting to reality. For many people, experiencing GYE as nature is their way to get to the bottom of things, at least in part. It is a way to find a deeper meaning to their life and for their own living through a felt sense of connection to people, place, and the natural world. It is a chance to move beyond the fast-paced lives we all live and to reflect on ourselves, who we are and who we want to be. Not everyone relates to the GYE in such a way, but many do. This is what nature immersion does for us.

We are arguably deep into a growing crisis as our social and environmental problems mount. There is a growing disconnect, or so it seems, between reality and our stories about ourselves, between our position relative to nature and what we are experiencing and seeing. We are caught up in a kind of ordinary, everyday living that assumes a given reality and that reality is coming apart. There are many reasons for the growing incoherence, but without an integrated thought system it is hard to accurately and adequately identify them. The widespread lack of philosophic engagement and appreciation is behind our growing self-inflicted disconnections, challenges, and crises. We do not consider our goals and values explicitly, nor do we sufficiently interrogate institutions and interactions, with great detriment. It is clear, or should be, that we desperately need to engage foundationally with psychological, sociological, philosophic, and historic matters in GYE, if we want to have a good chance at a healthy future for all. If we can come to agreement on what is true in the GYE and how we know, we can begin to set goals informed by our values and work toward them. I suggest that our highest goal is of healthy people in healthy environments. How do we get there?

Chapter 7

COHERENCE AND POLICY

The facts about humanity's contemporary damage—
and threats of even more perilous future damage—
to the environment are almost too much with us.
They scream in horrifying detail, not merely from the face of nature
but from every medium of communication.[1]

Policy is a word we hear often. There are at least ten definitions of the concept policy. For sure, policy is about decision-making, setting strategic aims, and implementation. It is typically political as well as empirical. Management uses programs to carry out policy prescriptions. Newspapers are full of comments about various policies in the GYE, especially the adequacy of those meant to conserve wildlife (e.g., Endangered Species Act and grizzly bears). Some people say that we need new policy. For example, the news is full of claims that better GYE-wide policy is needed for wolves, elk migrations, and visitor numbers. Claimants for new policy are frequently met by counterclaimants saying that current policy is fine (e.g., grizzly bear management claims by environmentalists vs. counterclaims made by some Wyoming citizens and officials). Elk management in Wyoming is another contentious issue, especially around feedlots and over chronic wasting disease (CWD). There are many problematic policy challenges in the GYE and there is always room to do better. Typically, we take on GYE policy issues one at a time and talk about them in concrete management terms (e.g., wolf numbers, grizzly bear deaths, deer movements). But there is confusion over policy as a concept and practice, and with its inherent politics, and therefore policy is often a messy process.

To me, overall GYE policy process raises bigger questions. Most broadly, the key question is what overarching policy is needed to ensure a long-term healthy future for the GYE? For example, what is being called for by claimants that new policy is needed and existing policy is ineffective? What is coherent policy? When we get bogged down in disputes over technical challenges and technical solutions, are we missing the forest for the trees? In this chapter, I examine the concept "policy," explore the policy process, analytic approaches available to us, and look specifically at GYE cases (e.g., elk management).

Coherent GYE policy should be logical, consistent, and justified. It should be clear, ordered, and integrated. The elements of coherent policy are carefully considered and each part of it connects or follows in a natural or reasonable way. In order to achieve our policy goals in the GYE, we need to start systematically asking hard questions, about policy, process, and consequences for both specific cases and higher-order, systems-wide problems. In this, we need to be as clear as we can about policy's relationship to management, the sciences, nature, society, values, politics, and more.

Management Policy for Greater Yellowstone

The task for people interested in ecosystem-wide conservation is to understand foundational components of natural and social systems. We need to harmonize those features practically—make them coherent. Functionally, we need to invent, accept, and initiate diverse ground-level citizen actions to address systems processes like long-range migrations, large carnivore conservation, and climate change. Simultaneously, we must attend to the high-order constitutive policy processes that promise to secure human-nature conservation in the common interest over the long haul. As I see it, we need more organizations that fully understand real management policy concerns, and that possess the capacity (perspective, knowledge, skill) to facilitate conservation outcomes toward the common interest—a healthy, enduring human relationship to wildlife and nature.

If we want coherent practical policy and management, we need to understand our values and the adequacy of the knowledge we have at hand.[2] We also need to be clear about our criteria for policy judging and action, as well as the policy process that we are using to decide. History and context are especially important. There are more questions and concerns, all of practical import in thinking about coherent GYE policy and management. Among these is whether officials and the public are adequately distinguishing among ordinary, systemic, and constitutive policy problems and processes in the GYE and responding appropriately. I explore these questions and issues below, as we look at the policy process conceptually and through concrete GYE examples.

The policy process

Policy is a word we hear all the time, but many people are not clear on what it is or what we mean when we use the word. Is *policy* referring to politics, law, science, government, implementation, values, decisions, advocacy, monitoring, or something else? What is the relationship between politics and policy, science and policy, and government and policy? How do we see policy content (e.g., wildlife biology) versus process (e.g., decision-making)? Is the policy arena organized to consider both content and process matters in an integrated, coherent, and ecosystem-wide way? Here, I use a functional definition of *policy* as a social process of authoritative decision-making by which members of a community clarify, secure, and sustain their common interests.[3]

Policy analytics

The policy sciences, the oldest and most comprehensive approach in the modern policy analytics movement, give us a practical concept of policy that allows us to better understand and participate in policy dynamics. The policy process is an approach for grappling with problems, small and large (Figure 7.1). We can understand this process through our heuristic from Chapter 1: *people* seek *meaning (values)* through *society (institutions)*

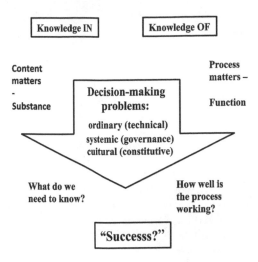

Knowledge needed about policy process

Knowledge IN   Knowledge OF

Content matters
-
Substance

Decision-making problems:
ordinary (technical)
systemic (governance)
cultural (constitutive)

Process matters –

Function

What do we need to know?

How well is the process working?

"Successs?"

Figure 7.1 The policy process illustrated here is a framework for clarifying, securing, and sustaining the common interest of communities. It recognizes interacting substantive matters and process dimensions.

using and affecting *environments (resources—natural and cultural)*. This heuristic is the simplest way to lay out the policy process, in general terms. Its four elements are the content of management policy and, as such, are open for investigation and empirical inquiry. In any case, we need to understand all four components and their interrelationship in real-world terms in any particular GYE case, and in the overall GYE system. Skillful use of the policy sciences framework enables people to realistically understand the problems we face, their relevant context, and what to do about them.[4]

Within the policy sciences framework, it is crucial to distinguish between special and common interests. Special interests are demands for values made on behalf of a person or group, and supported by expectations that the demand will be advantageous to them.[5] When people are looking to secure their special interests, conflict often abounds, as different people make demands that are personally justified, yet often contradictory, and not in the common interest. In order to truly maximize the efficacy of policy, however, we must identify the common interest—interests that advance shared values and goals. But, the common interest has to be discovered in each case as people work together. The common interest is not a rigidly defined set of prescriptive elements known in advance of that work. Rather, it changes given the conditions and participants in each situation. In the simplest conception, "interests are common when they are shared; special when they are incompatible with comprehensive goals."[6]

The policy process illustrated in Figure 7.1 is a framework for clarifying, securing, and sustaining the common interest of communities. It takes skill to use it well. It recognizes interacting substantive matters and process dimensions in all cases. If common interest policy is to be effective and enduring, it is necessary that people are clear about the nature

Table 7.1 Three classes of interactive policy problems in the GYE with descriptions, examples, and standards for each class.

Kinds of Problems	Description	Examples	Standards
Ordinary (technical)	Small-scale technical issues	Building new foot bridge in Teton Wilderness	Technically sound construction, cost-effective, minimize disturbance to wildlife
Systemic (governance, decision-making)	Decision-making level, small and large issues, a process	Cross-boundary issues of bison leaving YNP and entering Montana	Factual, comprehensive, timely, creative, rational, effective, open process
Constitutive (cultural, beliefs)	Cultural-level decision about how to decide, social rules for how all ordinary and systemic issues will be addressed	How should the entire ecosystem be managed, who decides and on what basis—public, experts, or officials?	Same standards as above, plus: • seek high-order common interest, • give precedence to high-order interests, and process should be controlling

of the problems they face; the utility of the conceptions, assumptions, and theories they employ; and the features of the situation they are trying to understand and manage.[7]

Most policy arenas present a complex mix of three types of challenges: technical (ordinary), governance (systemic), and constitutive (cultural), as shown in Table 7.1. We often overlook governance and constitutive problems, instead treating all challenges as though they are technical. However, this prevents us from thoroughly understanding and engaging with the challenges as they are, and limits our ability to construct effective management solutions.

Because the challenges that we face in the GYE are complex, it is often necessary for us to attend to all three interactive problem types—ordinary, systemic, constitutive—simultaneously. Additionally, we must functionally understand the policy process, as well as the technical content of the problems. To be successful, we need a practical way to gather, organize, and integrate information into an operational and practical framework for decision-making. The policy sciences provide a workable framework to do just that.

Participation in management policy

How can a person constructively participate in the management policy process, regardless of the issue? Most people draw on their everyday experiences and conventional understanding when participating in any policy process. However, while these experiences can be valuable, having clear concepts, technical and process knowledge, and analytic skills can increase the efficacy of policy processes. Commanding the concepts explicitly and systematically, and with contextual awareness and skill, can make all the difference in policy outcomes.

Often, people tend to simplify, misconstrue, or overlook key processes and content elements when participating in the policy process. This common problem or oversight occurs as a result of preoccupation with some idea, notion, or mental construct, such as ideology, experimental science, disciplinary boundaries, bureaucratic procedure, job descriptions, program boundaries, or policy preferences. Typically, individuals and organizations are unaware they are doing this. In contrast to conventional political participation, the policy sciences offer an analytic framework and method for empirical inquiry of policy process that is empowering by enabling people to move well beyond convention and into novel, functional ways of conceiving of and interacting with each other. The elk policy case is a rich example worth examining.

Elk management policy case

Take the case of elk management in the GYE as an opportunity to learn about policy process. Thousands of elk range throughout western Wyoming, and elk management policy in this region has been highly controversial for over one hundred years. Although elk are a common property resource, many special interests make claims on their use, including elk hunters, conservationists, ranchers, tourists, and businesses. The elk policy case study below from Jackson Hole, Wyoming, well illustrates the problems of governance, decision-making, ordinary and constitutive processes, and distortions caused by conventional perspectives and special interest. The case looks at concrete specifics as well as the concepts of policymaking and management involved.

Elk and their management

In order to sustain high populations of elk, the Wyoming Department of Game and Fish (WDGF) establishes winter feed-grounds to artificially feed elk. In Wyoming, there are 23 feed-grounds, most of which are managed at the state level, but the largest feed-ground, the National Elk Refuge (NER) in Jackson Hole, contains the largest and most renowned herd, of around eleven thousand individuals. These feed-grounds help prevent elk from starving to death when winter resources are sparse, but concentrating elk on feed-grounds leads to disease, crowding, and other unintended consequences, so conflict over elk management is intense.[8] Additionally, there is long-standing controversy between hunters and environmentalists over the elk hunt that happens each fall in GTNP. For example, historian John Ise referred to the park hunt as "an unfortunate concession to the selfish demands of Wyoming sportsmen" and "an insult to the Park Service and common decency."[9]

The primary problem in the elk case, simply stated, is that the government has seemingly failed to clarify and secure the common interest through the actions and decisions it has made. It is difficult—logically, analytically, and politically—to justify the positions and policies of the WDGF in this case. Instead of working for the public, the dominant state agency has served special interests, including its own, for the benefit of the few over the common interest of the majority.

There are three tests of the common interest that could be used to assess the elk case:

1. Procedural test (is there inclusion and responsible participation?)
2. Substantive test (are valid and appropriate interests recognized?)
3. Practical test (does it solve the practical problems?)

Addressing these questions specifically and with sufficient evidence is one way to determine whether the policy and management is coherent.

The GYE comprises a multifaceted political system that complicates the ordinary, systemic, and constitutive policymaking processes. There is a mix of private and public organizations and agencies operating at all levels, from local to regional to national. Each agency has its own authority, mandates, and jurisdictions for the resources that it controls. This leads to fragmentation and is part of the complexity that currently exists in the GYE. Often, because each agency is guided by different objectives and shareholder obligations, these various groups work in opposition to one another. County and state governments, as well as regional NGOs, lobby for specific economic, social, and environmental outcomes. This matrix of perspectives, values, and practices has created the elk policy fiasco discussed in detail below. As a result, the elk management arena is a conflict-laden, seemingly intractable policy dynamic—what policy scientist Dr. Jason Vogel calls a "persistent policy problem."[10] There are many persistent policy problems in the GYE.

A 2016 study of elk management in the GYE reported conflicting perspectives, problem definitions, and value orientations.[11] Opponents of the annual elk hunt felt their interests were not reflected in decision-making, while hunt proponents were dissatisfied with federal agency management. Agency officials focused on the technical aspects of elk management, such as overpopulation, and did not address the human conflict driven by underlying value dynamics. Decisions made by the NPS and the WDGF, through the implementation of these two different forms of wildlife management, failed to integrate diverse interests. As a result, many nonagency participants felt undermined by the decision-making arrangements and outcomes.

Analysis from multiple peer-reviewed published studies showed that organizational and institutional dynamics severely constrain the scope of deliberations, the production of practical problem definitions, and the search for improvements in elk management. In particular, one major problem is "allocation of competency" that is about who has the recognized proficiency and authority to be involved in elk management policy.[12] Officials usually determine competency, but how does this influence policy outcomes? Questions such as "who decides who is included in the decision-making process and who is not, and do the perspectives, power arrangements, and policy preferences of the dominant agency influence these decisions?" are questions about the constitutive policy process.

The Greater Yellowstone Ecosystem is recognized globally as a model for land and wildlife conservation, management, and policy, but as is demonstrated by the elk case, these often fall short of expectation or perspective. In the elk case, researchers concluded that high-level officials in the state government made decisions that favored their employees and excluded or discounted other experts. In short, the state favors its values, perspectives, and special interest policy over other valid and appropriate interests. The management process functioned poorly and special interests of groups, such as the state of Wyoming and hunters, are privileged, at the expense of other valid and appropriate common interests.

To address these shortcomings of the current policy process, I recommend including a focus on the constitutive level of policymaking (the underlying structure and functioning of policy processes). As well, I recommend increased exploration of how the four elements in the heuristic function in this case and how to affect processes. Finally, I recommend a programmatic emphasis on configuring the processes in ways that embody democratic principles, serve common interests, and resolve policy problems. The hard part is realizing the benefits of these recommendations in the current politized policy environment.

The common problem—a definition

The record of conservation and management policy in the GYE shows tremendous gains over the last century. However, it also shows deeply rooted persistent problems. There is a pattern of analytic errors stemming from disciplinary, programmatic, bureaucratic, organizational, and agency biases that go unrecognized and thus unattended to. One such pattern is in elk management as described above. Other examples of weak or poorly performing policy processes include recreational use of public lands, grazing and timber management, and local-state-federal coordination. We often only recognize the problem with these processes in retrospect—if we recognize them at all—and often only after the unintended, and often adverse, outcomes begin to manifest. Fortunately, such problems are preventable and corrigible, but only if we co-learn, increase our analytic thoughtfulness, and become farsighted through adequate policy process.

Typically, there are two groups of people involved in management policy: formulators (officials) and implementers (technically focused workers). Implementers include agency program managers, bureaucrats, administrators, team leaders, working groups, frontline biologists, technicians, and other people whose task it is to carry out management policy. Implementors must translate policy aims set by formulators—participants who draft and write policy—into actual programs, modes of thinking, and operations. In doing so, implementers make professional, programmatic, and organizational interpretations about projects, studies, and plans based on their own experiences and skills. Because of this process of implementation, the outcomes of a policy program often differ significantly from policy aims.

This is true, for example, with attempts to save the last population of the endangered black-footed ferret near Meeteetse, Wyoming. In the Meeteetse phase of the recovery effort (1981–86) researchers formulated a sound recovery strategy plan and detailed the needed implementation work, but the state of Wyoming failed to heed those recommendations, instead formulating their own special interest policy. The state's self-interested policy nearly led to the extinction of the species.[13] This multiyear case study showed that the state relied on simplistic, overly technical definitions of the conservation challenge, despite the availability of reliable, nuanced information. Further, the program was dominated by a single government agency (WDGF), failed to balance local and national interests, demonstrated a weak decision-making system, was not pragmatic, discounted reliable factual data, showed intelligence failures and program delays, was poorly coordinated, and demonstrated an overall pattern of the domination of special interests and incoherence. This case is an example of what not to do. Ultimately, the efficacy of management policy is determined by the activities undertaken by both formulators and implementers, and the efficiency of the translation between policy aims to on-the-ground application.

In the elk case, other cases in the GYE, many groups expressed concern with the official agency management policy, including ecological professionals, conservation group lawyers, and the public.[14] The primary analytic error in the elk case was the reduction of the conservation challenge into a set of social and decisional processes that favored convention and served special interest groups like the state of Wyoming and its hunter constituents. As such, the thinking, program, and process were highly vulnerable to domination by external and internal interests (e.g., states' rights ideology, hunting rights beliefs, economism), rather than common interest. With the arrival of CWD in 2018, this management policy took on an even more dangerous turn.

Further, why did elk management policy happen the way that it did? First, management policy officials reduced the complexity of ecosystem-scale conservation problem of elk management into a simple formula of inputs and outputs (e.g., feeding and hunting, respectively). By reducing the complexity of ecosystem concerns while trying to bolster elk populations, the program unwittingly generated vulnerabilities because it operated on a fallacy of simplicity. This program, significantly influenced by the WDGF, was constrained by special interest politics and value dynamics that have gone largely unrecognized and unattended.

Second, it happened by reducing the conservation objective to a simple formulaic framework; the program rendered the complex ecological and social challenge into an overly simplistic technical one. It failed to see that the case was really a complex policy dynamic. Misconstruing nontechnical problems, such as complex policy dynamics, as merely technical problems has caused delays and magnified the original conservation problem, an error that was compounded by a morass of bureaucratic rules and regulations. Clearly, the program used to problem solve was inappropriate or an insufficient process given the true conservation problem, which is highly complex, spans multiple jurisdictions, and contains biotic and abiotic elements, as well as human and nonhuman agents.

And, a third error in the elk case was ignorance to the array of forces and factors (social, ecological, bureaucratic, chance) that influenced the social, decisional, and policy arenas of conservation. As noted by Professor Steve Yaffee at the University of Michigan, "Sociology of the network of institutions that participates in its implementation [is key]. While policies are written in words on paper, they exist only in the form of the individuals, organizations, and agencies that implement them and the nature of the information, resources, authority, and incentives that flow between these actors."[15] Yaffee concluded that bureaucrats redefine or translate new challenges (biological, programmatic, or policy) so that they fit into the existing formal and informal operating goals. Therefore, the existing agenda and overall power setting largely determine whether the challenge will be addressed effectively in the common interest.

This brings us to the common problem, which is widespread in many GYE policy processes. The common problem is that we too often overly simplify complex challenges, reducing them to technical problems that we are able to address through specialization (veterinarians, wildlife biology, bureaucrats) and our conventional systems. We tend to rely on traditional prescriptive methods rather than attempt to understand the emergent properties of the actual complex, dynamic systems involved. Commonly, it is easier to take a conventional approach and call the problem "solved" than it is to undertake the nuanced discourse and critical cultural reflection needed to identify the systemic and constitutive challenges that gave rise to the issue in the first place.

How should we see or understand this recurring common problem in GYE management policy process? In the common problem, the emphasis is on content (e.g., elk biology) rather than the adequacy of the process (e.g., see wolf, grizzly bear, bison cases for other examples). Often, the problem with this conventional approach is not clear until the policies have been implemented, and sometimes not for long after. Dr. Archie Carr of the Wildlife Conservation Society captured this common problem when he concluded that biologists, managers, and policy makers should be "willing to use their training and analytic skills [well] beyond the confines of biology, reaching out to examine the cultural and sociological factors that bear on the survival of their favorite species."[16] This point was also made by Dr. Noel Snyder, an endangered species biologist, who said that professionals should guard against "too narrow a focus in any one direction; scientific knowledge is just not enough. We need a new breed of manager who can employ and understand biology without being trapped by it, economics without being trapped by it, and so on."[17] The common problem is a kind of self-imposed blindness, as well as a systems-level process problem. So, how can we best understand and address this widespread challenge?

Recommendations for elk management

Based on analysis of the elk management policy above, I make four recommendations to upgrade policy and manage responses to this complex problem. First, there must be clarity on the concepts of policy process, common interest, and problem orientation among those people involved with decision-making. The public wants an outcome that is understood to be in the interest of the community and other nonhuman entities (elk, in this case). The policy process used to find that shared interest must clarify, secure, and sustain elk conservation. Under the current governance format, the responsibility of securing the common interest falls on authorities and experts, although we are all participants. Importantly, the common interest is not a physical thing or single outcome. It is a fair, open, and effective process. There are many people interested in the elk case who are willing to interact to bring about a common interest outcome. As well, some interested people have recognized competence in these matters but are currently excluded from existing agency-dominated processes. Among those excluded are Native Americans, key citizens, and professionals knowledgeable and skilled in policy process. If we are to capitalize on the expertise and knowledge of these folks, we must develop a new, effective policy arena and process.

Second, individuals involved in the elk policy need to be systematically problem-oriented.[18] Problem orientation is a strategy of rationality, which helps us understand the underlying conditions that actually influence or cause complex problems in the first place. Many of these conditions can be hard to tease out, as they are about values, interests, and adequacy of established social and decision processes. Not everyone is knowledgeable and skilled enough to flush out these deeply rooted assumptions, special interests, and preferences. Rationality provides a strategy for problem definition and finding practical solutions in cases like the elk case. However, use of these proven concepts and tools, such as the policy sciences, is stifled by the agencies in the present arena. Agency experts continue to be trained in colleges and universities in outdated, ineffective ways of addressing problems. We still teach disciplinary-based, technical outdoorsman methods when in fact wildlife management is actually a complex social and technical matter. As a result, individuals focus narrowly on their own ideology, organizational loyalties, or policy preferences rather than effective decision processes in the common interest.[19] The elk case is only one among many cases where we are really being tasked with increasing our abilities to think in a pragmatic, systematic problem-oriented ways, yet we continue to rely on old, partial approaches that are deeply entrenched.

Third, improvements in this elk case will require changes in knowledge, skills, and the people involved. There is a plethora of knowledge and skill already in place about how to encourage effective governance in the common interest. However, these resources are seldom used, largely due to deeply institutionalized conventional ideologies that now dominate.[20] Change is needed in how we conceive of management policy, as well as how we use technical experts, authorities, special interests, and the public to discover our common interest. Workshops, courses, and cases studies on problem solving have been used to good effect to make some gains toward addressing such problems. These workshops could help conventionally entrenched experts and officials alter their paradigm of practice.[21]

And fourth, more generally, we must simultaneously upgrade our attention to the structures, process, and context of policy and management. An effective decision-making process should do the following:

1.   Arrange for common interests to prevail over special interests.
2.   Give precedence to high-priority common interests.
3.   Protect both inclusive common interests.
4.   Give preference to participants whose value position is substantially involved, without losing site of the overall common interest.
5.   Allocate resources to adequately control the situation.[22]

To achieve these goals in a coherent fashion in the elk case, changes are needed in the areas of resources, knowledge, funding, public attention, and media.

These four recommendations draw on the policy sciences to improve the elk management policy, and conservation policy overall. Because the policy sciences are committed to advancing human dignity, democracy, and effective problem solving, it is a useful tool for securing the common interest. On a large scale, the policy sciences approach can help society secure the kind of nature–human connection that we claim to want in the Greater Yellowstone Ecosystem and beyond.

Policy and Society

Conservation aims to maintain resources for future generations in a coherent, practical way.[23] GYE's public is concerned with Wyoming's elk feed-grounds, wolf predation, and transmission of diseases between wildlife and stock animals, all of which are conflicts that have been ongoing for decades and, in some cases, over a century.[24] Though a few large carnivores have recovered from near extinction, no states in the GYE have codified the "public trust doctrine" to protect them adequately, says geographer-professor Adrian Treves at the University of Wisconsin.[25] Public land management is changing as the current federal administration removes what it calls "bureaucratic obstacles" to resource and economic development on federal lands.[26] This will bring less protection for endangered and threatened species and to the complex ecological processes that we all rely on. Given this, what is the main policy problem and what should we do about it? The answer to this question requires that we move well beyond conventional, everyday views of politics and policy, and look at the bigger picture, functionally.

Securing the nature–human environment

The GYE's significance rests on much more than a collection of material things found within or outside the borders of the parks. Of course, YNP is composed, in part, by its material entities, including rivers, wildlife, and other resources that we use and exploit today. Fundamentally though, YNP is also a cultural entity that reflects our relationship with nature. Many of the challenges that we are currently facing in the GYE are rooted not in the material world but in the way we conceptualize nature and our relationship to it. If we are to address many of these challenges, we must first generate situations where common ground outcomes are likely to be found. Presently, we lack the appropriate spaces and tools to inclusively discuss effective conservation in the GYE, and this contributes to our inability to secure the common interest.

A good example of this is the conflict over elk, deer, and bison migrations throughout the GYE. As mentioned in earlier chapters, these large ungulates traditionally migrate throughout the region, causing "human-wildlife conflict" when they run into barriers such as roads, farmland, or fences. As present, there is no forum in which migration management challenges are addressed and coordinated by officials systems-wide to the common interest standards needed. In 1964, the Greater Yellowstone Coordinating Committee (GYCC) was founded to try and unify decision-making across the breadth of jurisdictions that have authority in the large-scale, regional problems facing the GYE. And while the GYCC has begun organizing on these subjects, there is much work that remains to be done. We need to bolster interdisciplinarity and public forums where people can come together to debate and decide on policies. Without such appropriate methods and arenas, it is impossible to secure the needed conservation processes.

So, how do we ensure a more successful human–nature relationship in the future— one that secures many of our conservation and other social goals in the GYE? How do we ensure that our ecosystems are not pressured by the threat of overexploitation. How do we ensure that wildlife populations are able to coexist with human populations in a way that enables the flourishing of both? We often denote these goals as "sustainability" or ecosystem "resilience," but these phrases are unspecific and thus are not particularly helpful when trying to lay out a set of clear goals and actions in the common interest.[27] And of course, it would be flawed to assume that achieving such goals is simply a matter of getting more data, doing more science, and electing better politicians. Such is not a recipe for success. While these things would certainly help in the GYE, the fundamental problem that we are facing is not technical but cultural and constitutive. In essence, we are experiencing a mismatch between our society, its mechanistic worldview, and the way that nature functions.

If we are to truly make progress on these issues, it is imperative that we transform our view of the human relationship with nature from a binary ("Society" is humans with our built environment, and everything else is "Nature") to a set of interconnected systems that are constantly shaping and informing one another. Again, this is the role of the heuristic I presented in Chapter 1: **People** seek **meaning (values)** through **society (institutions)**, using and affecting **environment (resources)**. Nature is not a static entity separate from ourselves. Rather, we are constantly influencing natural systems, while simultaneously being influenced by them. We are not distinct from nature, far from it. We are an integral part of its dynamic systems.[28]

Unfortunately, we do not presently have the concepts and problem-solving tools as a society to fix the fundamental problem. If we did, we could alter the dominant progressive cultural myth, attend more to the problematic myth, and ultimately develop a myth of self-restraint and responsibility.[29] Currently, we tend to "scientize" both social and environmental challenges, and proceed with "economism," despite overwhelming evidence that the challenges we face are not scientific or capitalistic.[30] If we are to transform our dominate view, intentional engagement will be needed from many, diverse members of society. However, presently some people are developing political fatigue or, more specifically, fatigue over arguing-about-politics.[31] People are becoming disillusioned with the process of policymaking and thus are alienated rather than incorporated constructively.

We are living presently inside the "Overton Bubble"—a political science concept coined by Dr. Joseph P. Overton that describes the range of acceptable views and opinions that can be held by respectable people.[32] In the GYE case, "respectable" refers to views that can be integrated into society without significant controversy (i.e., not upsetting the status quo), and it is the progressive myth that largely defines these views.

The concept of the Overton Bubble further refers to the inability of those inside it to hear or think about other viewpoints. It is a mechanisms for political control and maintenance of the status quo. If a troublesome viewpoint can be defined as outside the

Overton Bubble, then respectable society is free from having to consider it. The elites in government, business, and politics are free from its influence and those who hold that viewpoint are marginalized. The Overton Bubble is a social and political tool essential for the elites and the system in power to maintain themselves, allowing them to subvert views that threaten their dominance. Hence, the reimagined human–nature relationship that I am calling for in this book can only occur at the margins of our society presently. But, I hope that this new view will gain popularity and its efficacy for improving conservation outcomes will be proven. If so, we may be able to generate the kind of broad and deep discussion and collective action needed to transform our current cultural model of policy for the better.

Constitutive process of authoritative decision

For the GYE, environmental problems are systems-wide, as in large mammal migrations, large carnivore conservation, and climate change. However, the processes of policy and law operate at smaller scales. This incongruity yields high fragmentation in the region. At this time, the GYE's constitutive policy process has not risen to the level needed to ensure harmonious human–nature coexistence. Ideally, the constitutive policy process would stabilize expectations about authority and the degree of control needed. Presently, the current policy is way too rudimentary to meet this need throughout the GYE.

As it is now, the constitutive policy process provides only the basic features needed for the effective construction and implementation of sustainable human–nature policy and law in the GYE. Fortunately, the GYE constitutive policy process has been improving at an accelerating rate over the decades, particularly with regard to issues of conservation and environmental protection. With the GYCC, some attempts have been made to decrease fragmentation of policy and decision-making, but much of this improvement has been haphazard.

There are several key dimensions that must be considered when examining policy process in the GYE and elsewhere:

1. **Participation**: The GYE's policy processes have been somewhat democratized; individuals and groups concerned with the environment and wildlife today have more opportunities to participate in policymaking and law through citizen science, advocacy, implementation, monitoring, and appraisal. However, the system still preferences technical and bureaucratic authorities, so there is a long way to go before we have adequately addressed the questions of participation (who gets to decide?).

2. **Perspective**: The perspectives of the most influential people and groups are the basis for most of social process of decision-making in the GYE, and this depends on the stability and realism of these elites. There has been a perceptible movement toward demands, identification, and expectations needed for effective human–nature policy, and a public order for human dignity.

3. **Arena**: The policy process occurs within a structure of authority in which participants interact (e.g., YNP, national forests, federal and state relations, GYCC). In recent years, the variety and inclusiveness of these arenas has grown. However,

even these gains fall short of effective management policy. This is because they exclude many valid and appropriate citizens and their interests. Additionally, many of these arenas are simply short-lived responses to incidents and crises and do not have the temporal scope to address many GYE's foundational problems.

4. **Power**: The GYE is encompassed by three states: Wyoming, Idaho, and Montana, but also is comprised of a large number of federal jurisdictions. Do these states have the exclusive competence to determine wildlife management policy for the entire GYE? No, nor should they. What happens when state and federal management policies are at odds? Should criteria for competence to participate in management policy processes be more inclusive than states admit? What is the role of everyday people in policymaking when they may not have some of the technical knowledge or skills required to be competent in a policy arena?

These are among the many key questions that need immediate attention in the GYE.

Who has authority and control over the management policy process and resources in the GYE? Authority and control are two different things. Just because government has authority, they may not have control and vice versa. Currently, the relationship between authority and control varies in the GYE depending on the specific case and situation. We saw through the elk case that the primary agent and authority, the WDGF, had enormous influence (control) over the public policy process, its outcome, and its effects. The policy system in the elk case worked to concentrate exclusive competence in the state, excluding valid and appropriate interests in elk management from external, but equally legitimate interests. Thus, understanding how and why we allocate decision-making authority and competence is directly relevant for addressing GYE's environmental problems. Failure to include full competent participation is not in the common interest.

Over the decades, there has been a gradual movement in the GYE toward greater inclusivity and shared competence (i.e., those with the value outlook, knowledge, and skill to aid problem resolution in the common interest). This is helpful, but it is hard for authorities to recognize who has genuine competence. This inclusive competence trend must be accelerated. We need to come to grips with and fully understand the consequences of failing to include adequate competence and reliable knowledge in management policy processes. We can best manage GYE's collective problems, only if we can maintain an appropriate balance in authority and control, and in inclusivity and competence. We are not there yet.

Policy process—strategies

The current strategies for achieving goals in the GYE range from persuasive to coercive. In carrying out the functions that maintain society and the GYE, we use four key strategies: diplomacy, education, economics, and force (e.g., laws, sanctions). In the GYE, groups employ these strategies differently to meet their own objectives. First, diplomacy is coordination across high-level decision makers to influence leaders. For example, the superintendent of YNP may speak with US senators to achieve more funding for the park or NGO leadership may communicate with local business owners to change advertisement messages about appropriate land use. These are examples of diplomatic channels. This strategy employs exclusive competence and participation and is very common in the GYE.

The second strategy, education, attempts to increase individual and public understanding about technical information, and to selectively encourage participation in the management policy process. The targets of education programs vary widely, but the intended audience is often youth in schools.[33] Adults are also commonly targeted through public talks, citizen science, and advocacy. Examples of education programs in the GYE include the Teton Science School, Yellowstone Forever Institute, AMK ranch in Jackson Hole, local bird clubs, and university field trips. This strategy tends to be inclusive and could be significantly expanded to promote conservation and policy competence in the GYE.

Third, business and economic strategies seek to achieve goals through the direct control of resources and the management of wealth through markets and neoliberal capitalism. This strategy is dominant in the GYE at present and largely results in the exclusive cloistering of competence in the business community, which controls who is permitted to engage with policy. This strategy is reflective of the mechanistic or progressive myth of human development and our relationship to nature.[34] This strategy has serious implications and can impede effective policy process. However, given that we must work within the dominate paradigm until change for the better is manifested, this approach can be a key aid in conservation management policy throughout the GYE, if employed selectively and in the common interest.

Finally, force is the use of laws, social norms, and sanctions to structure and stabilize the way we live. Force is a necessary strategic instrument. Most people comply with laws and norms. Yet, some people choose to violate social norms and practices, as well as civic and public law (e.g., wildlife poaching). This force strategy (e.g., policing, suing, coercion) must be used selectively and wisely in the GYE to bring accountability to the policy and management process.

Over the decades, ordinary, systemic, and constitutive processes in the GYE have been modified and improved using these four strategies. This has generated much of the progress that we see today, although we have not yet achieved a fully functional policy process. Overall, policy decisions in the GYE are becoming more comprehensive, and ecosystem-wide management policy is slowly evolving. Yet, most ordinary, systemic, and constitutive processes are still too fragmented and parochial to secure the needed conservation objectives. Looking to the future, these processes must become more rational, employing the most up-to-date and reliable knowledge about people, society, nature, and resources to attain public goals, demands, and expectations. In short, we need to work toward incorporating potentially divisive and contradictory claims about common interest in the GYE, and pattern those claims into a common interest outcome that is widely supported.

Inherited public order in relation to resources

The policy and management we now have in the GYE was passed down to us from the work of society and many individuals over the past decades. We are the beneficiaries of their foresight, hard work, and effective participation in diverse policy processes. We inherited policy prescriptions (and policy processes) that determine how we should allocate and exploit resources in the GYE, both cultural and natural. In reality, the exigencies inherent in the human relationship to nature do, in fact, impose severe restraints on us that are reflected in what we are experiencing today—species extinctions, invasive species, and influx of new diseases. Only now are we coming to more fully recognize the need for reciprocal protection of resources in the common interest. We still have a long way to go for resource protection in practice.

The human community in the GYE has allocated resources among different interests to maximize inclusive competence. At its best, this process aims to regulate the behavior of individuals, groups, and communities to avoid the overexploitation or damage of resources.[35] Together, these objectives reflect a desire for a more rational relationship between people and the resources on which they rely. This requires the current policy process in the GYE to be updated, made more realistic, and focused on ecosystem-wide competence.

Yellowstone National Park and all other public lands are a great public good and a part of our historic policy, especially constitutive process. We need to protect that investment as we go into the future. As well, we need to mitigate the injurious use of resources that is happening throughout the GYE currently (e.g., harmful techno-adventure recreation, exurban development, large volumes of tourism overloading environmental limits). In the GYE, we must functionally achieve a continuous flow of authoritative community decisions to minimize harm and to maximize conservation and coexistence. This will require increased capacity and shared reciprocity of understanding and action throughout the GYE for many people and levels.

Decision-making functions

Conservation gains in the GYE only come about through decision-making processes. The decision process is the system made up of groups or individuals negotiating their desired expectations and demands. Decision-making outcomes can result in win-loss relationships, compromise, or, ideally, win-win (integrated) outcomes. Effective GYE decision-making processes will help individuals identify mutually agreeable outcomes, hopefully integrated, win-win outcomes where possible.

There are six decision process functions in all management policy process, each of which has standards that we should try to meet (Table 7.2). Though different people are typically involved in each functional phase, realistically the overall process should, to the extent possible, address problems in the common interest. In the GYE, we can work to upgrade these functions and the process as a whole for greater conservation.

The six functions are as follows:

1.   *Surveillance and Planning.* Also known as the intelligence function, this is the process of collecting, analyzing, and disseminating information to those involved in decision-making. It is imperative that the information gathered about the situation is accurate and unbiased. The information should also be widely available so that the entire community can legitimize it.

Table 7.2 Questions to ask when investigating a decision process.

Decision Process Functions	Standards	Questions to Ask
Surveillance and planning (intelligence)	Comprehensive Factual Selective Creative Open	Are data (intelligence) being collected for all components of the problem? Do these data cover all affected participants and the problem's context?
Promotion (debate, discussion, advocacy)	Rational Integrative Comprehensive Effective	Which groups support which courses of action (informal and formal)? Which groups benefit from the different courses of action?
Prescription (deciding, commitments)	Balanced Effective Inclusive Future-directed	Does the new course of action harmonize with existing rules and institutions? What rules are self-imposed (by the community)? Which courses of action are binding?
Implementation (invocation and application activities)	Timely Rational Dependable Effective Nonprovocative Unbiased Constructive	Is implementation consistent with the new course of action (prescription)? To whom do the rules apply? Who enforces the rules? What sanctions will be enforced and when? Are there currently resources to carry out implementation?
Appraisal (evaluation, monitoring)	Dependable Ongoing Unbiased Practical	Who does the program serve and not serve? When is the program evaluated? Who is accountable for successes and failures?
Succession (termination, adaptation, moving on)	Timely Comprehensive Dependable Supportive	Who terminates or changes the program? Who does the change serve or harm?

Adapted from Clark (2002), modified Table 4.1.

2. *Promotion.* The primary objective of this function is to identify possible policy alternatives that may be able to address the underlying factors that are driving an observed problem. Examples of this include media or advocacy. Ideally, promotion will formulate effective policy options through rational evidence that considers the multiple realms (biological, social, political, economic) that will be influenced by the policy once it is implemented.

3. *Prescription.* Once the proposed alternatives have been debated, the prescription function solidifies the community's expectations. This phase includes setting laws and regulations, developing new programs, or changing social norms that determine how the broader community should act. These regulations or programs are then enforced to achieve the goals articulated by the prescription. To be legitimate, good prescriptions must be fair, balanced, and widely accepted by the broader community.

4. *Implementation.* This is the function by which the community puts the new prescription into effect, ensuring that it is in concert with preexisting rules, thereby resolving disputes that arise. Prescriptions should be implemented in a timely and rational manner and should not antagonize affected participants.

5. *Appraisal.* Once a plan has been implemented, we must examine its practical outcomes. Are we achieving our intended goals? Are there other impacts that we did not anticipate? Are these impacts positive or negatively affecting our goals? Appraisal should be practical, cost-effective, timely, and unbiased. Rather than finding data to promote the success of a given policy, appraisal should aim to understand what does and does not work, why that is the case, and how the current prescription can be improved.

6. *Succession.* Once we have evaluated the prescription, we need to decide whether to continue with the current policy or move on. Succession is the formal process of modifying or ending a prescription that does not achieve the desired result, and replacing it with a more effective program. Succession should not connote a failure on the part of policy makers, but rather be viewed as necessary step in the advancement of nuanced and effective policy solutions. This function should be timely, factual, and supportive of the individuals that the succession harms.

These six functions have long-term consequences or effects, and together these are called the "management policy process." The creation of Yellowstone National Park is but one example of a long-term consequence of an earlier constitutive policy process.

Policy arenas

Constructing appropriate venues for discussion and encouraging effective social and decision process outcomes in the GYE are always challenging. In the GYE, we are struggling to organize the needed arenas to address large carnivore conservation, large mammal migrations, and many other challenges. There are challenges to building an operating arena adequate for GYE's challenges. One was noted by Murray Bookchin, a historian and political theorist, who claims, "As long as domination organizes humanity around a system of elites, the project of dominating nature will continue to exist and inevitably lead our planet to ecological extinction."[36] This sentiment reflects a growing awareness of some people in the GYE about how our social structures influence and inform our environmental challenges, including how we organize arenas.

Management policy arenas

All of human interactions and decisions take place in social and policy arenas. A policy arena is an organized situation or social process (e.g., courts) that shows certain features, if it is to be effective. This includes features such as whether or not arenas can be centralized or decentralized, continuous or short-lived, specialized or generalized, organized or unorganized, open or closed to broad participation.[37] From a policy perspective, an ideal arena is sufficiently flexible to balance these factors in a manner that allows the broadest common interest community goals to be realized.[38]

In the GYE, there are many people and groups participating in the decision-making arenas. For example, elected officials at the national, state, and local level interact with one another to create legislation. Bureaucrats in the agencies interact with the superintendents of Yellowstone and Grand Teton National Parks, as well as US Forest supervisors, to determine funding requirements for the upcoming fire season. National, regional, and local environmental NGOs in the GYE interact with local community leaders to develop projects that can protect parks and forests, while serving local needs.

However, using the everyday established arenas that exist now, conventional understanding, and accepted action to address GYE's current challenges is insufficient to address the more complex problems we face. I maintain that we need to be more integrative, practical, and effective going forward as we organize arenas and include citizens. Ideally, a management policy arena or situation should be coherent, realistic, inclusive, justified, and committed to the common interest.[39]

Given the multitudes of people now working as conflicting agents in the GYE, how do we secure effective policy arenas? Understanding the factors that distinguish and characterize arenas—special (self-interests) versus common interest arenas—could help participants in the GYE stabilize current arenas, create new ones, and develop the conditions necessary for better decision-making outcomes and effects, in any particular case and overall.

Organizing effective arenas

We must pay close attention to, and work within, the five key features of arenas shown in Table 7.3 in the GYE, and as discussed below. Let's examine these five aspects of arenas to see how we can organize more effectively in the GYE. Consider how these five elements played out in cases we have explored in this book (e.g., the elk case, ecosystem-wide policy, long-distance migrations). As well, look at own experience in these terms.

First, the degree of centralization of an arena describes the range from top-down to bottom-up decision-making. The two approaches are not in opposition but are complementary to one another. Using top-down decision-making, government can manage public lands and work for grizzly bear conservation in ways that bottom-up approaches do not permit. At the same time, environmental NGOs use bottom-up approaches for livestock management, capacity building, and others. Neither approach in this continuum is adequate alone. Instead, we must find a suitable balance between grassroots and bureaucratic management.

The second salient feature is whether the arena is continuous or short-lived. Some problems require attention over the long-time scales, such as grizzly bear recovery (which has been ongoing for nearly a sixty years) and climate change (which may take hundreds of years) in the GYE. Others require only short-term attention, such as a proposal for a new hiking trail or a system of mitigating of landslides. The degree to which an arena is

Table 7.3 Questions to ask when investigating and participating in any management policy arena.

Structure of Arenas	Questions to Ask
Centralized or decentralized	Which groups (organizations and individuals) have control over decision-making?
	Are these groups local, regional, national, or international?
Continuous or short-lived	What is the time frame of the issue at hand?
	Does the pattern of human interaction match the time scale of the issue?
Organized or unorganized	Is it an organic process or is it highly structured?
	What are the rules (formal or informal) on how a decision should be made?
Specialized or general	Does the problem affect social values and issues of social equity/justice?
	Is the problem a technical decision/narrow in scope?
Access	Who has the ability to participate?
	In what capacity?
	With what restrictions?
	What change can they effect?

Lasswell and McDougal, *Jurisprudence for a Free Society*; McDougal, Lasswell, and Reisman, "World Constitutive Process of Authoritative Decision," 191–282.

continuous or short-lived is determined by the scope of the issue it is designed to address, and should be adequate for the temporal scope at hand.

Third, arenas can be organized or unorganized, depending on how extensively an arena is structured. Organized arenas allow decisions to be made quickly and with economic efficiency, particularly when the issue is technical. In contrast, unorganized arenas allow for more organic and creative policy alternatives to develop. Just as with the dynamics of centralized or decentralized arenas, the goal is to find an optimal balance between organized and unorganized arenas. For example, an education program requires enough organization in the curriculum for administrators to ensure a consistent quality of education. However, it is desirable for the educator to be independent and sufficiently flexible to adapt to students' different learning styles.

Fourth, is the arena specialized or general? That is, to what extent is the arena guided by experts? Currently, many arenas are designed to include only experts, with citizens operating strictly as observers. One example is when biological experts are consulted for assessing and implementing conservation targets. In contrast, general arenas include broad democratic participation, such as the promise of effective community-based initiatives. An effective balance of these two kinds of arenas may use general arenas to determine conservation goals and specialized arenas to design and implement programs to achieve those goals. This strategy would mitigate the detrimental impacts of conservation programs on local communities, which often serve to alienate, rather than empower, the communities that rely on a threatened resource base. An effective management policy should uplift these communities as well as the environment around them.

Finally, arenas vary in accessibility, or the degree to which any particular participant can be involved. Access to arenas can be free and open, as in community meetings, or highly restricted, such as arenas for national security. Access varies widely in the GYE, depending on the issue at hand. Access is often limited by physical or geographical attributes that restrict individuals' abilities to travel and attend. In order to attain democratic decision-making, arenas should be open to those who have a legitimate interest in the outcome of policy decisions. Or, access may be compulsory if those deemed critically affected by the policy are required to participate. Of course, fully democratic arenas are not always favorable. Sometimes it is helpful or necessary to limit participation, particularly in specialized arenas. However, when management policy decisions affect the broader community, significant effort should be made to increase participation by those directly affected by the outcome. Policy arenas in the GYE today generally limit access through either formal restrictions by government agencies or a lack of infrastructure to promote geographic access.

Scientists in arenas

Many management policy issues involve technical information that scientists and experts command, but the impacts of these policies affect the broader community. Should the role of scientist differ among ordinary (everyday, technical cases), systemic, and constitutive issues? In our society, we know that the majority of beliefs, identifications, and attachments—including those to "nature"—are derived from the thoughts and words of people. For example, in YNP the dominant representation of nature is the one favored by scientists, one that asserts that there is a true order to the world that, if we study it closely and diligently enough, we will be able to ascertain it through data. Because of this widely accepted view of nature, today, scientists comprise the most influential epistemic community, a network of knowledge-based experts who help decision makers and the public understand problems and develop solutions.[40]

It is important to consider what kinds of scientists and what specialties are needed to help society and decision makers understand the challenges in the GYE. Currently, there are around 1,300 recognized environmental disciplines.[41] Disciplines in this case are the institutional mechanisms for regulating the relationships between consumers and producers of knowledge.[42] Typically, scientists and experts are knowledgeable in just one discipline, which promotes entrenched, fragmented, and conventional thinking. This impedes our ability to find integrative solutions in the GYE. Thus, when scientists convey their representations of nature and problems to each other and the public, they are perpetuating cognitive and methodological biases that permeate into our culture. Further, these scientific perspectives or stories differ from the personal and cultural stories that we hear in daily conversation and see in the media, examples of which are given in Chapter 1. It is important to note, however, that while scientists' narratives follow different standards than everyday storytelling, the purpose is the same—to make meaning of the natural world and our role and responsibility in it.[43]

Because the most common representations of nature in the GYE are those put forth by scientists, especially ecologists, we find ourselves in a situation of "epistemic dependence" on scientists.[44] We have internalized these viewpoints and rely on them when constructing our stories about the GYE and our place in it (see Figure 1.3).

We see examples of scientific narratives in *Yellowstone Science*, the official publication of the Yellowstone Center for Resources, which is the chief science arm for YNP.[45] The March 2015 issue, for example, focused on *Ecological Implications of Climate Change on the Greater Yellowstone Ecosystem*. It included articles from world-renowned university-based ecosystem and landscape ecologists about topics ranging from changes in snowpack to water balance to forest and wetland composition to wildlife to steam sediments, and more. These scientific narratives are authoritative and all had implications for the future of the GYE, but each was represented through the lens of climate change and its consequences on the GYE.

With the previous discussion in mind, I recommend that people concerned with the future of the GYE realize that scientific knowledge is important, but seldom enough. As Victor Hugo wrote, "Science says the first word on everything, and the last word on nothing." At the same time, I recommend that scientists, decision makers, and citizens use the integrative, empirical framework presented in this book to best understand and participate in the management policy process, regardless of the issue at hand. As well, I see that we need to better organize most conservation arenas and decision processes for better outcomes and effects. Asking the questions in Table 7.3 can help organize arenas to be more effective for conservation gains in the GYE.

Lastly, we must create and use better structures for the decision-making process, attending to all six functions and standards described in Table 7.2. Conservation decision-making in the GYE focuses heavily on surveillance, planning, and promotion to the exclusion of the other functions (e.g., systematic appraisal for learning and improvement). Considerable effort is made to collect, process, and disseminate biological information about the ecosystem. For example, detailed planning exercises, such as environmental assessments and impact statements, have taken place to recommend ways of proceeding in management decisions (e.g., Moose-Wilson Road upgrade in Grand Teton National Park). However, overlooking the context and social aspects of conservation in these functions has led to inadequate prescriptions, as well as an inability to effectively implement conservation programs in the common interest. The ability to draw upon past successful experiences—and understand why they worked—can also help participants develop better practices to apply in the future. To further advance effective management policy in the GYE, I recommend practice-based projects and prototyping as introduced in the next few chapters.[46] Prototyping is a continual process that helps participants learn from past experiences to improve decision-making functions and is particularly useful for the succession function. Lessons can and must be learned from project failures, too. The GYE offers many learning opportunities. However, at present, too many go unrecognized.

Integration and co-learning

There are practical alternatives for avoiding the common problem diagnosed previously that can greatly aid future management policy in the GYE. There are two ways to respond to the increasingly serious human–human conflicts. First is to seek more positivistic science and develop technical solutions—"scientific management." Second is to understand the basic human behavior that drives the problem at hand and find dynamic practical remedies. This includes understanding how problems are defined in the first place and why decisions are made—this is "adaptive governance."

The Jackson Hole Conservation Alliance and allies (e.g., Yellowstone to Yukon, The Nature Conservancy, Greater Yellowstone Coalition) use the first approach to tackle technical problems, such as car collisions with wildlife. These organizations push for more data on animal movements and traffic flows and propose engineering (technical) solutions, such as wildlife under- and overpasses on the highway. While this approach certainly makes sense, the underlying causes for collisions with wildlife (i.e., human values, distractions, speeds) go unaddressed fully. This dominate approach seeks to engineer a way around our problems. However, this approach is limited because it fails to address underlying human causes. Is the problem a lack of overpasses? Distracted driving? Or is the problem the choice to build roads along crucial wildlife migration corridors? I argue that it is the latter, which is further related to our exploitative understanding of nature, and thus that engineering technical solutions will only mitigate, not solve, the problem at hand.

What is a better approach? Some people promote adaptive management to improve on traditional strategies. Adaptive management policies operate under the assumption that positivist science can be optimized by combining it with experience and knowledge about context.[47] However, adaptive management often falls short because it is not well understood, and therefore little used.

I argue, however, that we need more "practice-based" management approaches. Practice-based management uses a systematic, problem-oriented, contextual, and interdisciplinary approach to address problem rather than simply employing a continuation of expert positivistic science or overly simplistic community-based initiatives.[48] Ronald D. Brunner, a professor of practice-based approaches at the University of Colorado, wrote two books introducing "adaptive governance": *Finding Common Ground: Governance and Natural Resource in the American West* (2002) and *Adaptive Governance: Integrating Science, Policy, and Decision Making* (2005).[49] Adaptive governance is a system of administration that encourages individuals and groups to gain the knowledge and skills for a practice-based approach. To date, there is only one organization in the GYE that explicitly applies this approach—the Northern Rockies Conservation Cooperative (NRCC).[50]

Adaptive governance (detailed in later chapters) takes a large step beyond adaptive management. Adaptive governance requires that we see problems as evidence of a need to fundamentally change our underlying notions of reality and truth for practical management policy (Figure 1.3). Our understanding of ourselves in the GYE wildlife-rich place needs to be changed, and this requires critical reflection, insight, and the development of new skills to help us understand who we are as biological entities, members of complex societies and cultures, and agents in dynamic environmental systems. By externalizing our problems and relying on scientific management, we fail to achieve these objectives. To move forward effectively in the GYE, we must build on the advantages of the positivistic approach, while encouraging progress toward the practice-based approach.

The Starting Point

One way to begin moving toward our goals in the GYE is by employing future mapping, also known as the "developmental construct" method. Future mapping was first proposed by Professor Harold D. Lasswell in 1951 and is a strategy that provides a way for us to move beyond simple experience toward broad understanding. Lasswell's future mapping method is problem-oriented and contextual, so it clarifies the challenges we face and encourages reasonable, responsible, and practical responses.

Future mapping

The future mapping approach provides data about ongoing change that may be systematically overlooked. Lasswell introduced the idea of creating developmental constructs to help leaders and citizens understand trends in ideology, public opinion, class analysis, social affiliation, and environmental conditions. This framework could be used to understand the overall GYE situation, and can be used by any group, organization, or agency.

A view of society

Without going too far from the GYE, many social scientists have noted that we are experiencing a gradual transition from capitalism toward socialism, although there is currently a huge backlash against this that is calling for a return to neoliberal economics to guide society, governance, and management.[51] The point here is that this transition is currently affecting all that we do, especially our relationship to resources and nature. It is certainly affecting our relationship with one another politically. Let's explore these very large social trends, so we can keep them in mind as we focus on any single GYE case. And, these larger social processes are the real context for the GYE. Under these social trends is the context inside of which we all live today. The following is a bit general, but this broad accounting can help us keep the GYE in perspective.

Under the current worldview or framework for our society, we have been riding the tide of success, both socially and materially for a couple of centuries. We were, and still are, confident that neoliberalism and scientism will give us a better life. These intermixed beliefs strongly dominate our culture today. However, we have become increasingly aware of the harmful global and regional consequences they have generated. This has both social and environmental implications, including the 2008 stock market crash, global temperature rises, and the compounded effects of climate change and resource overexploitation on global conflicts. And, with the rise of globalization, we are also seeing a homogenization of worldviews as the dominant Western civilization influences matters worldwide.[52] These big problems are well beyond what any individual or group can address. Yet they are important, always.

There are many theories of humans in nature that speak of some higher truth, a truth that is "more real" than the truth of everyday, conventional living that dominates now. The GYE ideal invokes this higher truth and semiotic symbols like "Yellowstone" or "Greater Yellowstone" evoke a sense of metaphysical or theological transcendence by confronting notions of Manifest Destiny, frontier ideology, and the pristine nature myth. Often, these symbols and ideas are rooted in religion that, as I discussed in both earlier and later chapters, often perpetuates our flawed progressive myth.

Futuring—criteria for making judgments

If we want to map and understand ongoing events and the likely future in the GYE, we need criteria to judge what is happening and what to do. How can we best understand our present human–nature relationship using our experience and empirical inquiry? How should we set rational criteria to understand our present, given what we know of the past, and at the same time have a focus and a way to make judgments on our future?

If we want to understand the future of the GYE, we can extrapolate from past and ongoing trends and the conditions behind those trends. For example, we can postulate that the trend of increasing visitation to YNP will continue and propose a mechanism that is driving this trend today and into the future (perhaps the increasing cultural cache associated with outdoor recreation or the increasing alienation of people from natural environments in their everyday lives leading to increasing desire for connection). In doing so, we are extrapolating the conditions that caused that trend, which will inform our imagined future outcomes. This method of futuring mapping is applicable for cases of wildlife, human use of resources, and broader environmental concerns.

There are also other grounds for projecting human activities and their effects on the GYE into the future. We have data on standards of living, longevity, and economic growth in the GYE, which we can use to gauge conditions of the human ecology in the region.[53] Then, an estimate can be made from these data, giving us insight into the future. To answer questions about GYE's future, we have to mobilize all the available knowledge on conditioning factors such as anthropology, psychology, sociology, philosophy, and policy. This is the role of integration and interdisciplinarity. So, the critical question: How do we support the continuation of utopian trends and alter conditions to mitigate dystopian outcomes?

Although many biophysical and social trends in the GYE predict a dystopian outcome, we can have pragmatic hope because we know that humans can and do alter their conduct and subjective and collective understanding. New awareness and knowledge can change our meaning of and relationship to places such as the GYE. This is the task in the coming years, if we are to secure a healthy future of the humans and wildlife alike.

Mapping the future

Again, speaking broadly, one of the key roles of scientific knowledge and empirical thought in this framework is monitoring trends (documenting history) and conditions (explaining history). These are two parts of problem-oriented policy mapping. Available scientific knowledge is highly relevant to understanding the past (history) and the present. It is also the basis for projecting policy constructs about possible futures. Without reliable data, views of possible futures will have blind spots and may reflect wishful thinking rather than reality.[54] People tend to suppose or presume "facts" in the face of deeply embedded cultural stories to the contrary. At present, there is no comprehensive, contextual map of the future of the GYE, though such a map would be helpful to all.

Goals

Future mapping (the making of constructs) is a tool for understanding complex and dynamic systems, including the GYE. Generating progress toward conservation in our future requires us to increase our capacities for critical thought and our problem-solving skills. We cannot assume that the future will look the way it is often presented in media today. To adequately estimate the future, we must consider the likelihood that future events will not be a continuation of the present. And to do this, we must appraise the degree to which conventional, ordinary thinking dominates current judgment. Are we ignoring the signals from the future that are now coming in (e.g., climate change) because it is comforting to stay within current beliefs—the Overton Bubble?

One key variable in our thinking about the future in the GYE is our present expectations for it. The function of management policy is to attain our goal values for the future, but it is incumbent on us to be clear about our goals and values in the first place. In my view, the unifying goal (the common interest) for the GYE is to maintain human dignity (justice) in a healthy environment. Given this objective, then, how do we go about inventing, using, and evaluating a contextual map of our present time? Officials and citizens should consider how values are manifested (or not) in present policy and in futuring scenarios. All people are concerned with their own dignity, sense of community, and security. All people want to be valued and respected.[55] Working for universal human dignity in the GYE or elsewhere ensures that these values are realized equitably for all people. Unfortunately, we have yet to achieve this goal as discrimination and other egregious forms of human indignity continue in various places worldwide.

We are now in a global environmental crisis that is a threat to human dignity and well-being. In order to effectively address the challenges this brings about today and in the future, we must be clear about our goals. Consequently, we must constantly reconstruct our future map in response to unfolding events, evaluating how well it tracks our trajectory. There are basically two possible future constructs. The first construct, a utopian society, operates under the assumption that historical trends and conditions will continue to lead us away from poverty, environmental degradation, and oligarchies toward free people in healthy environments worldwide. The second construct, which is dystopian in nature, suggests that history is reversing itself, moving toward more inequalities in wealth, well-being, and access to knowledge. To assess the shortcomings of each construct, we must consider the plethora of empirical data available and constantly assess whether these models are in line with the realities of the world. Assuming the present environmental crisis continues to grow, which is the likely outcome?

Time

If we want to map the future, a developmental construct for the GYE, what time scale should we use? What period should we consider for the beginning of our data collection?

We could choose any arbitrary date: the end of the Civil War (1865 onward), the closing of America's western frontier (1890s), or World War II (1945 onward). There is no universally appropriate time scale. Rather, it is dependent on the context of a given situation.

We need to place the current challenges in the GYE into a deeper perspective than we do at present. I recommend that the Greater Yellowstone Ecosystem developmental construct begin in the post–Civil War era (1865 onward), as this marked a key transition point in the internal balance of power in the United States, perspectives of the public, and international context. Starting future maps with data from a post–Civil War era offers two primary advantages. First, it allows us to concentrate on the study of a comparatively short period in recent history of America. And second, this period is very well documented. Our purpose for the construct is to contribute something new, and by choosing the starting point of 1865, we are able to apply methods that have been developed only recently (visual organization of data, remote sensing, GIS). Both the biophysical and social sciences offer a range of essential data acquisition methods and interpretative approaches that have, thus far, not been applied in the GYE.

The aim of the future construct is to make important historical data available to discussion of contemporary challenges and how to address them. Social trends, such as patterns of behavior, distribution of wealth, relative power of the elites, and social affiliations, are all examples of relevant factors that are seldom considered today in management policy of the GYE. Overall, understanding changes over time and across many variables is important for understanding the challenges we face in the GYE today and for deciding what to do about these challenges in the future.

Conclusion

In moving forward in the GYE, we want to be clear about our goals, our present challenges, and the conditions that are generating those challenges. From there, we should rationalize and unify our management policy strategies to address both short-term and long-term problems. To do so, we need to attend more thoroughly to the past record of successes and failures and reflect upon them to learn from experience. Using that record, we could deepen our understanding of what works and what does not. Presently, even though there is shared interest in the GYE and its future throughout the public and government, we need to work toward clarifying or harmonizing the various desires into a common interest that benefits from enduring public support. We now know that different groups use different worldviews or paradigms to inform their beliefs about how the GYE should be used and who should decide. Today, the management policy process in the GYE is too often fragmented, contentious, and incoherent. Elk management in western Wyoming well illustrates the challenge before us. The good news is that we have established sound ways of working toward coherent policy and management. We just need to use them.

In short, the management policy for the GYE needs to be upgraded, explicitly and systematically, with more attention to both policy processes and content matters. This will require that leaders, experts, and citizens come together to clarify actual challenges and options to address them. We know the functional steps and standards of sound management policy, and they can be used in the GYE. But doing so requires self-awareness, clarity of concept, appraisal capacity, and co-learning opportunities.

Overall, the GYE illustrates the diversity of worldviews that persist in society about our relationship to nature and what that relationship should be in the future. The way we view ourselves and the natural world is a cultural or constitutive policy process that needs explicit analytic attention and adaptation in the GYE. We know what the overall policy problems in the GYE are in ordinary, systemic, and constitutive terms, and we know what tools we might use to address them, at least in principle. Now, we must apply these in practice through cooperative policy and management. We—all of us—can work to create more sustainable management policy for the Greater Yellowstone Ecosystem, if we so choose.

Part 3

WORKING FOR ECOSYTEM CONSERVATION

Part 3 offers strategic, yet practical options to address GYE's multifaceted problems. It builds on and extends all preceding sections. It draws on my half-century's experience in the ecosystem, working in other ecosystems in North American and on other continents, and from broad research, teaching, and fieldwork worldwide.

This part encourages greater foresight and the development of new knowledge, skills, and enhanced self-awareness to best address GYE's problems. The options and examples offered provide a practical way to transition to better thinking and actions that lead to enhanced conservation. As I see it, we must culturally create a new conservation story that offers a sustainable way for nature and people to coexist— a new GYE conservation paradigm that becomes widely accepted and lived on the ground. It must be sufficiently realistic to provide pragmatic steps, actions, and process to make a difference.

In Part 3—on challenges, learning, and the work ahead—I offer options for greater effectiveness in conservation on the ground through cooperative arrangements and upgraded management policy. In Chapter 8 on "Challenges and Future," I revisit promises we made to ourselves about the special place we call GYE today. This chapter offers a series of actionable problem definitions and alternatives to address existing problems. If used, these can be transformative. What we do on the ground is the only place that really matters.

Chapter 9 on "Learning and Transforming" highlights the need to upgrade our ability to orient realistically to shared, widespread problems, to colearn a way to address them, and to transform our thinking and behavior to make a real difference. To be successful, we will need high-order leadership and new everyday skills for enhanced problem solving. This chapter offers practice-based, on-the-ground options.

Chapter 10 on "The Work Ahead" suggests a new cultural (institutional) story of meaning for GYE and ourselves in it. All of us need to see GYE as a whole, with ourselves as part of that holistic, evolving system. The sharp distinction between people and nature is outdated. The future of GYE depends on whether we choose to come to an integrated conservation story and appropriate follow-on actions or not. The new story or paradigm is slowly emerging in some circles in the regional society and well beyond.

Following Part 3 is Chapter 11 on "Creating a New Story, the Long View." This conclusion says that we are at a critical juncture in our relationship to nature and wildlife. This last chapter summarizes my case for a new GYE conservation story—paradigm.

Given our present social struggle, in the end, there a strong basis for pragmatic hope for GYE's future. Achieving that future will turn on us carrying out of certain cultural and policy transformations. Simply put, it requires that we individually and collectively lift our imagination and actions in GYE. Fortunately, there is a growing number of people and organizations supporting such advancements. This is good news.

Chapter 8

CHALLENGES AND FUTURE

Compromise is a difficult thing to find in Teton County
[…] there's so much money at stake.[1]

The Teton Mountains, which lie in the heart of the GYE, are one of the most widely recognized mountain ranges in the world. The glaciers that sculpted the mountains over their four million years ago largely disappeared around ten thousand years before the present, though some glacial ice and snow remain year-round in the heads of glacial troughs and are visible from the highway in Jackson Hole. Due to the rapidly changing weather and climate, these glacial remnants are disappearing quickly, as are the glaciers in the Wind River and Absaroka Ranges in Wyoming and Montana. Retreating glaciers are just one example of the effects of climate change on the ecology of the GYE. For humans and animals alike, big change is on the wind.

Humans are a very recent species in a very old landscape, yet currently, we are agents of dramatic change, and not only in the GYE. *Homo sapiens* is now a planetary force, and the impacts of human activity can be found in most planetary systems. Our widespread use of ammonium fertilizer, for example, has dramatically altered the global nitrogen cycle, and humans have transformed 75 percent of Earth's surface.[2] In recognition of the profound impact that human beings have on our planet, scholars have introduced the term "Anthropocene" to describe our current geologic epoch.[3] Unfortunately, many of our impacts are negative and uncontained. We are losing species and ecosystems at rate almost one thousand times the background rate. Tom Butler of the Northeast Wilderness Trust noted in his 2005 book *Overdevelopment, Overpopulation, Overshoot*, "The wild beauty, ecological richness, and cultural diversity [is] being swept away by the rising tide of humanity."[4] The Greater Yellowstone Ecosystem is not immune.

This chapter focuses on challenges that we face "out there" in the environment and "in here" in our minds, thoughts, and meaning making.[5] It offers a three-part, interconnected problem definition—conventional, systemic, and cultural—and offers pragmatic options to address these challenges. We must focus on understanding problems first, before we can find solutions to them.

Greater Yellowstone for Tomorrow

The Greater Yellowstone Ecosystem is a value-laden cultural icon in which many people have a stake. As such, the GYE provokes veneration, confusion, and political conflict. So, what is GYE's future?

Social and decision processes

The social process considers people and the ways that we behave. People are involved in every GYE natural resource problem, as both the cause and a potential solution.[6] The interaction of individual or group interests in society constitutes the context for resource and social problems. These interests must be accounted for in the GYE, if we are to be successfully problem-oriented. Neither the problems we face (and create) nor the decision-making processes we need to use to address the problems can be understood unless the human context is known and managed constructively.[7]

We have many indicators of our impacts on the GYE, for both the social and biophysical dimensions of change throughout the GYE.[8] We have good indicators for many of the hot-button issues, such as the status of wildlife, bioprospecting, biotechnology, renewable energy, climate change, growth, business, development, housing, transportation, recreation, and health care. We can use these indicators as data to make judgments and inform our decisions about how we manage the people and wildlife in the region, for example. Many individuals and organizations are working currently to understand these dimensions and ongoing changes.[9] These provide an invaluable picture of our context that helps us understand our present situation, including records of population growth, recreation use, and visitation to the GYE.

For example, in 1990, Dr. Ray Rasker of Head Water Economics estimated the total population in the counties surrounding Yellowstone National Park at 427,000 people and predicted that it would grow another 75,000 by 2017. In reality, it grew another 170,000.[10] And another 25,000 people are expected to move to the area in the immediate future, according to reports by Dennis Glick and Hannah Jaicks of Future West. The phenomenon of mushrooming population growth in the region has been reported by journalist Todd Wilkinson of Mountain Journal. Population growth is a major concern for many who are working to reconcile human development and conservation challenges.[11]

In addition to population growth, total annual visitation to the region is also increasing rapidly. In 2016, over four million visitors entered YNP—a 17 percent increase from 2014—which now puts the park at a historic crossroads. Dr. Ryan Atwell, a social scientist in Yellowstone National Park who studies park visitation, is scoping the challenges for the future and proposing changes that we may undertake to adequately address them.[12] The COVID-19 has accelerated visitation rates in the GYE.

To best understand and work with GYE's growing social complexities before us, we need to better understand people, their perspectives, and their values, including how all these come together in our decision-making. This requires collective and intentional examinations of the ordinary, systematic, and constitutive decision-making factors that are at play in the region.

We need to close in on effective problem solving as the core concern of our decision- and policymaking. To do that, we need to understand the processes by which people and organizations make decisions. The process of decision-making involves many people engaging in one or more decision-making functions: problem identification, information gathering, debate, implementation, evaluation, and restarting new processes (see Chapter 7). Of course, these processes vary widely depending on the issue at hand. Some cases are highly controversial.[13]

The GYE is a complex and dynamic system, as we know. The scientific study of system complexity was made popular with James Gleick's 1988 *Chaos*,[14] followed by Roger Lewin's 1992 *Complexity*, among others.[15] These books were the origin of our modern "systems thinking" and interest in complexity in theory and practice. One common feature of complex systems—such as the GYE—is that they can gradually drift into failure.[16]

Failure in systems is not necessarily catastrophic or sudden. Rather, failure can come inevitably as a by-product of the normal functioning of flawed systems, often driven by slow-acting environmental and social pressures. We may be experiencing just that currently in the GYE. For example, we are seeing a "governmentality problem" (different from government) in many social and decision processes today in the GYE. Governmentality stifles efforts to upgrade the decision-making process because efforts are deflected or absorbed by government agencies and other organizations and ignored or reduced to a narrow technical problem.[17] The elk hunting program in Grand Teton National Park is an example.[18] The governmentality pattern is also seen in large carnivore conservation and in large ungulate migrations in the GYE.[19]

Drift into failure can occur as an organization, town, or community uses its resources to achieve a mandate (e.g., recreation), then gradually invests in capital improvements without allowing for maintenance or feedback. From such, it gradually borrows from the margins that were previously available to protect from failure.[20] Thus, the carrying out of normal functions of government and business can create the very conditions for erosion of the community. The Greater Yellowstone Ecosystem is just one complex system that is following this corrosive pattern. This is cause for considerable concern because the hard consequences are not forgiving, neither are they reversible. So, is the GYE dipping into such a failure? How will we know?

Possible futures

From my home in Jackson, Wyoming, I can look out across the National Elk Refuge in winter to see thousands of elk. My backyard is full of mule deer tracks and occasionally a moose spends the night on my back porch. Bighorn sheep winter a short distance away on Miller Butte. Coyotes are common, howling almost every day. I have seen wolves out of my window and mountain lions hunting deer and elk on the slopes to the east. This world has been there for the more than the forty years that I have lived in this house. And now, it is the same as it has always been, a seemingly unchanging and spectacular landscape full of abundant wildlife. Federal lands protect this wildlife, for which I am eternally thankful. Many people in and out of government have worked for decades to protect the public good that I see and experience through my window. However, this seemingly perpetual present cannot be guaranteed, given the significant ongoing changes in the landscape and in our culture. How much longer will this environment be as it is now? We know it is changing.

We are inclined to assume the future will always be the way it is today—automatically utopian. This type of thinking is called living in a "perpetual present." Despite the abundant science that tells us the future is likely to be significantly different from the past, we often avoid these emerging realities and hold onto our bounded views, utopian hopes, and the self-important stories of ourselves in the here and now.

Our material progress is taken for granted. It is assumed that there will always be a forward progression and that our lives will necessarily be more comfortable in the future. Most of us are secure in this belief. However, the things that I experience by looking out of my window do not reflect the entire picture. The timeless view of the present is misleading. In fact, it misses a lot. There are unprecedented levels of anthropogenic change underway locally, regionally, and globally that are happening on scales not visible through a stationary window. I know that my imagined, utopian future is unlikely to be a given considering those changes. Ongoing changes, perhaps imperceptibly slow, carry novel risks and threats to the GYE, and we have little capacity to predict their impacts.[21]

As we think about a utopian future, we need to remember that all humans share the fundamental human values.[22] In recent decades, we have grossly exaggerated individual and cultural differences that obscure the many features we have in common. But, as demonstrated in previous chapters, our core functional values are nearly universal: respect, enlightenment, well-being, affection, power, wealth, skill, and rectitude.

These human values are inherent in our evolutionary history, too, stemming from our interactions with the physical environment and with each other as social primates. These values comprise what we often refer to as "human nature," and are preconditions for society and culture in its varying forms today. The sum of these values is often stated as human dignity, and their commonality makes human cooperation possible. And, of course, we must also consider the value of nonhuman life. When all of these values are produced and shared widely in a society, we call that society democratic, but this is seldom achieved fully.

Some believers in the utopian future do acknowledge the present and future challenges in the GYE. However, they argue that we are resolving those challenges and that there is no cause for concern. Under this model, all we need to do is stay the course with our current beliefs and formulas—our stories—and all will be okay. Utopians argue that the same science, technology, and markets that led us to this level of materialism and living will solve our problems of biodiversity loss, pollution, and waste now and into the future. We just need more time and patience for this to happen. But, is this utopian view realistic?

In contrast to the utopian view, dystopian scenarios of the future are dominated by visions of irreversible loss of nature, environmental destruction, potential for societal collapse, and a greatly diminished human quality of life.[23] Social scientists, international scholars, and global organizations have been tracking major trends in our social, political, and environmental situation for decades. Many of these trends are not encouraging. One key concept is global "environmental debt."[24] We all understand a personal or even a national debt, but few appreciate the extent and consequences of our environmental debt (e.g., loss of top soil, deforestation, overgrazing, pollution, extinctions, and more). As with any debt, it will come due, and the environmental interest we will have to pay in the meantime may be severe. Professor Herbert Bormann of Yale University, who coined the term, called for scientists and the public to "emerge from their cocoon" of convenient thinking that avoids payment.[25] Before his death in 2012, Bormann was one of a growing number of people investigating our various environmental debts. For the Greater Yellowstone Ecosystem, we do not yet have an accurate idea of the size of this debt, much less a widespread acknowledgment of its effects.

Professor Bormann debated whether we could design a sustainable society in harmony with nature. He noted two major accomplishments of the twentieth century in the environmental sciences: (1) the realization of the biosphere as a highly interactive ecosystem and (2) humankind is a geological force capable of altering global, biogeochemical processes in destructive ways. We now see the earth as a dynamic and self-regulating ecosystem, powered by solar energy and characterized by millions of species, including humans. We also know that humans possess the power to alter biospheric processes dramatically and irreversibly. In the past 300 years, humans have used resources and altered such systems to support rapid material and technological improvements. This has contributed to a higher quality of life for people around the globe but at a significant cost to global environmental health.

As noted by Professor Bormann, a "global economy dedicated to relatively unrestricted growth seems on a collision course with the goal of a sustainable world based on harmony with nature."[26] Progress is not a foregone conclusion in our society, despite people believing in the perpetual present. History shows that societies do sometimes take steps backward into darker periods.[27] We must not take our forward progress for granted.

Problem Definitions

Defining a problem is much more than finding something you do not like about a system, or finding someone or something to blame for a troublesome situation. Professor Janet Weiss of the University of Michigan describes problem definition as a "package of ideas that includes, at least implicitly, an account of the causes and consequences of undesirable circumstances and the theory about how to improve them."[28]

Defining problems

Without a clear and practical problem definition for our "problematic situation" in the GYE, there is no basis for talking about solutions, much less selecting an appropriate course of action. Consider the wolf example.

Wolves and wolf management are highly contentious in the GYE.[29] Wolf management should be thoroughly grounded in systematic problem solving, but what exactly is the problem with wolves, as some people claim? Some people think there are too many wolves, in the wrong places, and behaving badly. Other people think there are too few and that people in the surrounding states need to be more tolerant. So, is there even a single problem?

We have erected systems of management policy to address the perceived wolf problem, but it has become fragmented, cumbersome, and a target of public criticism.

The problem-solving approach we are presently, conventionally using is in itself part of the complexity we face and, in turn, is part of the overall problem. Conventional problem solving, folk knowledge, and standard operating procedures seem to be failing us, as we struggle to find sustainability or coexistence in the wolf case. So, a key question remains: what is the wolf problem?

As discussed previously, three kinds of interrelated problems are evident in the Greater Yellowstone Ecosystem: ordinary (conventional), systemic (governance), and constitutive (cultural) (see Table 7.1). All three are evident in the wolf case and also in my own decade-long study of elk management in western Wyoming.[30] As discussed earlier, elk conservation was considered an ordinary problem by 46 percent of respondents and a governance problem by 37 percent. Only 17 percent of the people interviewed recognized that elk conservation is also a cultural or constitutive problem. One can presume that these outlooks are probably fairly consistent across all conservation issues in the GYE.

What does this say about how different people view and understand problems? Even agency representatives, wildlife biologists, and other technical experts saw the elk problem differently than their colleagues. So, what is to be done when the public, experts, and officials cannot agree on what the problem is? Who, if anyone, is responsible to work with the diversity of people and their different views of the problem, in order to develop a practical problem definition that is actionable? Who is knowledgeable and skilled enough to provide this problem orientation?

There are problems with how we see and talk about problems. Those who professionally study differences in how the public, experts, and officials see problems note that many of us do not have the concepts or language to express the problematic content or context of the issues that concern us.[31] Ecologists Dr. Claudio Campagna and Dr. Daniel Guevara, for example, noted that our critical concern is not only our inability to articulate a shared understanding of nature and problems but also a confusion of language, reflected in how the problem at hand is depicted, thought about, and understood in our attempts to solve it.[32] So, if we do not have the concepts and tools to fully express our concerns, how can we address them? For example, the growing confusion over long-distance elk, deer, and antelope migrations is but one set of challenges. We need to ask, what can be done to address the holes or blank spaces in our concepts, language, and thinking about problems and needed responses in this case? One way forward is to ensure open, grounded problem-oriented work that exposes those holes and blank spots in the first place. And, then fill them, cooperatively.

Kinds of problems

Most problems in the GYE today are a mix of ordinary, systemic, and constitutive problems. We know now that to understand problems fully requires us to look not only at ordinary and concrete issues but also look well beyond those issues to the underlying systemic and cultural processes, which can be truly problematic (Figure 8.1).

Ordinary problems are evident all around us. In the Jackson, Wyoming, area, we see deer and moose killed on roads, animals caught in fences, water pollution at the Jackson Hole airport, and pesticides in irrigation ditches in South Park. From the volume of

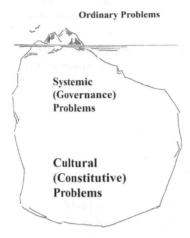

Figure 8.1 Three kinds of interrelated problems in the Greater Yellowstone Ecosystem.

newspaper coverage, one would think that these are the only kinds of problems there are in the GYE. These problems appear to be external and are typically about material things.[33] For example, how many domestic cattle do wolves kill annually? What is the status of whitebark pine blister rust? How many tourists came to the GYE this year? These questions reduce overall problems to questions of numbers, implicitly demanding a judgement over whether the numbers are too great or too small. In conventional understanding, we objectify and materialize ordinary problems and provide engineering solutions, thus rendering them technical.

Ordinary problems, and their technical solutions, are isolated, tangible, and easy to understand for most people. If the state wildlife agency wants greater elk numbers and better overwinter survival on the National Elk Refuge, they simply engineer an irrigation system to grow more forage material for elk feed. By doing so, the agencies engineer an extended farming model to manage elk and, in turn, increase hunting opportunities and revenue for the state.

Environmentalist and the highway department want to engineer new roads or over- and underpasses for wildlife. With ordinary problems, the objects of our interest are talked about as targets of physical manipulation/engineering (e.g., instituting a hazing or hunting program, trapping or killing problem animals, fencing, feeding). People have the expectation that traditional, technical management can fix ordinary problems. Solutions are expected, if only we give the agencies, scientists, and technicians more support. Fixes may also require educating the "unknowledgeable" public by printing brochures, putting up signs, and writing press releases. This view of problems, it is argued by environmentalists and officials, is quantitative and objective, so it keeps politics out of management policy. However, it is also inaccurate.

Ordinary problems and solutions are based on a litany of assumptions about people, nature, reality, ways of knowing, and patterns of behavior that facilitate certain conclusions while excluding others (see Figure 1.3). These assumptions divide the object (elk) from the observer (people), perpetuating a human–nature duality. As such, these assumptions perpetuate a utilitarian view of nature as a set of resources that exist for humans to exploit for material gain, rather than a sustainable view that imagines humans and natural environments as different facets of complex, interconnected systems. Ordinary problems keep the observer outside the realm of inquiry, immune from being considered a part or even a source of the problems. This immunity blinds the observer from themselves, their thinking, and their values. If we look deeper into ordinary problems, it is clear that they are really problems with our conventional system of thought—our governance and culture.

Ordinary problems exist because of deeper systemic problems with our societal system of decision-making or governance. A governance problem exists at the level of organizational or institutional systems. These problems are not due to a specific individual, isolated factor, or missing piece of information. They are not caused by a scientific error in data or user error on some method. For example, if many elk have diseases, it is likely not a problem with each single elk, but rather a systemic problem of crowding on the elk refuge caused by official decision-making systems. A systemic problem affects an entire system, be it an illness that affects a body through the digestive system or a problem in the decision-making process overall. Systemic errors require a change to the structure and culture of an organization, institution, or decision-making system.

When discussing systemic or governance problems, bear in mind that governance and government are not the same thing. Governance problems derive from inadequate organizational systems, both structurally and culturally.[34] Politics, the negotiating of differing value positions, will always be with us, no matter how hard we try to render all problems into technical problems. Systems give us a way to address these value differences and see problems more realistically. The challenge for us is to be aware of this fact and create and operate systems to manage ourselves responsibly.

Keep in mind that objects (e.g., deer, rivers, forests) are all managed through organizations that are systems of people, information, and operations—social and decision processes. Sets of organizations make up institutional systems. The National Elk Refuge is managed by several federal and state organizations. Systemic problems originate out of the way we organize ourselves in such programs.[35] In addition to the elk case, there are many other examples of systemic problems in the GYE today. One such problem is the high volume of deer killed on roads and in old, broken-down fences entangling migrating mule deer. No amount of fence building or removal will permanently keep deer off roads, and signs for drivers to slow down will never address the systemic problems at hand because these challenges are fundamentally problems with how we make decisions about such matters.

Because we tend to see matters in a piecemeal, ordinary problematic way, we often misdiagnose our situations. We seldom think in terms of systems, so we experience problem blindness on a case-by-case basis. Consequently, we are seldom able to see or address our systemic or cultural problems.

In order to function, systems require five conditions: self-interests, technique, market factors, organizations and institutions, and objectifying our exposure to issues. First, because individuals function only through what they can imagine, possess, measure, describe, control, fear, or rationally use, systems directly reflect the self-interests of those who create them. Second, systems are dominated by standard operating procedures that are based on rationality, technology, and a linear ordering of things. Techniques and methods usually demand the use of even more techniques and methods. Third, all systems are subservient to the thinking and needs of a profit-oriented, property economy, where profit is typically put before all else. Issues of market expansion, dominating market share, and more system operations are sacrosanct. Fourth, systems are made up of sets of organizations using system thinking and operations and they rely on each other to function. Consequently, reputable information, opinion, and authority can only come from organizations and institutions, not outside or from others. Usually, other external sources for these are considered corrupt and biased, and rejected. Finally, the task of a system is to rationalize its existence to its members and the target audience. Hence, individual behavior within a system is often under continuous bureaucratic surveillance, a phenomenon called "objectifying exposure." These features are always at play in challenges in the GYE.

Cultural or constitutive problems are the most fundamental of all problems, as the paragraph above notes. Both ordinary and systemic challenges are direct outgrowths of our constitutive (cultural) outlook and machinations. Cultural problems result from how we as a culture understand ourselves and our relationship to others and the world. These cultural problems are seldom talked about directly, likely because they involve grappling with complex and difficult questions that lack immediate or singular answers.

Cultural problems come from our persistent patterns of conduct and the rules underlying those patterns that our society uses to address its ordinary and systemic issues. We sometimes have to infer these constitutive rules from how our institutions are organized and operated, and from the outcomes and effects of our system of governance. The US Supreme Court was created to address constitutive problems, not ordinary ones, which are addressed by the lower courts.[36] The constitutive process played out through the courts and, in many other ways, specifies, changes, or adjusts societal rules about how everyday decisions about ordinary things should be made. Constitutive rules are the rules for making rules about how governance functions and society should be structured and operated. These rules are not always clear or explicit. In fact, we take the cultural dimension of our living so much for granted that most of us are not even conscious that it exists.

Several authors have written books to help citizens see the cultural process in real time. For example, Professors Robert Healy and William Ascher of Duke University, both policy specialists and economists, examined how new knowledge is used in natural resource policymaking.[37] They noted that new scientific knowledge seldom produces improved management policy, but it does impact ordinary decision-making by altering the intensity of demands.[38] Aldo Leopold, a US Forest Service employee and professor (often called the "father of wildlife management"), recognized the cultural process in his 1949 essay *Land Ethic*,[39] which considered how ethics about natural resource use have changed over the past 3,000 years and how our sense of morality around such issues functions today. Leopold called for new "rules" for interacting with the environment that he referred to as a new "land ethic."

The most visible part of cultural problems is in the system of public order that we all live under and depend on for predictable, secure lives. It is much easier to see ordinary problems conventionally, and even systemic problems, than it is to appreciate subtle but vitally important constitutive dynamics in society. And to make things more difficult, not many people are trained to conceive of cultural processes, as such that it is the means of adjusting constitutive rules or norms.[40] Nevertheless, the cultural process guides our management policy in all aspects of society. Without an understanding of the cultural process, it is impossible to clarify a comprehensive and workable means to influence outcomes in the GYE.[41] What we are seeing in the GYE today, I believe, is the perpetuation of our fragmented authoritarian system, a focus on ordinary understanding and problems, and underattending to systemic and constitutive challenges. And, importantly we are seeing the present cultural systems function to deflect challengers who are striving for new arrangements in the GYE that may be more inclusive, dialectic, and effective in adapting public order to changing ecological and social conditions. As such, the present system prevents the discussion of and implementation of necessary changes all too often.

Persistent problems

Many of the problems we face in the GYE are "wicked" or persistent problems.[42] If we are interested in solving these kinds of problems, we need to understand their nature. A persistent problem, according to Dr. Jason Vogel of the University of Colorado, is a "policy that fails to meet its formal objectives despite the sustained attention by decision makers."[43] This failure occurs even though there has been a focused effort and significant resources allocated to the problem's resolution. This kind of problem often shows goal displacement, among other troublesome issues.

The Greater Yellowstone Coordinating Committee (GYCC) represents, perhaps, the best, single example of a persistent policy problem in the GYE.[44] The organization is now over sixty years old, yet many of the problems it is mandated to address—such as meaningful coordination at relevant spatial, temporal, and complexity scales, large-scale evaluation and implementation of new transformative management, protecting large mammal migrations, addressing invasive species, climate change, and conservation of the large carnivore guild—persist as subjects of fragmentation and conflict.

Government, all the agencies, and GYCC could better address most of these challenges in two ways: (1) by dedicating themselves to understanding problems more fundamentally, comprehensively, and functionally and (2) by reorganizing themselves in a fully problem-oriented, contextual way to address these challenges, especially enabling themselves to attend to systemic (governance) and cultural (constitutive) dimensions. Persistent policy problems are irreducibly complex with fuzzy contours.

Persistent policy problems are not an exclusive or exhaustive category. There are infinite combinations of nonlinear causes and complex conditions involved in any particular issue. The very act of human intervention in any problem changes the nature of the problem. There is no simple checklist to comprehensively define or address a persistent policy problem. Science is needed, but we should not prioritize biophysical science over the social, applied science, or integrative sciences. We must leverage what we know of persistent ordinary, systemic, and cultural policy problems to improve matters in the GYE in the common interest. This is certainly a challenge for the GYCC, even under the best of circumstances.

The term "wicked problems" was popularized in 1973 by urban planners Horst Rittel and Melvin Webber, who detailed the concept and contrasted it with "tame" problems, which are complicated but have a clear solution.[45] If one looks at management across the GYE, many of the issues are assumed to be tame problems, when in reality they are not. Wicked problems are typically resistant to resolution under current conventional thinking and agency arrangements. In such cases, the problem itself maybe not be fully formed and as such it is open to misdiagnosis due to contradictory and changing symptoms. And, because of complex interdependencies in conditions, addressing one part of the problem may exacerbate other aspects. Further, new parts of a problem that were overlooked may be revealed as actions are taken to remedy the symptoms. This kind of problem typically has interrelated political, environmental, and economic stressors. Wicked problems are multifaceted, thus our conventional structures, thinking, and operations are often unable to effectively understand much less address them.

Wicked problems do not have an enumerable set of potential solutions, neither do they have a well-described set of permissible operations (standard operating principles) that may be incorporated into a plan to address the problem(s). Consequently, every wicked problem is unique and each problem should be considered a symptom of another problem. Those tasked with solving wicked problems, whomever they may be, should be liable for the consequences of their actions.

Metaphors

It is difficult to describe these interconnected, persistent, and wicked problems in ways that people conventionally understand in ordinary language. As a consequence, scientists and policy makers have created metaphors and stories to facilitate broader understanding of these problems.[46] One such metaphor proposed by biologists is the "Sixth Great Extinction."[47] Today, global species extinction rates are found to be 100–1,000 times higher than background rates, largely due to human activities. Today, one in five vertebrates is threatened or endangered.[48] In the past 42 years, we have witnessed a 58 percent decline in the known vertebrate populations worldwide, and scientists estimate that we could lose another 30 percent by 2020. The rate of extinction we are experiencing today rivals that of the extinction event that wiped out the dinosaurs—the late Cretaceous extinction. Throughout the history of life on Earth, we have experienced five major extinction events, wherein we see a dramatic increase in species loss in the fossil record. These extinction events, however, were driven by environmental processes and catastrophes, not by human actions. Thus, by equating the current extinction event with the previous great extinctions, scientists are able to implicitly demonstrate the extent to which humans have affected the biosphere—a degree comparable to catastrophic geologic events.

The concept of the Anthropocene is another metaphor used by scientists today to express the magnitude of the impact humans have on the environment. This metaphor is inspired by a recognition that humans affect biotic and abiotic systems to an extent that their mark will be preserved in the fossil record. The human species is overwhelming the natural functioning of the planet, affecting the atmosphere, oceans, soils, land, and life. While the magnitude of this change is difficult to grasp, the Anthropocene metaphor defines it in terms people can understand.

Persistent and wicked problems, presently understood dimly by many people through metaphors, are leading us to likely unpleasant outcomes, unless something is done.[49] Fortunately, as I have laid out in this book, there are logically comprehensive frameworks and propositions that could be used to help us to address challenges. Many scientists and others, some cited above, argue that our future depends on our ability to recognize and confront these persistent and wicked problems, especially the systemic and cultural ones. We need to approach problems not with panic, despair, outrage, or denial but with thoughtfulness, insight, and pragmatism—pragmatic hope.

Common Problems

Thus far, we have failed to promote a practical unity between social and ecological systems in the GYE and throughout most of the planet. This is an everywhere problem. Perhaps it has to do to with the way we have tried to address complex issues within limited short-term conventional, technical, and self-interested ways. If so, we will continue to fall short, unless we dare to address the foundational drivers of these persistent and wicked problems that we are inadvertently creating.

Description

We are the ones who created the present "unsustainable and unbalanced patterns of resource appropriation, production, and consumption, unplanned economic growth, and social inequity."[50] Most fundamentally, the cause of our common problem lies deeply buried in cultural (constitutive) processes, including our individual and collective psychology, philosophy, and policy capacity.

One alternative description of the common problem, other than the one I present here, is that people lack satisfactory concepts, theory, or self-awareness to address our social and environmental problems. If so, improvements in management policy inquiry depend on diverting attention from applied, ordinary understanding to higher-order theoretical research and work. Other definitions of the common problem allege that management policy officials simply lack adequate funding or access to information, or are too deferential to an unknowledgeable public. If this problem definition is accurate, then any improvements that may come about depend upon new self-awareness, knowledge, and insights in the GYE arena—education and ethics.

Problems exist when actual events move us away from our promises and goals, as is happening now in the GYE. Several key conditions allow the "everywhere problem" to manifest. First, our present capitalist system creates an inequitable distribution of wealth and promotes an exploitative and materialistic view of natural resources. In turn, this system promotes competition and maximization of profit, rather than cooperation among people, and thus it encourages people to overexploit natural systems for economic gain, regardless of diminishing returns. This system is, of course, upheld and reinforced by the dominant neoliberal myth, which exalts markets and nationalism as key mechanisms of improving our lives.

The interaction of these conditions is considered by Professor Michael Fotos, a political scientist at Yale University, in his definition of the everywhere problem:

> The modern nation state maintains its monopoly to transform (i.e., monetize) the human and natural capital at its disposal into taxable commodities or events. The most efficient means for doing so is by building a consumer economy organized as a partnership between competent public authority and corporate enterprises. The consumer economy, so-named because consumption is its central purpose, draws more people into the taxable wage economy than any known, feasible alternative and creates bottomless demand for converting natural resources into taxable commodities, a service that corporate enterprises are uniquely suited to provide. The nation-state system and the consumer economy are two halves of a single social order that happens to be the dominant form of order in the world today. Corporations are not uniquely perverse entities. They are the key enablers of the system's critical processes of consumption, wage labor, taxation, and the distribution of economic rents to buy political support.[51]

Dr. Fotos concluded that "the older I grow, Greenpeace gets more reasonable," and I wholeheartedly agree.[52] Greenpeace is an international, activist environmental group.

As I explored in the previous section, causes of the common problem are buried in our present, dominant paradigm that rests on three elements: our dualistic relationship to nature, our overreliance on positivistic science to understand the world and ourselves in it, and our propensity to arrange the world through economism. These three elements lay at the very heart of our "modernist humanism" myth.[53]

Problem diagnosis

Given the three problematic dimensions that I refer to throughout this book—dualism, scientism, and economism—and their overwhelming domination in the GYE, we see that the present model of management policy is failing on a number of grounds. Current forms of science, government, business, and even environmentalism are poorly equipped to cope with the magnitude of changes underway or their consequences. The GYE arena is increasingly rife with conflict and the real and symbolic outcomes of political battles serve to entrench perspectives, interests, and destroy trust among various factions too often. There is a polarizing effect from the compounded accumulation of nonresolved management policy conflicts in the GYE. Frequently, the losing faction attempts to reverse the authoritative decision, and if successful, the other side does the same through endless, future iterations. This makes the conservation of the GYE more arduous, if not seemingly impossible.

And, as we know, our understanding of the problems we face is limited to ordinary problems typically and thus cannot adequately address the complex systematic and constitutive elements. And, our systems of education are shaped by the political, cultural, and economic logic of scientism and neoliberalism, which represents a strong force perpetuating these problematic worldviews.[54] This trend is occurring at a time when we need graduates from our colleges and universities who can (1) understand the basic and foundational issues at play in dynamic systems such as the GYE and (2) work to address actual problems in all their complexity. The policy sciences as I introduced them in this book are uniquely suited to this task.

Realistic problem definitions

First, we must come to view our diverse problems as evidence of a need to fundamentally change our underlying notions of the "good life." We need to revisit and modify our understanding of ourselves in wildlife- and wildness-rich places like the GYE. This requires us to reframe the discourse (and our stories) over the GYE's future and come to realistic problem definitions. Clear problem definitions will offer new ways to see our challenges in the GYE and beyond. The methods for doing so that I present here are grounded in the relational insights of ecology, geography, sociology, psychology, philosophy, policy, and pragmatism. They come together in the policy sciences. To implement the skills, knowledge, and methods proposed in this book, we must build co-learning networks, undertake more practice-based work, and organize collectively for pragmatic action.

On small-scale levels, people are already implementing these strategies in pieces as they seek to conserve flora and fauna, strengthen human relationships, and enable people to gain a greater voice in the communities and management policy processes of importance to them. One particular area where we see such effort is in education.[55] Currently we educate people about the dimensions of the GYE in a variety of ways, from superficial entertainment-like engagements with plants, animals, and scenery to a few instances of deeper investigations of ourselves, our culture and its practices, and the central role we play in the ecosystem. Fortunately, this array of approaches is slowly changing toward the latter educational approach, for the better. Accelerating educational transformation is one important task as we move into the future.

Learn and transform

Second, we can use active co-learning networks to harvest, learn, and rapidly accelerate our trajectory out of the mess we are currently in (see Chapter 9). I believe we all want a healthy life that is full of meaning in a world of inclusive human dignity and biologically rich environments.[56] To quote the physicist and philosopher David Bohm, we need a new system of thought and action that captures this needed big, overriding goal.[57] To meet this goal requires us to embody and use a different way of making meaning for ourselves than we are doing now. The needed new understanding for the GYE is in marked contrast to the story that we currently employ.

This new learning-based and cultural story should be one that recognizes an "implicate order" in the world. A new cultural story should be about unity, wholeness, and coherence. Our striving toward this end requires all the knowledgeable and responsible attention we can muster. If one thing is clear to me, after half a century of living and working in the GYE, it is that the future is arriving much faster than ever before. This oncoming future, whose contours are becoming clearer by the day, signals that we need to change our old ways and replace them with healthier tools for rapid co-learning and adaptive governance practices.

Practice-based work

Third, we should use proven practice-based problem-solving approaches (see Chapter 10). We can work pragmatically to create new experiences, understanding, and practical actions through our projects. This practice-based learning approach tends to surface in conversations and site visits with some resource managers and other experienced practitioners. However, this approach rarely manifests in policy for the GYE, though there are exceptions. Practice-based approaches do not offer universal guidelines. Instead, each particular case and operational map of that case is based on practical prudence and a knowledge of how to confront real problems.

In practice-based problem solving and learning, the leadership group (and all workers for that matter) should skillfully command tools of controlled problem solving, deep contextual knowledge, and a capacity to cooperate with others. Overall in the GYE, we need to create a track record of successful on-the-ground demonstrations of successful practice-based work for all to see. Then, we can more easily show people the common interest goals that we are trying to achieve throughout the Greater Yellowstone Ecosystem.

Conclusion

We know that the world-famous Greater Yellowstone Ecosystem is under growing stress and that interactive problems are compounding. The ecosystem, along with our social system, is experiencing dramatic change, including shifts in weather, vegetation, and loss of wildlife migrations, to mention only a few. Currently, we are struggling to grasp the magnitude of the change itself, although the personal discomfort and uncertainty that we feel alerts us to the magnitude and consequences of the change that we may experience, as well as the more immediate concrete problems it presents to us. These ongoing changes—signals from the GYE's future—are clear and getting clearer day by day. At present, we are embedded in a complex sociocultural and governance system in which problems are too often misdiagnosed and left unaddressed. We need a forward working strategy to get out of our current mess.

In thinking about GYE's problems, this dialogue from Lord of the Rings comes to mind. Let's take a lesson from *The Fellowship of the Ring*. Frodo said to Gandalf, "I wish it need not have happened in my time." Gandalf responded, "So do I, and so do all who live to see such times. But that is not for them to decide. All we have to decide is what to do with the time that is given us [...] And already, Frodo, our time is beginning to look black. The enemy is fast becoming very strong. His plans are far from ripe, I think, but they are ripening. We shall be hard put to it."[58]

In the GYE, some of our conservation efforts have done better than others in controlling, reducing, or avoiding problems, but few have done well enough. We live in a participatory democracy that requires citizens to be responsible and responsive to collective problems. As I argue throughout this book, we must all be well informed with systematic knowledge of present and future thinking and with skills to respond adaptively to the challenges we face in the Greater Yellowstone Ecosystem and beyond.

Chapter 9

LEARNING AND TRANSFORMING

*It is only through labor and painful effort, by grim energy
and resolute courage, that we move on to better things.*[1]

In May 2015, I attended my 40th Greater Yellowstone Coordinating Committee (GYCC) meeting in Cody, Wyoming, as an interested member of the public. Founded in 1964, the GYCC is the highest-level government body in the GYE. Because of its important administrative role in the region, I spent 16 years dutifully attending meetings observing the committee, and in 2008, I published a book-length analysis of the GYCC—*Ensuring Greater Yellowstone's Future: Choices for Leaders and Citizens*. I concluded that GYCC's formal leadership, organization, and behavior were problematic because of its self-limiting thinking, bureaucratic rigidity, and contextual challenges. I was curious to see, years later, how agency leaders in the ecosystem think about the challenges they face, how the group is choosing to address those challenges, and how they engage the public in their meetings.

I was, first and foremost, interested in GYCC's guiding strategy for problem recognition and resolution. I discovered that their strategy was to focus on the least controversial, most concrete issues in the region, such as doing more data collection. A few large issues were forced on them, such as grizzly bear conservation, but they deferred to existing program models and interagency interactions—status quo approaches—to address them. The Cody GYCC meeting reinforced that the co-learning at individual, group, and organizational levels needed to most effectively address Greater Yellowstone's growing challenges is not taking place at high levels of leaderships and governance. In fact, I observed that opportunities for co-learning were actively blocked by conventional thinking and bureaucratic structure and procedures. Meeting authorities blocked productive conversations, perhaps not intentionally. As well, fruitful avenues of inquiry were not followed up. The meeting came to people talking past one another without adequate facilitation or integration. This was largely due to rigid agency arrangements that presented a prescriptive method for addressing challenges, a method that rendered issues technical and limited interaction to that "box" largely.

That being said, I am supportive of GYCC's mission and sympathetic to the individuals involved. Practical co-learning is challenging, given the context, bureaucratic prescriptions, constraining incentives, and limited time for appraisal and reflection, both in and out of government. Yet interconnected appraisal, reflection, and co-learning are the only way we can bring about constructive change for more effective problem solving in the public interest. In this chapter, I will delve into these subjects and make recommendations about how we may implement them in the GYE. I also offer options for addressing the challenges presented in the preceding chapter—and throughout this book—to move GYE conservation forward in sustainable ways.

Virtually all recommendations in this chapter, and the one to follow, rely on learning strategies that require us to learn our way out of the diverse challenges that we face in the GYE. The kind of learning that I am calling for is not about finding new ways to engineer our way out of problems with technological advances or rigid, ineffective modes of "collaborative" work with the public. These things are important, but they do not capture the depth of the change needed. Instead, I am calling for transformational co-learning to change our frame of reference, our understanding, and ways that we behave. Needed learning is about transforming our being, doing, and knowing in ways that are commensurate with the biophysical and social world and that are capable of addressing the multitude of problems that we have created for ourselves. In this chapter, I go into the theory and practice of learning and transformation. I understand transforming is a tall order. However, the reality of the situation we find ourselves in today requires such a strategy. It is the only way out of our current problematic trajectory that we are on and toward a promising path to sustainability in the GYE.

Learning

In order to move conservation efforts forward through learning strategies, we must first understand the realities of our present work and management policy, its effectiveness, and the actual contours of the future, as best as we can. The Cody GYCC meeting was a good opportunity to see how GYCC addresses these subjects and tasks. It was also an opportunity to demonstrate the co-learning needed, as I try and do in this chapter.

Cody meeting: A case study

Part of the day-long Cody GYCC meeting was open to the public while other parts were closed. In the open sessions, the GYCC ran discussions where they asked the public to respond to a litany of issues, soliciting public views on management issues and agency performance. Did this generate meaningful dialogue? Was it a co-learning experience? If so, what was learned?

The public participants, myself among them, were divided into groups of eight, each moderated by an agency official. In the group I was assigned to, our moderator was the supervisor of one of the big national forests in the GYE. My co-participants came from a variety of contexts and included a preacher from Worland, a rancher from the South Fork, an outfitter, a county planner, a weed and pest eradicator, a university extension agent, and a Bureau of Land Management representative, all from northwest Wyoming or southwest Montana. The experiences and knowledge of these people at my table were certainly varied and grounded. The rancher from the South Fork had learned of the GYCC only an hour before the meeting, whereas I had been attending public meetings for 25 years. Overall, the social dynamics were not representative of people in the region—of the nine people at my table, for example, only two were women: myself and the US forest supervisor.

Before the small group table exercise, several agency specialists gave presentations in which they scientized issues, turning complex social matters into objectified science problems and rendering cultural challenges into technical concerns. Agency experts from nine technical working groups reported on their subjects: climate change, invasive species, fisheries, wildlife conservation, whitebark pine beetle, fire management, air quality, hydrology, and operations. As these presentations were given, I observed reactions from the audience. Overall, the audience seemed overwhelmed by the disjointed technical reports and varied PowerPoint presentations. There were no goals offered, few trends mapped, and no efforts made to integrate the subjects toward a conclusion or practical problem definition on which we could act.

And, all of the GYCC meetings that I have attended have begun this way: with hours of technical reporting to an audience with varied backgrounds, areas of expertise, and diverse interests. The meetings are organized on the starting premise that scientific management, expert technical operations and data, and the authority of bureaucracy are the place to begin. The challenges the GYCC presents for itself and the GYE are therefore informed, bounded, and limited by such sideboards, assumptions, and operations, which manifest in the organization and possibilities of the citizen forums it set up.

When our small groups were asked questions about management and conservation in the Greater Yellowstone, there was a strong bias toward technical responses. We were asked about the key management issues in Greater Yellowstone and about what a successful partnership would look like, although we never received clear direction about what kinds of arenas and participants we should be considering. Were we to consider the public or private sphere? Individuals or political groups? Locals or tourists? Conservationists or businesses? Nevertheless, we began a general, wide-ranging conversation about the issues at hand, although its broad range and unclear motivations made it rather fruitless.

As we engaged in this "directed" discussion, I noticed that most individuals were far more concerned with their personal issues—how livelihoods were to be maintained in a changing economy, how citizens can and should have a voice in the public sphere—and, as well, everyone showed a deep concern for the future of the region. Given the strong predilections of the citizens to discuss these kinds of specific questions, I was curious why the GYCC chose to focus on scientific and technical reporting when they curated their discussion questions we were asked to speak to. I soon realized that the agencies felt they had a public relations problem and were trying to fix it. They believed that a key barrier to their effectiveness (and public support) was a lack of communication or understanding on behalf of the public about a wide range of technical matters. The GYCC seemed to believe that by merely doing "educational" outreach and increasing interactions with the general public, they could increase their public favorability and bolster support for their initiatives.

Because of our varied backgrounds, the individuals at my table were all on different pages, and none of us were on the same page as GYCC officials. Each person seemed to be experiencing a growing sense of incoherence about how events and issues, such as endangered species conservation, were playing out in the region relative to their views. While no one at the table had the skills to talk specifically or explicitly about his or her values, participants did offer statements that provided indirect clues to their value concerns. First and foremost, the people in my group were making demands for respect, reflected in a desire to have their concerns recognized, acknowledged, and included in decision-making processes. As I observed these conversations and implicit value demands playing out, I realized that the GYCC had a built-in inability to hear citizens' value interests or grasp the context or social process playing out in front of them. This, to me, was the core problem of the day. The functional value dynamics in play were not recognized or addressed, and the GYCC mode of operation strongly biased conversation to technical matters that could be arrayed in lists.

Overall, the citizens at the table felt marginalized, alienated, and hostile toward government. These were manifest in different ways. While some of those feelings may have been misguided, they were clearly rooted in participants' real, though unspoken, value concerns. People were finding it hard to make meaning of their lives amidst all the change underway in the GYE. They looked to forms like this GYCC meeting to address their concerns. However, officials of the GYCC seemed unable to hear this fact, and instead of engaging in these potentially rich and productive discussions, they redirected interactions toward listing the technical challenges facing the region and participants' conventional comments.

I tried to engage the central topics (values, perspectives, change, and pragmatics) and talk about needing to build trust within the community. I wanted to find a way to get at the foundational value concerns people wanted to discuss. I wanted an effective democratic process to address people's concerns. In doing so, I pointed out that simply creating lists of symptoms, observations, and opinions did not constitute a respectful discussion or practical problem definition, and thus was unlikely to generate trust or real change in the region.

However, the supervisor who was leading our discussion appeared uncomfortable and did not respond to my comments. Her default assumption seemed to be that comments should be taken literally, at face value and listed in her notes. She did not consider that the things people were saying and asking reflected deeper values (e.g., demands for respect). Thus, the golden opportunity the meeting provided for insights, communication, trust-building, partnerships, and joint problem solving went unseen and not acted upon.

The meeting was a lost opportunity to move toward the GYCC's mission—to work with the public toward conservation in the GYE. The community meeting could have been designed to address the underlying personal and social problems people were concerned with, clarify common ground, and identify practical next steps. It could have been used to stock community trust and to engage people in our shared concerns over GYE's future. But in the end, it was a hollow exercise in listing technical problems and symptoms, and maintaining the status quo through conventional, in-the-"box" thought, dialogue, and operation. Those in attendance may have hoped for two-way communication, hands-on problem solving, and meaningful involvement, but most participants left feeling unfulfilled.

In my assessment, the GYCC meetings lacked a problem-oriented process, one that is sensitive and focused on real concerns—value and otherwise. Consequently, local people misattribute the government's inadequate response to their expectations and demands to sinister motives on the part of officials and powerful outside forces (e.g., environmentalists). To me, the exercises we were tasked with neither clearly answered what a workable partnership might look like in general nor succeeded in building common ground and trusting relationships. I believe that GYCC community meetings would have been more successful if instead of harvesting statements and making lists, they had employed a respectful co-learning strategy.

People's knowing, doing, and being

If the GYCC's Cody meeting had been structured to use a respectful, co-learning strategy, we may have been able to develop a practical outline for a partnership and actionable problem definitions. GYCC meetings are also an opportunity to develop cooperative skills within the community, if leaders so chose to do so. There are many ways to think about encouraging diverse people to work together on problems of shared concern, but these require first stopping and thinking fundamentally about who people are, their motivations for attending the meeting, and how a cooperative partnership might be constructed. This advanced work provides a foundation to learning and hopefully realizing gains on the ground for citizens and officials. In the spirit of trying to make positive recommendations to GYCC in this case, I suggest that the writings of one of our greatest educators and pragmatists might help.

The American educational reformer John Dewey laid out a practical system for progressive learning and education that functions for different "modes of experience"—knowing, doing, being.[2] *Knowing* is about being informed and knowledgeable, *doing* is about action and engagement, and *being* is about existing in the present. This system has implications for problem solving in the GYE. I suggest that the GYCC meeting would have been more successful if it had used this progressive mode of education, one that focuses on method, process, and participant-centeredness. This model would enable us to not only clearly identify the problems we face but also learn to work together to address them. There are other co-learning approaches that could also be used by GYCC to address shared problems, but in the following sections, I will examine Dewey's work in the context of challenges, participant diversity, and value goals in the GYE.

Knowing—a conscious knower

What does it take to be a conscious knower in the context of conservation of the GYE? A conscious knower must demonstrate facility not only with concrete facts but also with abstract ideas (Table 9.1). Conscious knowers are open to the task of learning in groups as a systematic cultural practice. Knowing is a way of being a fuller person, and knowing is not independent from doing and being. If we are to generate the changes we are hoping for in the GYE, we must embody these three fully—we must know, do, and be to the highest extent we are able.

The GYCC meetings would ideally be used to bolster all three of these tenets in people in attendance. Meetings would be educational opportunities—a chance for

Table 9.1 Summary of John Dewey's modes of experience (being, doing, knowing, including both aesthetic and reflective modes) with implications for successful partnerships. It is important to combine and align these two modes of experience with how we live our lives (occupations).

Aesthetic Experience (Affective/emotional feeling)		Reflective Experience (Cognitive thinking)
Nonreflective	Trial and error	Regulate
Experienc*ing*	Experienced	
Pre-reflective	Method	Subject matter
Post-reflective	How	What
	Process	Content
	Concrete	Abstract
OCCUPATION	OCCUPATION	OCCUPATION
As holistic	As continuous	As organizing
	purposeful	principle for
	activity	knowledge
Being **WHO?**	*Doing* **HOW?**	*Knowing* **WHAT?**

Table is modified from Quay and Seaman, *John Dewey and Education Outdoors* (2013), 87.

participants to learn more about the social and biophysical context, as well as gaining skill development and action-oriented works. As it exists today, however, GYCC meetings offer little, if any, opportunity to consciously know, gain insight, or advance skills and cooperative working relations.

We know that people learn most effectively when they participate in activities and gain salient experience. And experience is understood as knowing, doing, and being at the same time. When we formulate our educational systems in these ways, we employ both reflective (thoughtful, insightful) and aesthetic (appreciation, emotion, beauty) learning together. Perhaps aesthetic experience is the key aspect to learning. Conscious being is the origin of growth, development, and assumed responsibility. Knowing these things, the GYCC and others in the GYE could adjust their leadership, thinking, and practices to help people become cognizant knowers. This is the clearest road to consensus and constructive transformation in the common interest.

A *knowing* person is able to engage in a reflective appreciation of the aesthetic experience of existing in the world. High-order knowing people are able to deal with concrete and abstract matters at the same time, and are more effective problem solvers. Not everyone has the time or interest to be a high-order knowing person, but certain frameworks, such as the policy or integrative sciences, actively encourage this kind of growth. Becoming a high-order knower takes time, reflection, and insight, and not everyone is organized to undertake these activities consciously or systematically. Thus, most people who operate according to convention carry out continuous purposeful activity—doing and being— rather than knowing in the sense I talk about it here.

Doing

People spend most of their time *doing* something. We are focused on goals and actions, although we often engage in activities without conscious awareness or reflection about what we are doing or why. Consider the professional world, which rewards technical expertise and skill above reflective thought. We are built, culturally and evolutionarily, to do more and not necessarily to know what or why.

In the past, the functions of knowing and doing were divided between the elites and others. Consider, for example, the distinction between what we refer to as "blue-collar" versus "white-collar" work. Today, however, mass education provides a more democratic access to these functions. We should continue to work toward encouraging critical reflection about the things we do—enabling us to use our doing as a way of knowing. Perhaps this is what people mean when they talk about "intuition." This not only will increase our awareness of the realities we face in general and in the GYE but will also encourage and foster grounded meaning making through action.

Being

Being is central to human life. We dwell most often in our nonreflective aesthetic experience. It is our life in an immediate, tangible, and holistic sense as we go through each day. Perhaps this is the way that many people experience YNP? Our lives unfold in real time—the present. In life, both aesthetic and reflective forms of thinking are continuous, whether we are aware of it or not, and both modes of experience are necessary.

Any opportunity for learning and education should be conceptualized to capture the inherent relationship between reflective and aesthetic modes of experience (*being*, *doing*, *knowing*). Learning is about encountering various ways of being a person—aesthetically and reflectively—in different contexts over time (e.g., YNP, GYE, beyond). Learning increases our ability to adapt our ways of being to our specific context through our doing and knowing.

I believe that employing these kinds of learning experiences in meetings could greatly aid GYCC's future work and public support. Experiences are effective not only for individuals but also for groups and organizations. The field of organizational and policy learning is very active and I believe there are many opportunities to use this knowledge and these vehicles in the GYE today.[3] The GYE is ripe for a more sophisticated and effective projects to enhance reflective experience—knowing, doing, and being for conservation in our common interest. Individual learning is key in all situations, and from individual learning, more complex learning at organizational and policy levels becomes possible. Levels of co-learning (individual, organizational, and policy) are interactive and should be fostered throughout the GYE.

Reflective experience

From a reflective view, one of the most critical elements that I felt was missing from the GYCC meeting in Cody was open discussions to define the relationship among people, the agencies, and the ecosystem in Greater Yellowstone. Perhaps this omission occurred because meeting facilitators felt it was too big an issue or for other reasons. Likely, they simply did not think this way or know-how.

Typically, reflective dialogue is teacher-centered, but the GYCC meeting was government-centered. However, reflective dialogue should be centered on concerns of students and citizens in the GYE. The tendency is for people to remain in the confines of conventional thought and prefer aesthetic over reflective experience. One way to address convention is through deeper reflection as a student-centered or, in this case, as GYE citizen-centered activity.

Learning and education are a social institution that is uniquely able to advance problem-solving capacity and enhance individual lives. However, these functions were not done well at the GYCC meeting in Cody. Dewey and others have been critical of the distorting aspects of scientism (as ideology and method) and bureaucracy (as ritualized standard operating procedure and power relation) because of their harmful effects on democracy, public process, and individual awareness and responsibility. Yet, the GYCC meeting was heavily focused on scientific data and bureaucratic process, the things that Dewey argues are problematic. Thus, the GYCC meeting missed the opportunity to reimagine the relationship between people and nature, citizens and government, and problems and solutions. There was no effort made to orient to the content of citizen comments or value differences, grasp the context or social process of the meeting, or to identify an actionable way forward. Other GYCC meetings have proceeded similarly to this one I describe here.

For Dewey, education is the cultural practice of organizing and distributing opportunities to learn. GYCC meetings can and should be a place where education happens in this sense. In the future, I contend, we must focus not on new curriculum content or instructional methods but rather on individualized cultural-historic natural learning. The GYCC, as important leaders in the GYE, would do well to take heed of Dewey's philosophy and employ it in their public gatherings and internal deliberations along with the analytic approach of the policy or integrative sciences. This would enable individuals, groups, organizations, and institutions in the GYE to move beyond the confines of convention and elucidate clear and actionable goals, invent solutions, build trust and cooperation, and make more genuine conservation gains.

Learning what?

In the more than twenty-five years that I have been attending GYCC meetings, I have seen little active reflection or co-learning, as I speak of it here—conscious, systematic learning at higher cognitive, organizational, or policy levels. The GYCC has stayed well inside the confines of conventional and bureaucratic thinking. The single most important element of learning is reflective experience.

Reflection typically requires finding the barriers to learning and overcoming them, though this is often a time-intensive task. The GYCC's operations seem ripe for this kind of reflective exploration as a way to make their work more salient and effective. This requires that we look at the behavior of ourselves and our organizations (e.g., government, business, and NGOs), as well as the social, cultural, and policy systems of which we are a part. We must consider in each context whether we are limited by convention, standard operating procedures, and unseen psychological and philosophic assumptions, or if we are operating in adaptive and effective ways. However, if we conclude that we are, in fact, constrained by convention, what do we do?

Learning

Learning is the process of using information to adjust our responses to the environment. It is a process of detecting and correcting errors or mismatches between our goals and the actual outcomes, intended or not. Learning to meet value goals in the GYE requires much more than holding on to positivistic, conventional bureaucratic procedures, formal meeting structures, and recycled ritualized dialogue on a few narrow issues. It requires more than reading technical reports or listening to experts and others. It is not simply a question of presenting scientific knowledge to the public under the assumption that they just do not know enough. Instead, we should begin to implement an emergent co-learning network through which individual transformation can occur in perspectives and actions (adult development), as we attend to actual challenges in the GYE.

All learning and education proceeds, as Dewey said, "by participation of the individual in the social consciousness of the [community]."[4] Learning is not merely knowing about the social consciousness of the community but also participating in it and experiencing it in meaningful ways. How do we do this in Greater Yellowstone given the current polarized, distrustful, conventional situation?

Currently, debates about learning and education remain inside conventional thinking way too often, even in our educational organizations, whatever the subject may be. By this I mean within the ordinary, averaged off (conventional) way of thinking about learning and education. In contrast, I suggest that as we seek to move beyond those limitations, we must consider how to engage people in doing and being, while conveying vital content knowledge in ways that promote conscious reflection and insight. This is a tall task.

Education programs that exist in the GYE today are biased heavily toward content (e.g., biological facts) and a certain kind of interaction (e.g., field identification of plants and animals). These programs promote memorization of facts and acquisition of certain skills. As important as these are, they do not foster critical thinking, reflection, or problems-solving skills that are needed as tools to move conservation forward in the GYE.

There are ongoing debates about the subjects and methods of education in the GYE that indicate that there is not yet agreement on the educational problems that we face.[5] Without a clear problem definition, we also have failed to identify any concrete course of action for improving the educational systems, as yet. We have successfully avoided the basic question: What education and learning is needed in the GYE in the public or in GYCC? To get a practical answer, we need to maintain our focus at a deeper level of reflection and dialogue than simply prescribing more conventional management and educational actions (Figure 1.3). The basic question is what kind of education is needed to equip us to effectively address real challenges in the GYE?

Experience

John Dewey called for finding an educational "plan of operations preceding from a level deeper and [that is] more inclusive than is represented by the [current] practices and ideas of the conventional contending parties."[6] Going to this deeper level in the GYE is not about compromise between different interests or views, nor is it some eclectic combination of existing conventional thinking and standard program elements. Instead, it is about finding innovative (integrative) ways forward. The needed GYE-level innovations cannot be achieved by selecting the most attractive parts of competing educational and learning approaches now in place and configuring them into an incohesive patchwork. Gains require a new and comprehensive framework that includes the controlled problem solving of the policy or integrative sciences.[7]

Dewey said that learning consists of three elements: (1) observation of surrounding conditions, (2) knowledge of what has happened in similar situations in the past, and (3) judgment that puts together what is observed and what is recalled to see what they signify.[8] This view promotes *reflective experience* in education and problem solving. As long as we act intentionally to consider the implications and foundations of our experiences, we are able to engage in this kind of reflective experience. This is what is needed in the GYE.

Reflection

Most people have some propensity toward reflection, but with varying degrees of insight, self-awareness, and contextual knowledge. Even at the GYCC meeting, there was a wide range of insight at my table. These varying levels manifest in two types of reflective experience.[9] The first is a method of trial and error, which involves less deliberative thinking. The second approach involves much more intentional, focused mental effort and analytic thought. The first method is most common and is related to the earlier stages of adult development, while the second method is rarer and corresponds with self-transforming and self-authorizing stages of adult development.

Dewey noted that problems could trigger reflective experience, but not necessarily (Table 9.1). He adds that reflection is found in "perplexity, confusion, and doubt,"[10] and if "no problem or difficulty in the quality of the experience has presented itself to provoke reflection [...] inquiry does not take place at all."[11] In pragmatic terms, feelings of doubt are rooted in a desire to know and are thus a partly emotional experience. Dewey labels the doubting phase *pre-reflective* and the resolving phase *post-reflective*.[12] The GYCC meeting in Cody stayed well within the pre-reflective phase and did not adequately engage participants in either kind of reflective experience, let alone resolve questions in the post-reflective state.

Thinking in the mode of reflection requires that we use our firsthand experiences and make those experiences an object of our thought. Dewey's original description of this kind of reflection is enlightening. He refers to five elements:

1.   Perplexity, confusion, doubt, due to the fact that one is implicated in an incomplete situation whose full character is not yet determined

2. A conjectural anticipation—a tentative interpretation of the given elements, attributing to them a tendency to effect certain consequences
3. A careful survey (examination, inspection, exploration, analysis) of all attainable considerations, which will define and clarify the problem at hand
4. Taking on consequent elaboration of the tentative hypothesis to make it more precise and more consistent, because squaring with a wider range of facts
5. Taking a stand upon the projected hypothesis as a plan of action that is applied to the existing state of affairs; doing something overtly to bring about the anticipated result and thereby testing the hypothesis[13]

For high-level (positional) leaders, such as agency officials in GYCC, learning and education should involve active reflective efforts directed toward shaping the future in Greater Yellowstone. Work should focus less on ordinary tasks of management—technical details about soils, rivers, plants, and animals—and more on putting practices, thinking, organizations, institutions, and society on a trajectory that is sustainable and open to continual co-learning, reflection, and adaptation. This is the needed mission of the GYCC. For many people in the GYE today, this would mean shifting the focus of their work away from case-by-case technical and content matters and toward understanding values, reflection, and learning. This work, if undertaken, should lead to significant gains in the processes of governance (decision-making), controlled problem solving, and cultural or constitutive dynamics toward sustainability objectives.

Aesthetics

We are all dominated by our aesthetic experiences (Table 9.2). They are our lived reality, our truth. Aesthetic experience is nonreflective and it captures our immediate sense of a situation. In life, we only retrospectively apply words to our experiences. Experience is emotional, first and foremost, and it will always be so. Emotion gives us a qualitative, integrative experience, a conventional unity of sorts. It can be pragmatic and helpful when we feel and respond appropriately given the situation, but disconnected and destructive when not. It all depends on who we are and what we do with our aesthetic experiences. Perhaps aesthetics is what most people visiting YNP experience.

We should appreciate the fact that emotions are part of all human experience, including educational experiences where the focus appears to be on knowledge and rationality. But, the GYCC meeting, as a learning experience, failed to address the aesthetic (emotional, appreciation) dimensions of being, which are equally important to facts in learning. As such, the meeting demotivated people and failed to uplift its participants or recognize shared interests. The meeting was meant to advance learning and support for government management policy, but it failed to do so because it offered little pre-reflective and post-reflective experiences or opportunities. In fact, the meeting was facilitated and structured to preclude these.

Learning and education in the GYE should be focused on addressing the serious environmental, social, and psychological conflicts that we are experiencing. It should address foundational issues in our individual lives and in society that take us away from

Table 9.2 A comparison of the present (conventional) model and the transformational (integrative) learning model.

Feature	Present Traditional (Status Quo) Learning	Transformational (Integrative) Learning
Goal of learning	Better concrete outcomes are sought	Systematic, adaptive changes are sought
Purpose of evaluation	Efficiency, effectiveness, impact, stability	Performance support for greater learning, innovation, and responsiveness to context concerns
"Modeling" the learning approach	Positivistic—causality is hypothesized, predicted, tested, often using statistics	Post-positivistic—causality is determined by pattern recognition, constructed in real time and retrospectively from direct observations
Accountability	To "higher ups," authorities, funders	To fundamental values and making a pragmatic difference
Idea findings	It is assumed that validated "best practices" can be generalized across contexts, time, and space	It is assumed that principles can be discerned that can inform practice and be adapted to local contexts
Method of reporting	Expert, scholarly, or authoritative voice often as third-person, passive voice	Involved persons who are engaged in the learning process and the actual project as a first-person, active voice
Evaluators standpoint and characteristics	Objective, independent, rigorous, credible/standing Authority, expert, official	Teamwork, creativity, flexibility in approach, tolerance for ambiguity Co-learner, interested, knowledgeable, unofficial

Modified from Patton, *Developmental Evaluation* (2011).

our idealized goals (Figure 1.3). My basic recommendation to all officials and citizens in the GYE is that we must pursue co-learning and education at a deeper level than we have so far. We must engage transformative co-learning rapidly and successfully.

Transformation

We need a shift in how we operate (do), think (know), and live (be) in the GYE away from historic patterns—"in-the-box" thinking and status quo approaches (Table 9.2). We cannot continue on the same trajectory that we are on currently (the one that created the problems we are struggling with now), if we expect to address them adequately in the future. Because of the magnitude and immediacy of our present challenges, we need transformational, not incremental change.

This immediately raises two questions: What is transformation, and how does it take place? In partial answer: We must engage transformation at the individual level before we can manifest transformation in organizations and institutions, and we must engage transformation in ways that are more complex than simply learning new information. Transformation is about changing the very form of your mind. It occurs when you are able to step back, reflect on an experience, and make a new decision or judgment about it. We must make efforts to transform ourselves and the ways that we interact with and make decisions in the GYE.

Transforming ourselves

We all have an intuitive sense of transformation from our personal lives, a vague awareness of or hope for changing our frame of reference and living to be more reflective and rewarding. One key role of an effective leader, activist, or policy analyst in the GYE context is to function as a self-transformative agent. Effective leaders model transformation and learn from transformation in others. Good leaders can engage with other people and complex social systems in reflective ways and adapt their ways of thinking and doing in response. Fortunately, this is a skill that any one of us is able to develop simply by being knowledgeable of the transformational process, being self-aware, learning the analytic skills involved, and engaging deliberately with the wicked problems in the GYE, without discarding the best of what we are doing now.

Deep shifts

For transformation to occur, it is likely to require "a deep shift in perspective, leading to more open, more permeable, and better justified meaning perspectives."[14] In turn, this requires transformation of knowing, doing, and being. The focus for transformation is the individual, starting with ourselves. Transformative learning and change are explicit processes. They can be taught, learned, and acted upon, and they are a focus of study at some colleges and universities. Transformation can be a rational or emotional endeavor, depending on an individual's specific context.

Importantly, individual transformation drives social transformation—it always has. We can be encouraged by desperately needed transformation in the GYE. We can use workshops, informal discussions, scholarly courses, or self-motivated learning to advance matters.[15] The approach to transformation (and skill set) that I propose is described fully in *The Policy Process: A Practical Guide for Natural Resource Professionals.*[16]

Perhaps the clearest account of individual transformation is of Dr. Mike Gibeau, who served as a park warden, conservation biologist, and carnivore specialist for Canadian Mountain Parks for 33 years. In 2012, Dr. Gibeau published a thoughtful article in which he detailed his transformation from relying on the epistemology and mindset of wildlife biology and positivism toward an integrated work strategy (detailed in Chapter 10).

From Dr. Mike Gibeau's years of experience, he came to understand that positivistic science, although necessary, is not sufficient to solve problems in the real world. He observed that most biologists and other technical experts, when challenged by a problem, would simply collect more data—render problems technical and miss what was really going on in terms of social process and value dynamics. Through his experiences and conversations with others, Dr. Gibeau came to realize that he was conducting science in an a-contextual, self-unaware way. Through his self-transformation, he adopted the integrative policy science framework that allowed for a more comprehensive and realistic approach to problem solving. He applied this framework to his work on grizzly bear management in Banff National Park.[17] This was one of the first and only times that the policy sciences framework has been implemented in wildlife, parks, and conservation arenas.[18] Dr. Gibeau used reflection to transform his thinking and work. On this, he said,

> Something changed during my career that has fundamentally rearranged the way I think—not just about biology, or bears, but about almost everything I do in my life. I've come to understand a more comprehensive way to solve problems. By sharing it, I hope to provide an alternative to the way fellow biologists approach problem solving.[19]

So, I ask here: are GYE leaders, professionals, activists, and the public ready for transformation of this kind, via the integrative opportunity that Dr. Gibeau used successfully?

Transforming capacity

Transformation at the individual level requires both an epistemological shift and a cognitive restructuring. Transformation of mind is a fundamental change in a frame of reference, culture of mind, or system of thought. The process and outcomes of transformation may be either affective (feelings, emotions) or cognitive (rationality, thinking), or some combination of these two. Both forms involve multiple dimensions of being, doing, and knowing. But which is the form that transforms? This is not an esoteric question.[20] The answer has real, concrete consequences. Professor Robert Kegan of Harvard University showed that the specific transformation formula determines if it is a transformation of one's being or a transformation of one's thinking.[21] These forms are interactive (Table 9.2, Figure 5.1, Table 5.1).[22]

Transformational learning can happen in the context of working for social and environmental change in a community that one cares about, such as the GYE. People interact, most often in personal ways in such situations, as they work together to affect desired changes. As they do so, they challenge the status quo, reimagine systems of advantage and power, and act. This is emancipatory learning.[23] Transformational learning can alter our very core identity, group loyalties, and personal and social responsibilities. It can change our self-awareness, critical consciousness, and relational empathy to other people and to nature.[24] Importantly, it can change our knowledge, skills, and the ways that we apply them. Transformational learning may give us insight into the foundational questions about what it means to be human, our purpose, and our responsibilities to loved ones, communities, and nonhuman life (Figure 1.3).

Learning models

We see a clash of thinking styles, knowledge, and values across all issues in the GYE today. We see people interact from very different positions of adult development, self-awareness, and capacity for problem solving.[25] It seems clear, though perhaps it is not, that we have not learned our way out of the problems of fragmentation, competitiveness, and reactiveness that we are experiencing. We are not moving toward more acceptable outcomes for all concerned, except in too few cases. So, what is to be done?

Learning is the demonstrated ability to connect knowledge across disciplines and other knowledge domains, including varied contexts and perspectives. It is about integration of knowledge and action to upgrade our lives and, as such, our relationship with the nature—the environment, especially wildlife. Active learning teaches us how to serve the public trust and the GYE goals that we laid out for ourselves in the past, for the present, and for the future. The dominant learning mode today is not transformational, although it should be.[26]

The present conventional model of learning dominates almost all deliberations in the GYE, both at individual and organizational levels. But this learning is directly opposed to, and in fact, necessarily inhibits, transformational learning (see Table 9.2). To secure a healthy future for the GYE, we must learn to live responsibly and ethically in a fragile and complex biophysical and social landscape.

For integrative learning to occur in the GYE, leaders must have a genuine interest in and motivation to engage fully and clearly with integrative exercises and co-learning opportunities. This requires reflection, attention to process and content, and capacity in both concrete and abstract concepts. Integration is directly related to synthesis, and it occurs toward the end of progressive education, which builds as a sequence, such as with knowledge, comprehension, application, analysis, synthesis, and evaluation. Few people complete the full progression of education, but this kind of thinking is desperately needed in the GYE at all levels from citizenship to high levels of management policy. How might integration help us learn about nature in the GYE?

Learning about nature

Our growing concern over the GYE's future against the backdrop of a rapidly declining global environment invites us to reappraise our individual and collective actions. Are there opportunities for progress in the GYE that go unseen and uncapitalized? What assumptions built into our modern way of living are causing problems in the GYE? How do we learn to see them? We know that conventional assumptions structure how we relate to the natural world, whether we are consciously aware of them or not. Recognition and appraisal of these assumptions and their consequences is a critical step toward adapting them and achieving a more sustainable human–nature relationship in the GYE and elsewhere.

Moral education

Over the last few decades, our environmental concern for the GYE has grown. As I have laid out throughout this book, there are two basic ways to respond to the environmental problems that we face in the GYE. The first, and most common, is to use technical solutions to minimize our harmful impact,[27] and the second is to see environmental problems as a chance to change our underlying conceptions of a meaningful life. The first approach does not require moral education. The second does.

What is moral education? First, morality is "a set of principles to govern human life."[28] We know that ethical concerns do not arise out of some pure form or abstract universal principle. If that were the case, our moral obligations to the GYE's many life forms would be clearer. Second, morality is an involvement in a place. This recognizes that our moral life is concerned with issues in a particular context, and through some aspect of our own participatory role in that context—our *doing* and *being* in the GYE. This view of morality highlights the intimate reciprocity between our ethical concerns and the GYE's environment reality.

Both approaches to environmental and moral concerns—the technical and the contextual—call for change in our present being, doing, and knowing. The moral approach calls for a new understanding of our place in nature and a reconfigured moral commitment. The contextual, ethical approach calls us to critically examine the postmodern period in which we live and interrogate how it influences our thinking, understanding, and actions relative to nature and nonhuman lives.

Thus, we need a transformation and clarification of our moral relationship to the natural world, if we are to move toward a sustainable GYE. We need a reimagined land ethic, so to speak, that defines our responsibilities and goals for the environment today and in the future. In order to attain this, we need a moral education centered on transformational learning. We need to understand that moral education arises from certain perspectives in society, and that those perspectives must be understood as combinations of aesthetic and holistic endeavors. We must account for the aesthetic (nonrational) and reflective experience (cognitive) in our moral judgments, as we interact with the GYE and all its life.

Reflective experience does have a significant role to play in making the needed changes, but I would argue that the aesthetic experience is the most important. People form intimate relationships to place that is typically felt as emplacement, the situational feeling of connection or belonging to a physical place.[29] Emplacement has both local and transcendent dimensions. For example, it can manifest as pride in living in the GYE, or in the feeling of personal restoration when spending time in wild spaces.

If we are to engage in the needed moral education and transformational learning, we must have a suitable platform or structure for leading these. One example of a place where I imagine implementing the needed co-learning is at the annual SHIFT conference in Jackson Hole, Wyoming. Each year, the SHIFT conference is advertised as a place where conservation meets adventure, as a forum where people can come to engage with basic and foundational issues about the role of wild spaces in leading meaningful lives (see Figure 1.3). Is wildness and wilderness meaningful because of the skill and pleasure (e.g., adrenalin rush) that we gain through recreation? Is it meaningful for the rectitude of knowing we can exercise self-restraint and preserve aesthetic places in the face of our harmful resource demands? Is it meaningful because of the potential for wealth and power that comes with adventure recreation or resource extraction? Or, is it meaningful because of the well-being benefits of exercise, stress reduction, and clean air or water?

Unfortunately, these questions are never addressed in practice in detail at SHIFT. The SHIFT conference does not consciously or systematically engage attendees in transformative co-learning by encouraging exploration of human–nature relations and responsibilities. Instead, the SHIFT conference leverages techno-outdoor recreation and commodification for gains by the Wyoming tourism industry and adventure recreationists themselves. This is an example of goal inversion, goal displacement, and goal substitution. Some people recognize the missed opportunity that SHIFT offers. Thus, SHIFT has come under increasing criticism as a conventionally shallow, self-promoting recreation group of diverse individual and business special interests.[30] Despite criticism, I believe SHIFT does hold the possibility of serious, in-depth discussion of recreation, business, and environmental and ethical matters—a possible transformative experience. SHIFT, if adjusted in focus and content, could be a leader in facilitating the needed changes in attitude, morality, and behavior. However, realignment of purpose can only come about with a deeper understanding of the real challenges we face, a high degree of self-awareness of our being, doing, and knowing, and a well-grounded and justified knowledgeable and ethical stance.

What to do?

Let us delve into moral education in the GYE as a key element of transformation. Attending to moral education can create space for the much-needed reappraisal of our current mechanistic or progressive worldview, myth, or paradigm. It can also create space for each of us to become more empathetic and responsible people. Experience is a critical component of moral change and transformative learning. As we begin to reimagine what our responsibilities and relationship toward nature may be, we should spend more time attentively and curiously exploring the natural world. We need to have a critical eye, not only for the biophysical experience of nature but also for the ways that it makes us feel and react. These opportunities are plentiful in the GYE. We can go to the Lamar Valley and many other places to experience nature's powerful presence.

When we immerse ourselves in nature, we can gain a breadth of ecological knowledge, and as well we can also gain knowledge of the implications of our own action in the ecological crises we have created. And, we can gain much more, including a sense of responsibility and agency for constructive change. One way to do this is to use some aspect of a local environment, such as Flat or Cache Creeks in Jackson Hole, to build attentiveness, knowledge, and relationships. We can do this on our own, in school systems, and overall in the community. This strategy is currently being implemented in part at least in Jackson Hole by Tom Segerstrom, a Teton County Conservation District leader.[31] His work is effective and well supported by landowners and the public. It is important to integrate nature into all learning and education in the GYE. Any moral education that recognizes the need to challenge the current dominant cultural paradigm must create opportunities to reveal, dislodge, and replace that paradigm with something more appropriate.

Many people have benefitted from places where nature is front and center, as in the GYE. Some of them have come to understand, quite implicitly, that morality requires us to live so that nature and its many processes can be preserved. This is not just a cognitive or intellectual concern for some of us. Rather, it is an important concern about our character that we can come to possess and manifest in our daily lives. We must come to live and operate with embodied attentiveness, an enhanced awareness of self and environment (and interaction) at the same time. This is a tall order and requires directed learning and education for most people.[32] Proper education would sensitize us to possibilities of more rewarding and sustainable living in the GYE than what we are doing now.

Any new human–nature relationship and ethic needs to clarify our sense of both *emplacement* in the present and *transcendence* in the future in the world. My recommendation here is that we can actively work to enhance our attention to our place in nature. The GYE invites us to reconceive our community, not as a closed system of humans in a fixed place but as a broad and dynamic system of interactions between human and nonhuman lives. What is the role of dialogue in accomplishing these goals?

Missing dialogue

Our collective, current public dialogue about the GYE is incomplete and overly selective, leaving out much of importance about our current situation. Thus, I recommend that we move toward comprehensive public dialogue that directly addresses problem orientations and definitions, and includes a full discussion of options about what to do about the many intermixed ordinary, systemic, and cultural challenges we face. At present, we routinely materialize, scientize, and marketize the discourse and management policy without thinking about what we are doing or whether there is a better way to go forward. Agencies, environmental groups, business, and elected officials produce numerous reports, plans, and actions under conventional assumptions and frameworks. We need to come to see that the current public dialogue about the GYE and its future is not adequate for us to either fully understand the challenges that we face or to fully explore our options to address them.

Looking ahead

The public discourse in Jackson Hole, other gateway communities, and even at the highest levels of civic or state and federal leadership (e.g., GYCC) commonly avoids foundational questions (Figure 1.3). Yet, ongoing thoughtful public discussion on these matters is prerequisite to a healthy future, successful transformation, and relative sustainability. Our thinking and dialogue must be predicated on our pragmatic successes and the latest knowledge and skills available to us.[33] We must have our heads in the clouds and our feet on the ground—high pragmatic hopes, realistic expectations. Fortunately, transformation is occurring in some individuals and groups through small-scale, practice-based projects in the GYE.

One way to cut through the present, conventionally blocked learning and education is to imagine education organized around occupations, as noted in Table 9.1. Occupations, as defined by Dewey, are the "socially meaningful units of cooperative work in which method and subject matter are inherently integrated and arranged carefully by a teacher to parallel [...] some form of work carried on in social life."[34]

Addressing the GYE's problems requires action through occupations, as I note. Yet, how we presently educate for occupations in colleges and universities and in society at large is coming under criticism.[35] Conventional education seems inadequate to prepare young adults to address the complex problems that we face in the real world today, and those that are foreseeable. How can we best organize public learning and education to address problems in the GYE, elsewhere in society?

Occupations

Occupation as defined by Dewey is the key unit of analysis when thinking about learning and education. The concept of occupations should encompass a range of engagements, including work across aesthetic and reflective experience. When we think of education for occupations (both for living and working), we need to realize that both aesthetic and reflective experiences are basic conditions of human existence.

The framework in Table 9.1 is not meant as a prescription but rather a description of human occupation in a Deweyian sense. Using it might help us upgrade our performance. Reflexive learning emerges from and returns to aesthetic experiences in all instances. Once we understand this, we can use this relationship to make prescriptions about how to organize education and occupations to best address the GYE's problems. For example, what occupations do we need in the agencies in the GYE or in environmental groups or local government and business to realize gains? In the end, what perspectives, knowledge, and skills are needed to attend to GYE's challenges, how do we educate about them, and how do we translate these into action that counts?

These are questions about education. Organization of education, according to Professor Dewey, is "nothing but getting things into connection with one another, so that they work easily, flexibly, and fully."[36] Organization in traditional education is centered around subject matter through isolated disciplines and jobs. However, problems in the real world are not singular or fall into either of those two. Problems arise from a multitude of dimensions, and thus effective problem solving requires an interdisciplinary, systematic, critical thinking, and integrative approach.

Education for the real world should be directed by policies and managing personnel that recognize the interrelationships among organizations, society, and the environment (i.e., *people* seek *meaning (values)* through *society (institutions)* using and affecting *environment (resources)*. This kind of education (and heuristic) should be useful for integration across complex problems and fields.

Thus, education and technical expertise on only subject matter is not sufficient in today's world to address such challenges. Instead, education should prepare individuals for needed occupations and life in general through integrative thought and skills. Additionally, education should be based on community context, social life, and the particulars of the environment. These matters are critical to the relationship between education, experience, and action. This educational arrangement should be manifest at all levels, from primary schools to colleges and universities. We should create educational organizations and learning environments that capitalize on the insight of Dewey, the heuristic offered in this book, and ideas from others to bring forth a generation of open-minded, creative, and interdisciplinary thinkers and integrative problem solvers in the GYE.

Education

Much has been written on outdoor education and connecting with nature leading to transformation.[37] Nearly all adults in the GYE have been educated using the subject matter approach deeply institutionalized in current education, without the value-added problem-oriented integrative approach that I recommend. The subject matter approach gives the learner substantive technical knowledge but not necessarily aesthetic knowledge, or functional, integrative understanding and skills at scale or self-awareness. It certainly does not provide the interrelated reflective experience or analytic capacity needed in the GYE today, given existing problems.

In thinking about a better educational strategy, we must recognize that our lives are constituted by occupations. As we mature and grow, we take on new occupations, modes of living and working.[38] Growing, maturing, and developing are not about preparing oneself to occupy a technical or substantive occupation. Neither is it about acquiring knowledge that we might apply in a typical professional job.

Instead, adult development (and existential maturity) is about taking on occupations that are significant to us at all phases of our lives. This is a different kind of occupation than the standard usage of the word. As Professor Kegan of Harvard noted in his work on adult development, this is what it means to transition from the socialized mind to the self-authorizing mind to the self-transformational mind—a changing of occupations (Tables 5.1 and 9.1). How can we facilitate this sequence of adult development through co-learning and education in the GYE?

Consider the following GYE example of a transformative education and learning program that I developed. This program is focused on the foundations of natural resources policy and management and is designed for any person in a subfield of environmental studies or other interests.[39] The program is designed to be delivered as a course, workshop, informal conversation, application, or through case study analysis (e.g., Cache Creek, Wyoming, citizens science efforts, cooperative group work).[40] The approach is designed to help people examine their perspective, gain knowledge, and improve their skills to think more effectively and act more responsibly in complex management policy cases. The approach explores comprehensive and integrated (interdisciplinary) concepts and methods for thinking about problems in natural resources policy and management and proposing solutions to them. It relies on case studies and applications, as well. In short, it uses the four-part heuristic in this book.

You may note that although the subject matter of this curriculum is ostensibly natural resources, the real goal of the course or application is about learning and transformation. First, this approach to learning and education introduces attendees to how to find comprehensive and integrated solutions to problems in management. Second, it introduces attendees to foundational concepts that underlie problems in management through diverse exercises designed to promote transformative learning and adult development. Third, it enables attendees to gain proficiency with these methods and concepts by having each student or citizen apply them to a particular management problem. And, finally, it embodies the concepts and methods of the course by creating an environment conducive to individual and collective learning.

This educational design is organized around the occupational question: What do environmental or natural resource managers manage? This question illuminates the concept of sustainability in ways that may be new to attendees/students or established professionals. For example, finding sustainability in policy and management is a matter of comprehensively identifying, responding to, and engaging with the changing conditions that shape our communities and their environments at different scales, and guiding dialogue and policy, while engaging respectfully with other people.

One challenge for scholars, practitioners, and citizens is to identify rigidities in both outlook and material conditions in any policy or management system that might inhibit transition to an adaptive, contextual, learning-oriented approach. Regardless of the subjects of concern—biodiversity, human populations, land and water use, justice, or human rights—every discipline and perspective has something to contribute to the examination and creation of alternative ways forward. In the GYE, variations to this approach can be used widely by governmental agencies, environmental NGOs, public and private educational institutions, or individuals should they choose to do so. There is a huge potential to use this approach widely in the GYE to good advantage.

Action and policy

Given all that we know about the problems and limitations we are currently facing, what are our first steps? Perhaps as Professor Louis G. Lombardi said, we should consider the welfare and freedom of wild animals to be wild.[41] Certainly, this is an important matter needing ongoing discussion. Given this, I and others think we should extend our basic moral considerations to all animals. Other animals are not moral agents. They do not understand the difference between right and wrong in any human sense of the notion. However, it is beyond question that nonhuman animals share many or most of the needs and desires of humans as physical beings. At a basic level, most share a capacity to experience thirst, hunger, and pain. And some species are semi-cognitive or even self-aware. As such, they deserve consideration and care from us.

Of course, we can and should make distinctions among animals as a basis for unequal treatment. Yet, as Robert Wright, journalist and Princeton scholar, noted, the "hopeless illiteracy of a chimpanzee is a fine reason not to admit it to Yale, but if we're about to strap it down and cut into its brain, the question isn't whether the chimp can read, but whether it can suffer."[42] Perhaps this is the kind of question we should be asking ourselves as we interact with animals in the GYE. We need to seriously reconsider our relationship and behavior toward animals. How do our actions directly or indirectly cause harm and suffering to other life forms? Currently, these and other questions are seldom considered in the public dialogue about the GYE's living forms.[43] We can rectify this omission.

We also know that the fate of the human species is closely tied to the health of the environment, nature, and other life on the planet. This is true in the GYE too. Given this fact and our present state of relative ignorance about the complex biotic and abiotic systems of which we are a part, it is prudent for us to be very careful when interacting with other life and natural systems. In the end, we should consider how can we contribute to and shape these interactions and our own thought and moral processes for a healthy life for all beings. By doing so, we might avoid the mounting problems in the GYE. This admonition is possible only through a transformative co-learning process toward a new ethical outlook, should we decide to take that route.

Farsightedness

A recent article shows that "every 2.5 minutes, the American West loses a football field worth of natural area to human development."[44] Both the GYE and the larger Northern Rockies ecoregion are being eaten away, piece by piece, by accelerating human development, intensified human uses, and resource exploitation.

As we imagine the future of the GYE, we need to draw on the field of future studies. This field seeks a systematic pattern-based understanding of the past and present as a guide for the future, and the outcome of this is referred to as a "developmental construct."[45] Developmental constructs seek to determine the likelihood of future events and trends. Constructing likely developmental pictures of the GYE is about postulating possible, probable, and preferable futures. Developmental constructs enable the decision maker, analyst, and citizen to find his or her way in the total situation in which he or she operates (e.g., the GYE system). The purpose of using this approach is to address complexity in an orderly way by revealing the significant contours of reality that we hopefully recognize and address.

Greater Yellowstone's future

What might the future of Greater Yellowstone look like? Are there clear signals about the future of Greater Yellowstone already visible on our radar? How do we find the important signals among the noise and confusion of daily life, overwhelming volumes of information, and the confusing dialogue in newspapers and social media? Are the signals strong or weak? Is there a way to connect the signals into a larger, clearer picture of what is likely coming to the GYE?

Future thinking

Greater Yellowstone is influenced by global social and biophysical systems (e.g., climate).[46] So, we must consider how these larger complex systems may function and interact in the coming decades. Concerning the future, Wendell Bell, a founder of the field of future studies and Yale sociology Professor, said,

> I live in a decade during which some of the most important choices in the history of human civilization will be made. I happily join others in facing the heroic challenges of this

decade—to move from our present catastrophic path to a new path that will dramatically improve our prospects for a flourishing future.[47]

The future is not just an extension of the past. It will most likely develop in surprising ways. Future studies concern the big picture and the context of our complex human and natural systems. It requires a deep look at existing human and cultural worldviews, myths, and paradigms that underlie them. Future studies as a field requires analyzing the sources, patterns, and conditions behind change and (in)stability. It is an integrative task, an interdisciplinary job.

We need to organize systematic knowledge for future thinking about the GYE. This needs to be done in ways that are widely shared, discussed, and acted upon. Taking stock of where we are in meeting our goals for the GYE requires an honest appraisal of where we came from, what we are doing now, and what we are planning or hope for the future. How does one take stock of all this in real life, relevant ways, and from a view of our collective society?

Where does one find future signals for the GYE? Perhaps they are in the stories people tell of Yellowstone, maybe in fact it is in their expectations and current demands, at least in large part. There is certainly scientific, social, and economic data at hand for us to use about the GYE. We can collect and organize all this data into a coherent picture of the whole that can be shared with the public and officials. A project of this size can only be carried out through a partnership of universities and government. I recommend that these organizations, regional and national, take the lead and produce a "futures map" for the GYE. Future studies, however understood and practiced, is a coherent field of thought and its empirical results are well established. We can use it to project the GYE's future and adapt our systems and actions accordingly.

Future signals

When talking about finding signals from the future, consider the metaphor of RADAR, a vital technology for sending signals across large physical distances. RADAR technology works only if we pick up on the returning signals and separate them from excess noise. This is what we must do with the data and signals about the future in the GYE. Unfortunately, we do not have a single technology or organization to read GYE's future presently. However, we do have signals from the past and we are accumulating new signals each day that we can use. The recent visitor and transportation report out of Yellowstone National Park is just one example of incoming signals, as it provides vital information about trends and conditions that we are presently experiencing.[48] The signals we need to pay attention to are all around us, not simply as single data points but as directional maps that reveal change through time. And, all this signals that data is waiting to be aggregated into an overall, comprehensive picture of GYE's likely future.

In our discussion here about the GYE, what are the signals we need to pay close attention to? How do we take data from the present day and the past and use it to project and address problems that do not yet exist? Is anyone in authority making a future-oriented, developmental construct for the whole system? Is GYCC? Assuming this is done, who decides what to do about the picture that presents itself? Future mapping and developmental constructs are keys to our farsightedness in the GYE. I recommend authorities and academics get right on this important mapping task, immediately.

Need for farsightedness

Now that we have the additional tool of future studies, I present four more recommendations to enhance conservation in the GYE. First, we need to recognize future studies as a legitimate and identifiable arena of intellectual and practical work and begin implementing and applying those tools and skills on subjects such as large mammal migrations, fires, human development and population growth, and large carnivores.[49] Already, some of the people working in the GYE have made important contributions through models on weather and climate, fire, vegetation, wildlife movements, and more.[50] Yet, collectively, much more is needed.

Second, we need to create greater capacity for future studies. This will require regional and national universities to draw on what already exists and build core courses and programs in future studies, as they learn to work closely with government and the public. Currently, much of the work and knowledge products about the GYE are widely scattered, disorganized, and of little use in giving us an overall picture. Agencies, environmental NGOs, and others need to focus on this vital task and aid one another in creating an overall cadre of skilled practitioners. Presently, there is a great opportunity to organize and contribute to this overarching need.

Third, many of the specialists and public officials in the GYE need to adopt future studies in their thinking. This futurizing will benefit everyone as we expand the horizons of attention, policy, and planning to address not only immediate problems but also those that extend well into the future. It will benefit scholars and scientists, as well as the public and private leadership, involved in decision-making. This will require reorganizing our current organizations and institutions. Perhaps the Yellowstone Science Center (YSC) could organize the overall effort to learn about and prepare for what is coming in GYE's future.

Finally, this futuring fieldwork can empower ordinary citizens, participants in the democratic process, and leaders at all levels to hone their skills and focus their attention on the needed real work before us. One vehicle to do this is "citizen science," which is now popular. However, it must be conducted with rigor and quality control. Futuring can help us move toward our long-term goals for ourselves, our communities, and for the GYE and its wildlife. Without the application of futuring skills, we cannot work effectively toward a sustainable, healthy future for the GYE. Without using them, we are handicapped.

Strategy

Greater Yellowstone is a national and global treasure, but evidence is accumulating to suggest that we are experiencing an accelerating system failure in the region. We must act quickly to address the existing and foreseeable problems as the system comes under more and more stress. There is only one Greater Yellowstone Ecosystem. It is an invaluable and potent symbol for what sustainability might mean in practice. In the end, it is up to us to learn our way to sustainability. The GYE is a great place to start with a model that can be helpful to people everywhere.

Using our resources

We need a strategy for how to use our limited resources (natural and cultural) to achieve our goals in the GYE. Our unfinished work in the GYE is a task for people who want to control their future rather than merely accept it. We need a strategy for understanding the past and present, as a basis for our likely future. The needed strategic work on our future has local to global implications and payoffs in terms of conservation, enhanced cooperation, education, and leadership development. Finding a strategy to get us there requires a deep look at existing worldviews, myths, and paradigms that underlie our current attitudes and behavior in the GYE, as I have introduced previously.

Many thinkers see that we need to adapt to rapidly changing circumstances in the GYE and in the world, and that means transforming ourselves to a new system of thinking, living, and individual and cultural stories.[51] Instead of seeing a future full of risks to be studied and minimized, individuals, groups, organizations, and institutions will need to address a future of unknowns. Adaptation, although important, will not be enough. Transformation is urgently needed.

Adaptation

Adaptation seeks to accommodate ourselves to changing conditions rather than contest the sources of the harmful change. The adaptation strategy is often viewed as making a series of technical and local changes. Adaptation schemes are generally viewed as apolitical and assume a system that is inextricably tied to perpetual economic growth as a driving force. Of course, this understanding maintains our current worldview and is insufficient for accomplishing our sustainability goals in places like the GYE and elsewhere.

The adaptation approach has implications for the whole environment, natural as well as human, including social justice and class and economic stratification. Adaptation tends to emphasize top-down forces for local resilience projects, often with a technical focus. Conventional adaptation assumes this will be enough for us to cope with foreseeable change. Such an approach does not call for large-scale systemic change or shifts in worldviews (constitutive change), which I believe are necessary. As such, this kind of adaptation is not what we really want or need given current circumstances in the GYE.

Instead, we need to question the underlying cultural and economic assumptions that are driving our current social and environmental decline.[52] It may be easier to engage with these harmful assumptions as it becomes clearer that our current thinking and actions have pushed us well beyond GYE's and the earth's biophysical limits.

The transformation needed should be understood as a process rather than a one-time shift. Transformation can happen through fast, catalytic events or through slower, deep shifts in thought and behavior over a long time. Fundamentally, both physical and political changes are energy-intensive in terms of innovation, individual shifts in thinking and doing, and in management and policy. All are needed in the GYE.

There are many challenges to transforming. For example, needed changes in infrastructure will become more difficult as energy becomes increasingly costly. The time lag between cause and effects of the current environmental and social decline may be fairly long, even generational. Any transformative actions will create a psychological and political obstacle for needed policy action that must be addressed. The basic question to us presently is can we make transformation in the GYE where the need is already clear and resources are readily available? Is co-learning and transformation possible?

Transformation

Our ongoing social sciences and humanities can help inform the best strategic approach to both instigate and guide transformation. Psychology and the behavioral sciences provide insight into how individual people's behavior can change in profound and durable ways, thus motivating change and creating a vision for a new status quo.[53] Even so, transformative change may not be feasible unless triggered by larger disasters or shocks.

Ultimately, the greatest barrier to an informed transformation is a lack of precedents for conscious change on regional and other large scales. The traditional literature in transformative theory does not explicitly apply to environmental problems on the scale we needed in the GYE, much less at the national or global level. The success and consequences of any true transformation will depend on the speed and openness to change of individuals, communities, and institutions. It will also depend on our understanding of how best to guide a transformation of such scales. This will require effective high-order leadership to be successful.

Actions

I recommend that we need a future studies unit for the GYE at the Yellowstone Center for Resources, Yellowstone National Park, as noted earlier, in close cooperation with many other organizations, universities, and individuals. This unit could clarify both utopian and dystopian visions of the future for the GYE. Such a project offers a way to think about our future and ways to sharpen our understanding, as a kind of farsightedness. It can help us explore practical ways to bring about the kind of future that we want to live in. It can produce realistic, problem-oriented problem definitions and help us understand how best to practically address them.

Judging preferable futures

How can we judge the kind of future that we want for the GYE? In the past, we have used religion, ethics, and professional knowledge to decide. These approaches appeal to some authorities and the public to help us discern and justify what is good and bad, but none of them alone is adequate.

For all of human history, humanity has relied on religions, superhuman powers, or a god or gods to organize their lives and societies. Religions prescribe values and formulas to live meaningful lives, and many religions encourage a stewardship of natural resources. Pope Francis's pronouncement in his Encyclical Letter *Laudato si: On Care for Our Common Home* speaks to the heart of the matter. He addresses the human roots of the ecological crisis (globalization of the technocratic worldview or paradigm), the effects of modern anthropocentrism, the principle of the common good, intergenerational justice, and the shortcomings of our contemporary dialogue on the environment.[54]

However, many societies are now secularized and religion does not have the hold it once did. The overall trend in human philosophy, society, and living in the past 10,000 years has been toward secularization, and this comes with an increasing reliance on pure scientism as a way of understanding the world.[55] This has implications for how we understand the world, how we define truth, and how we will go about responding to the large-scale changes that are underway.

Secularization has spawned two counterresponses from religion. One is a revival of religious groups in opposition to secularization and a return to less worldly and fundamental metaphysical concerns. The other counterresponse is the creation of new religious groups that accommodate and partially incorporate some of the symbols of secularization, yet do not change their basic religious beliefs or practices. As the relationship between religion and science continues to change, so will trends in reliance on scientism and other ontologies, epistemologies, and axiologies (Figure 1.3).

Appeals to collective judgments or law

The justification of goals and value commitments is commonly determined by making appeals to the collective judgment of society. The argument that law is based on the outcome of a democratic process is a powerful one. This approach goes back to Aristotle and Kant and is a founding principle of the US system of governance. It rests on the principle that values and norms should reflect the culture in society. We currently use this approach and structure it through elections, the three-branch governing system, academia, and the free press.

Some people make appeals to history and earlier charter myths. One example of a charter myth is the Declaration of Independence. Other people appeal to a survey of public opinion. In the United States today, polls are commonly used to see what preferable futures might look like. We can also appraise the actual results of our judgment, if we are introspective and analytic, to determine whether they meet our goals. Some people use neoliberal capitalism (markets) to decide. Still others want to appeal to future generations to help make decisions today. Appeals to beliefs in conventional society will continue to dominate our thinking and decisions about the GYE's future, for a while at least.

Can we judge the GYE's future based on law? Humans have constructed law and policy to address matters for group living, even at large geographic scales like the GYE and the nation-state. Formal policy and laws are clearly human constructions to organize individual and social life. How do laws and policy interact with other systems of moral designation? Religious beliefs have long dictated right from wrong and are incorporated today into our current law and policy. Political myths and religious beliefs are often similar because they share a reliance on presumed universal truths and value statements. Today, however, authoritative state and federal laws and policy largely replace religion as an arbiter of morality in modern society.

Currently in the GYE, we see four ways in which laws and social norms are interrelated. Social norms may become laws, laws may become social norms, laws and social norms may conflict with each other, or laws may facilitate the creation of social norms by other people. Each of these four interactions is mutually exclusive logically, but in real life they are not separate. They are in constant interaction all the time.

In today's changing multidimensional complex context, some laws, norms, and policies are in conflict with each other. This happens more often as the environment and society change. These changes bring old laws into conflict with new social norms. Sometimes, new social norms are destructive and so new laws must be developed to address the conflict. For law to be more constructive and authoritative in GYE's future, we must work to harmonize social change with policy changes at a high level.[56] In particular, the extent to which a law reflects prevailing social norms influences its effectiveness. People are more likely to comply with a law or norm if they believe it is fair and its outcomes are worthwhile.[57] Thus, all laws and norms in the GYE should be focused on common interest outcomes and longer-term concerns, while recognizing both special and short-term interests. Even so, it is likely that political struggle will continue to dominate GYE's future.

Professional ethics

Finally, another possible moral guide to the future is a new professional practice and ethic. Professionals are often involved in the process of deciding our future actions and management policy, as they do now in the GYE. We know that professionals participate in a whole range of activities, including research, teaching, publishing, consulting, and advising government and private organizations now. They work in government, environmental NGOs, and businesses, too. Who are the current professionals and experts in the GYE's science, management, and policy? How should they be engaged in deciding the future of the GYE? These are open questions and deserve serious attention and answers.

We can turn the question inward and ask: How should professionals themselves think and work in the GYE? What do we think they would say? These questions involve ethical considerations as well. Some professional societies, like the Wildlife Society, have a code of conduct, a kind of moral structure and shared value system, but others do not. As I have discussed, the things we do are not separable from who we are and what we think. Thus, professional ethics and the values upheld by others with similar occupations influence our understanding of what is right and wrong and what we should and should not do. If we are to derive a new moral structure toward sustainability in the GYE from a professional foundation, we must fully understand these interrelationships.

Conclusion

Securing the needed conservation management policy for the GYE's future requires that we learn rapidly, exchange lessons learned, and network ourselves to be most effective. Finding a sustainable place for ourselves in nature requires that we teach moral education and attentiveness, among other things. It also requires that we use co-learning insights and appraisals and rapidly turn lessons of hindsight into practical applications and foresight. Further, it requires transforming current operations to a new mode of learning, thinking, and action for sustainability. To be successful ultimately requires transformation of individuals, groups, organizations, institutions, and management policy processes. We know of practical designs to successfully achieve such outcomes. We need to use these widely.

Finally, future thinking—farsightedness in our judgments and management policy— is a vital way to encourage the needed learning at all levels. It requires deep shifts in our being, doing, and knowing. I recommend that when judging our preferred future and identifying methods for achieving our stated goals, we should draw on collective judgment, legal structures, and a new kind of professional ethic to secure a healthy future for the GYE. Most immediately, we need extensive dialogue throughout the GYE and educational spheres about foundational matters to our living so that we coalesce a vision for the GYE's future that is widely supported and identifies strategies to get us there.

Chapter 10

THE WORK AHEAD

To live successfully in a world of systems requires more of us
than our ability to calculate. It requires our full humanity,
our rationality, our ability to sort out truth from falsehoods,
our intuition, our compassion, our vision, and our morality.[1]

Grizzly bear #399 and her family are the most famous grizzly bears in the world, and their story provides a snapshot of our changing relationship with wildlife. It brings to focus the work needed to conserve grizzly bears, other species, and ecological processes for the benefit of people and all other life. Over the last few decades, #399 and her offspring have lived in and around Grand Teton National Park. In 2020 this 24-year-old mom had four cubs. She has become an easily recognizable symbol for conservation issues in the GYE. This is why she and her family have come to personalize wildlife and the Greater Yellowstone Ecosystem to millions of people worldwide. This fosters concern for this families welfare, and sparks a deep interest, as we watch the effects of social and environmental change unfolding all around them. People interact with these bears by viewing them from the safety of their vehicles and through social media. Grizzly #399 has come to exemplify a decades-long battle to protect grizzly bears throughout the GYE, which is itself really a metaphor, a symbolic story, about trying to conserve the GYE itself. The stories that we tell about #399 reflect who we are as humans, how we view wildlife, and what we are doing in nature.

We have created a powerful narrative about grizzly #399 and her family over the years. This narrative came about as photographers and journalists shared their observations, images, and story for the public, such as the beautifully documented photographs of Tom Mangelsen's book *Grizzlies of Pilgrim Creek: An Intimate Portrait of 399 (The Most Famous Bear of Greater Yellowstone)*.[2] Her playful cubs engage people. What does this story really mean to us? Is there a deeper meaning than just seeing a bear? In 2015, a headline claimed, "399 is the poster bear for all the bruin clan."[3] The author, Todd Wilkinson, said, "I think about the gift she represents in waking up millions of people to the miracle that Greater Yellowstone's grizzly population was rescued from the brink of annihilation. And I think about the fragile reality of recovery that still lies ahead."[4] He ended the article with, "Yet every day she is out of the den, 399 navigates a field of landmines. Most of the bear's offspring have already died at the hands of humans." Again, this raises questions about work ahead for us to address this case and many other of GYE's problems successfully. This chapter explores that needed work.

According to the 2015 NPS report, we are now in a "complex time." Yellowstone National Park and the entire Greater Yellowstone Ecosystem are experiencing massive increases in visitation, use, and harmful impacts. As the 2015 NPS publication concluded, "Although change and controversy have occurred in Yellowstone since its inception, the last three decades have seen many issues arise. Most involve natural resource conflicts."[5] Of course, as I have laid out in this book, the conflict is not simply about the natural or cultural resources themselves but rather about how people want to use and make meaning in YNP and the GYE.

In this chapter, I focus on the needed work ahead, including overcoming fragmentation, competitiveness, and reactiveness, and the need for a deep change in our system of thought or culture of mind relative to bears and nature. In doing so, I look at the #399 bear story as a referent of nature and offer a cautionary tale. I also look at our record of managing the GYE as an open-ended experiment and ask these questions: What is to be learned from this experiment? How can we apply those lessons of the past to oncoming challenges? How do we come to see the GYE as a "whole," in systems terms, perhaps for the first time ever? Bear #399's life story makes these subjects concrete and real for people.

The Bear as Our Story

Historically, people have viewed nature and animals as the nonhuman world—a collection of material things for us to use as we saw fit. But today, that paradigm is beginning to break down as the dualistic boundaries dividing humans and nature become ambiguous. As we learn more about the ways humans both affect and rely on the environment, we increasingly realize that we are part of nature, not outside it. In many ways, to many of us, bear #399's story is a metaphor for these emerging realizations and as such, it speaks to the need for us to work toward a better future for all concerned. Our connection to this bear and her family gives us a human way to connect to this larger paradigm shift, at least indirectly.

A cautionary tale

Ultimately, the story of bear #399 serves as a warning about the impending changes and threats to the GYE—wildlife is killed or harmed frequently in interactions with humans and populations are threatened by climate change, sprawling development, and industrial-level adventure recreation. But it is also a story of promise and pragmatic hope, of redemption and recovery of the North American grizzly and sustainability of Yellowstone as nature.

Bear #399 and family

Grizzly #399 first began generating press in 2006 when she was reported with her three cubs "digging for roots and rodents along [the roads]" and enthralling park visitors.[6] Since then, many more stories have been written about the trials and tribulations of these bears. In 2016, one of her cubs was killed in a hit-and-run car accident, which garnered widespread public outrage. But what are these stories really about, bears as real animals or our imagined relationship to them?

Significance of the bear story

We created the cautionary narrative around bear #399 as "a window into the drama of grizzly conservation."[7] One article about the famous grizzlies stated,

> 399's journey of survival provides a lens for pondering the deeper value of having grizzlies in Greater Yellowstone. And as well, for it to serve as a reference point in the debate over removing this iconic bear population from federal protection. And finally, to show what's at stake as Wyoming and other states consider recommencing a trophy sport hunt of grizzlies.[8]

The conservation struggle in which we are enmeshed in the GYE is a contest over the future we want for ourselves with wildlife and in nature. In the case of bear #399, our questions about her value and her future are not simply about what we hope for bears in Yellowstone. Rather, our stories are an attempt to clarify why we value wildlife and what relationship with nature we want to have moving into the future.

Bear #399 has been a heroine of our overall grizzly bear conservation story for the last many years. But, much like the future of the GYE, her story is unfinished. As we change bear #399's tale in the public dialogue and through management policy, we make meaning for ourselves and work to clarify our place in the natural world.

Stories of personalizing animals

Most people do not get the chance to interact with wild animals in their habitat. Hence, we have little understanding of their true nature or behavior.[9] Because of our limited interaction, we are constrained in our ability to accurately convey the reality of wild animals as independent entities. To compensate, we often filter our stories about animals through our own human worldview and project our realities onto them through anthropomorphization. Anthropomorphizing is the process of ascribing human characteristics to nonhuman beings or objects. Originally, anthropomorphizing referred to deities who were given the form of humans; however, anthropomorphizing applies to any nonhuman entity, including animals such as bear #399, plants, and inanimate objects. Some anthropomorphic entities represent specific human concepts, such as love, wisdom, power, or beauty. They may also show human weaknesses, such as greed, hatred, jealousy, and uncontrollable anger. Often, we imagine animals as lesser versions of humans, as morally undeveloped creatures that act entirely on impulse in matters of life and death, good and bad.

We use stories about animals as a dramatization of our inner conflicts. Many officials and wildlife ecologists oppose the anthropomorphization of animals because the official narrative is about the objectification of animals and nature, and they want to enforce that view on the public. Yet, for the public, anthropomorphizing provides a way to make meaning of nature and wildlife. It often fosters empathetic feelings and concern about wild animals, particularly when the animals are perceived as acting with moral consciousness—like a story of a mother protecting her offspring in a hostile world, a story of loyalty and dedication. Bear #399 is the subject of a powerful utopian story for the GYE and, in a larger sense, for people and nature. It reflects our hope that nature, in the form of bear #399 and her family, can survive and thrive well into the future.

Greater Yellowstone Ecosystem's story

On August 25, 1916, President Woodrow Wilson signed the Organic Act, which created the National Park Service, a federal bureau in the Department of the Interior:

> The Service thus established shall promote and regulate the use of Federal areas known as national parks, monument and reservations [...] by such means and measures as conform to the fundamental purposes of the said parks, monuments and reservations, which purpose is to conserve the scenery and natural and historic objects and the wild life therein and to provide for the enjoyment of the same in such manner and by such means as will leave them unimpaired for the enjoyment of future generations.[10]

Ongoing experiment

The history of Yellowstone and the GYE can be told through a series of progressive stages. It began in the precontact times when nomadic native tribes inhabited the area (today, 26 Native American tribes are associated with YNP).[11] Then, the landscape began to shift once Osborn Russell, a fur trapper, arrived in the Lamar Valley in the 1830s. Yellowstone National Park was formally established in 1872 and was managed by the US Army from 1886 to 1918. The National Parks Service was formed in 1916 and the service centralized parks management. Other federal agencies were created along the way to help protect public land, including the US Forest Service and the Bureau of Land Management. Settlements were developed in later decades, up to the period of rapid growth that we see today throughout the Greater Yellowstone Ecosystem.

Modern management of YNP began in the 1960s with the goal of "maintain[ing] biotic assemblages." Park management policy underwent significant shifts in 1963 after the publication of the Leopold Report and again in 2002 with the updated report by the National Academy of Sciences. Both documents called for enhanced wildlife conservation as a primary purpose of Yellowstone National Park. The 1960s and 1970s saw a host of new environmental laws, including the Wilderness Act, the National Environmental Policy Act (NEPA), the Clean Air Act, the Clean Water Act, and the Endangered Species Act (ESA). All supported greater environmental protection.

A precarious story of GYE's future

As President Theodore Roosevelt said over a century ago, "We have fallen heirs to the most glorious heritage a people ever received, and each one must do his part if we wish to show that the nation is worthy of its good fortune."[12] Yellowstone National Park and the GYE (both as an idea and a place) emerge from our civilization and represent our restraint, foresight, and vast nostalgia for the era of frontier individualism. Given our current context, now is the time to recognize the opportunities available to us and to capitalize on them for constructive changes. If we are successful, the foreseeable future will be a healthy GYE.

We say that Yellowstone National Park was established on March 2, 1972, but in reality, we have never stopped establishing Yellowstone. While a lot of people still view it as little more than a physical place, a growing number of people recognize that the park is the site of something much more dynamic in human culture, a kind of perpetual experiment that will never end.[13] As noted in the 2015 National Park Service publication,

> The years have shown that the legacy of those who worked to establish Yellowstone National Park in 1872 was far greater than simply preserving a unique landscape [...] This idea conceived wilderness to be the inheritance of all people, who gain far more from an experience in nature than from private exploitation of the land.[14]

The challenge is clear: we must work quickly, cohesively, and effectively if the GYE is to have a healthy future. We know what to do, we know what changes we must make if we are to manifest our goals in the GYE. In the words of Andre Guide, Nobel Prize Winner in 1947, "Everything that needs to be said has already been said. But, since no one was listening, everything must be said again."[15]

A Working Strategy

In the GYE, we need to develop and sustain an effective problem-solving process that can function in a democratic context—a practical process for addressing ordinary, systemic, and constitutive problems as interconnected challenges. Unsystematic thinking, which is common today, is not serving us well. Consequently, we need a proven workable strategy to think about cohesive and interrelated processes in a systematic way. Such a strategy requires that we develop and use many mental tools in an orderly fashion.[16] However, before we can implement such an approach, we must identify an overriding goal, a value-oriented rationale, a framework for analysis, categories of analysis, indices of values, and component operations of problem solving. We can do this by grounding our thought in practice-based projects that are currently underway.[17] This requires finding and describing relatively successful conservation efforts or, at least, efforts with great potential, adapting and diffusing them widely, and opening up new opportunities to build further successes.[18]

Practice-based problem solving

The multiple, reinforcing environmental and social signals from the GYE all show growing problems. To address problems, we must take on a problem-oriented, contextual approach as our working strategy and ground our work in pragmatic experience (Figure 10.1). We learned earlier that the current system in place materializes, objectifies, and instrumentalizes wildlife

**An integrated, problem oriented approach
to science, management, and policy**

Goals	What's happening?	Why?	Future?	Does a problem exist?	Solutions? alternatives
↓				↓	
Indices?				Problem definition?	
1.Environment					
water, air, ... ?	?	?	?	?	?
2. Social process					
people ... ?	?	?	?	?	?
3. Decision process					
management ?	?	?	?	?	?

A problem orientation

Figure 10.1 The problem-oriented approach and questions that should be answered, using evidence-based, reliable information.

and nature. In turn these are often monetized or commodified. Further, the assumption of human separation and dominance over nature aggravates problems. Finally, a host of scale (space, time, complexity) issues make integrated, coherent management policy highly problematic. Together, these generate many of the problems we presently face. And, the more we rely on those views and assumptions, the more we seem to be moving away from genuine problem solving for sustainability. Conventional problem solving (only partially problem-oriented and contextual) is institutionalized now in our system of management policy today. To put it most simply, we must transform this system by shifting paradigms from scientific management to problem-oriented adaptive governance.[19]

Addressing problems

Albert Einstein once said that if he had only one hour to solve a problem, he would spend 55 minutes on understanding the problem and 5 minutes on offering solutions. In short, he was explaining the need to orient to problems and understand their context and content before we charge off with presumed solutions that, if misguided, may only compound problems. To carry out sound integrated research, management, and policy, we need to avoid the temptation to be "solution oriented." Instead, we need to be fully "problem oriented," but what does that mean?

   Problem solving is largely about a set of logical steps about rationality (Figure 10.1).[20] It is both a process of known thinking and tasks or operations for dealing with the matter at hand. There are five tasks in problem orientation:[21] (1) clarify goals, (2) describe trends, (3) analyze conditions, (4) project developments, and (5) invent, evaluate, and select alternatives. These tasks are often overlooked, but when employed, they can help us orient our work and maximize our efficacy. They can clarify which alternatives will best achieve conservation goals. We need to ask: What outcomes do we want? What are the problems given our goal? What alternatives are open to participants to solve problems? Would each alternative help solve the problem? Did it work or not work when used for similar scenarios in the past? Why, or under what conditions, did it work or not work? Would it work satisfactorily under existing conditions? And repeat these questions on an ongoing basis within limits of time and resources.

I recommend this approach be used in all work throughout the Greater Yellowstone Ecosystem. The five tasks direct us to ask questions and to think critically and comprehensively about the information that we know in a fashion conducive to efficient problem solving. First, clarifying goals is simply a matter of asking the question: What do I want? Describing historical trends is about identifying what has been done to address or mitigate an issue now and in the past. Analyzing conditions means understanding the factors that are driving the trends that we observe, both positive and negative. These conditions should be the focus of our problem-solving work as they are the things we need to change (if they produce negative outcomes) or maintain (if they produce desired outcomes) through management policy. Next, making a projection based on past trends and conditions requires us not only to extend past trends or conditions into models for the future but also to think about how they may interact and influence each other, and how they may change given the complex social and ecological context of the GYE.

Finally, we are tasked with finding options to solve the problem(s), based on the previous forecasts and our understanding of the conditions. It asks us to identify approaches that we think will help change the conditions driving undesired trends and, in doing so, help us realize the goals we set for conservation. What alternative will we undertake to change conditions so that future trends will be favorable for animals and humans in the GYE?

The benefit of contextual problem orientation is that we can better understand the problem(s) or solution(s) in their context(s). This strategy may seem off-putting to some people who see this recommendation as a lot of tedious academic thought, when what we need is action. However, this problem-oriented process can be performed quickly and it dramatically increases prospectus for long-term gains of conservation programs. It is important to make an honest attempt to place all the facts, options, and potential consequences in context before we propose solutions. Doing so helps us ensure that our solutions yield the desired effects. We do not want to compound or create additional problems in other areas. Problem orientation, recommended here, is a strategy to be practically rational, as opposed to using conventional approaches.

Human factors

Conservation is always a human concern. Humans not only create the conservation problems we experience but also create the ideals and expectations that we work toward (recall our heuristic: *people* seek *meaning (values)* through *society (institutions)*, using and affecting *resources (environment)*). Yet, conventional conservation typically overlooks or ignores the fundamental importance of human social processes, and especially values and institutions. It overly focuses on resources as conventionally understood (i.e., things). This is one of the most significant problems with our current conservation process.

So, I recommend that we employ social process mapping around any problem to help us understand the conditions at play—the context.[22] The social process is the way that human social interactions proceed. It includes our core needs and ways that we try to manifest them. Typically, we fall short of getting a complete social process map about most all of the issues we are trying to resolve. As such, our incomplete maps limit our ability to fully understand the conditions at play or interact with one another constructively.

We all know that our social interactions affect the environment. Social process mapping describes the interaction among people in the context of a specific conservation challenge.

Clearly, we must use and affect resources (the environment) to help fulfill our needs and accomplish our goals. The way that we use resources can either help or hinder conservation management policy. Understanding human social processes in practical terms is vital because conservation can only take place through the social and decision process. We need to ask questions about the social process derived from observation, deep thinking, and sound social and integrative science.

These core questions, as noted below, delve into the essentials of human social interaction and come out of functional anthropology, psychology, and the policy sciences. Asking and answering these questions can greatly enhance our understanding of complex conservation management policy problems. This model offers questions that focus on participants, who have perspectives interacting in a particular situation. People draw on their core values, including knowledge, skill, and power, and use these values in their strategies to pursue preferred outcomes. Finally, the outcomes have long-term effects on all future interactions. The social process model is a mapping method for cooperative problem solving, regardless of the case at hand.

Question include the following:

1. Who is participating (identify both individuals and groups)? Who would you like to see participate? Who is demanding to participate?
2. What demands do participants (or potential participants) have in terms of values and organization? On whose behalf are those demands made? What are the perspectives of those who are participating, you would like to see participate, and making demands to participate? What would you like their perspectives to be?
3. In what situation do participants interact? In what situations would you like to see them participate?
4. What assets or resources do participants use in their efforts to achieve their goals? What assets or resources would you like to see participants use to achieve their goals?
5. What strategies do participants employ in their efforts to achieve their goals? What strategies would you like to see used by participants in pursuit of their goals?
6. What outcomes are achieved in the ongoing, continuous flow of interaction among participants overall and by phase? Who is indulged in terms of which values? Finally, what new practices are put in place over the long term? Are old practices maintained? What forces restrict innovation and adaptation? What are the long-term

effects on the social and decision processes involved? What new practices have been put into place? Were there any innovations? And, how were innovations diffused or restricted?

Some of these questions are technical, others less so. Most biophysical scientists, including wildlife biologists, have been trained to see conservation management in technical, material, and objectified terms rather than human social and decision process terms. Thus, the human social and decision process is easily overlooked, assumed, ignored, seen as irrelevant, or viewed as a political constraint to the central, technical biological task of conservation. From that standpoint, human interactions are often seen as "politics" and outside of serious conservation attention and work. This partial view often proves fatal to public trust and to really conserving wildlife and natural resources.

We must come to deal with the human factor realistically, if we are to resolve the conflict and challenges in the GYE. In most instances, people are unaware of others and their values involved in any particular case, thus they are unable to see why they are in conflict. Typically, conflictual situations are understood in terms of the "good" guys versus the "bad" people. Social process mapping is a method to realistically map the social process as a basis for finding shared interests. Successful conservation is only achieved when human social process is made to effectively support common interests. The social process (and questions) can be addressed fully in terms of the five interrelated, systematic tasks of problem orientation.

Some trends in things (e.g., bears) and processes (e.g., fires) are improving in the GYE. For example, there is growing awareness that the GYE is threatened. The recent hiring of a sociologist by YNP is laudable, but we need the integrative sciences in the mix. The integrative sciences possess the capacity for problem-oriented, contextual, and multimethod problem solving. Our overall inattention to human social dynamics in systematic ways can be addressed. Data suggests that problems in the GYE are currently outpacing our ability to understand. We need to move on to fully problem-oriented contextual paradigms as our best strategy to address those problems.

Value transactions

Doing conservation is a technical task, but most importantly, it is a value-laden task. I spoke about human values earlier in this book because they are central to conservation and the social and decision process. I will touch on them briefly again for emphasis. Values are the things and events in life that people—including scientists, managers, and decision makers—desire, want, or seek. Values are the basic medium of exchange in all human interaction. As stated in previous chapters, values can be understood functionally as eight interactive categories: power, wealth, skill, enlightenment, affection, well-being, respect, and rectitude. The sum of the eight value groups when enjoyed amply is what we collectively refer to as "human dignity" (i.e., justice, fairness, the good life). Value dynamics must be mapped realistically as part of understanding the social process in any and all conservation efforts, and attending to them requires an explicit, conscious, and self-aware effort. These values are at the very heart of conservation and all that we do.

Conventional approaches to conservation seldom attend to these fundamental value processes, explicitly or systematically. Attending to value dynamics in a practical way requires that we become much more aware of our own values at play (our own standpoint) in work and living. It requires us to develop a broader, practical, and perhaps theoretical conception of our values well beyond conventional, everyday understanding. Armed with this value conception, we can work to find our shared interests, and hopefully, it will allow us to clarify, secure, and sustain our common interest.

Making partnerships work

Partnerships are being used in the GYE to improve effectiveness. They range from passive to interactive participation and "self-mobilization."[23] Using partnerships does not guarantee successful or improved outcomes, especially if they are conventionally conceived and carried out. In some instances, partnerships can compound the original problem if the partnership design, its problem-solving strategy, and the problem to be addressed are not matched. A social process, combined with a sound problem-oriented approach, can increase cooperation, maximize resources available, and improve chances for high-quality decision-making for successful conservation. Ideally, partnerships are united by a shared interest, civility, and evidence-based approaches to conservation. Thus, for partnerships to be effective, they must devote considerable attention to productive social process and making the decision-making process work well.[24]

Effective partnerships

Partnerships are comprised of individuals. Participants in partnerships are likely to differ in their motivations, knowledge, and skills. They are likely to be interested in different values and outcomes. The relation of these individuals to one another and the ways that their viewpoints interact to shape conservation strategies influence the likelihood of success. Often, partnerships are unsuccessful because of inverting, displacement, or substituting goals, the act of turning goals inward toward the partnership itself (getting along), rather than focusing on the conservation problem to be solved.

The goal of a partnership should be to understand and solve the problem at hand, not simply for the individuals to get along with one another. There is often a tendency to subordinate the goal of conservation problem solving to other interests represented in the partnership effort, such as "collaboration" itself. Goal displacement or substitution guarantees failure. For example, pressure to get along with fellow participants may force members to be superficially friendly and present a positive public face at the expense of orienting to the problem realistically. This can reduce the likelihood that individuals speak up when they disagree or have ideas that are contrary to the prevailing group sentiment. In this way, goal substitution or inversion can trap partnerships in special interest goals and misdirected strategies. If this happens, the partnership fails to adequately understand the problem, create a realistic problem definition, or explore and offer options to address the problem. Thus, the conservation problem goes unaddressed.

Given this, creating effective partnerships is not an easy task. Partnerships too often are focused on getting *more* input ("inclusion?"), rather than *substantive* input. Of course, it is important to involve the public in deliberate and democratic ways, but effective partnerships require knowledgeable and skilled problem solvers to address real problems.[25]

There are diverse considerations in partnerships, public participation, and decision-making and all these are all seen in use in the GYE. The most effective partnerships and decision-making processes are able to avoid goal inversion and other conventional traps. They do so by understanding the social and decision-making processes in a problem-oriented, contextual way.

Addressing weaknesses

Partnerships must make decisions to bring about successful conservation (Table 10.1). However, there are common, recurring weaknesses that diminish the efficacy of these efforts. I will quickly review weaknesses as a reminder. Recall that the decision process is a means of reconciling conflict (over values) among people, whether differences are about rationale, political interests, or morality. A realistic understanding of the decision process can help us evaluate and improve the critical functions of problem solving, regardless of the issues at hand.

We know from experience that there are common weaknesses in decision-making.[26] For example, key information is missing, debate is restricted, wrong decisions are made and implemented, efforts are delayed, and harmful practices are not ended. It is important to recognize these weaknesses early on in the partnership process so that they can be addressed for improved effectiveness (Table 10.1).

Problem solvers

Conservation work is carried out by both professionals and citizens. The problem solver is any entity (a person, partnership, or institution) that comes to understand the problem and works to resolve that problem. Problem solvers who are very aware, self-knowledgeable, and technically skilled in the integrated social and decision processes will be most effective.

For our purposes, a professional is a person with specialized education who participates in a community with standards of practice and shows commitment to public service. Professionals typically implement management policy. The role of professionals who carry out management policy work is a key variable in how well problem solving unfolds. As scientists and citizens, we need to look at this variable in some detail.

The role of professionals has changed significantly in recent years (Table 10.2). There are huge differences in the identities, roles, and effectiveness between the traditional and policy-oriented models of professionalism. The traditional standpoint is often not as effective in today's complex social and decisional contexts as it needs to be. Being

Table 10.1 Common weaknesses or pitfalls to be avoided for each of the decision functions in the decision-making process.

---

1. Planning (intelligence)

Delayed sensitivity. Perception of a problem comes only after the problem has developed and harmful effects are widely felt.

Biased initial problem definition. One interest sets out a problem definition that favors its own interests and view of the problem. This definition fails to capture the full and true nature of the emerging problem.

2. Debate (promotion)

Inadequate analysis of the problem. Needed data and analysis of trends, conditions, and projections are lacking or only partially carried out.

Study the problem to buy time. This form of delaying is a common tactic of people who oppose the emerging policy picture and problem definition. These people do not accept the problem definition or want to take action on it.

3. Selection (prescription)

Poor coordination in government decision-making. Often, complex problems are addressed by several groups simultaneously that may not be aware of each other or communicate well in developing a common understanding of the problem and what needs to be done to solve it.

Overcontrol. Groups may respond to problems by automatically imposing greater controls on everyone involved. This leads to bureaucratization and sometimes paralysis or gridlock.

4. Initial Implementation (invocation)

Benefit leakage. Certain socioeconomic interest groups may seek to capture and benefit more from the policy than other intended recipients.

5. Final implementation (application)

Limitations of state enterprises as natural resource managers. The size, slowness, political interests, conservative, and bureaucratic features of governmental organizations can all be limitations.

Poor coordination of implementation. Often bureaucratic overcontrol, rivalry, and exclusion of key parties can lead to muddled policies and programs.

6. Evaluation (appraisal)

Insensitivity to criticism. Critics may try to improve a policy honestly; but government often simply ignores their input, regardless of its merits.

Failing to learn from experience. Organizations can fail to learn and repeatedly respond to new conservation challenges using the same programs, approaches, and techniques.

7. End/exit (termination)

Pressure to continue unsuccessful policies. Even unsuccessful or poorly performing policies programs that have outlived their usefulness may benefit someone, who then clamors for the policy to continue.

Failure to prepare for termination. Groups may fail to appreciate and prepare for the difficulties of terminating even a bad policy, early in the overall policy process.

---

explicitly aware of a professional's or citizen's standpoint is an important way to improve performance.

The traditional standpoint often includes views such as the following (see Table 10.2): (1) Participants know what they want and follow a prespecified plan or project design; people tend to be rigid. (2) Participants are generally aware of a single, tangible reality, which is

Table 10.2 A comparison of the traditional and policy-oriented professional standpoints in conservation.

Conventional Professional (Scientific Management)	Policy-Oriented, Professional (Adaptive Governance)
Participants know what they want and follow a prespecified plan or project design; people tend to be rigid.	Participants do not know where projects will lead, so work is an open learning process; people tend to be flexible.
Assumption of single, tangible reality, which is generally known to participants; "correctness" is clear, and "right and wrong" actions known.	Assumption of multiple realities; reality is partially socially constructed and must be discovered by participants; "correctness" and "right and wrong" to be decided by participants.
Method of participation tends to be singular, disciplinary, reductionistic, positivistic, and narrowly ideological (cause and effect, prediction), often with a special interest focus; thought and actions "bounded."	Method of participation tends to be holistic, interdisciplinary, and broadly ideological, with a common interest focus (empirical, systematic); thought and actions "unrestricted."
Policy and information are extracted from situations that should be controlled; authority, control, and dominance are at issue.	Policy understanding and appropriate focus of attention emerge from interaction with context; authority and control are important issues, but focus is on solving common problems fairly.
Problem solving is blueprint-like; a "formula" is known and should be used to address problems.	Problem solving is process-like; guidelines are known to address problems as well as general standards (e.g., reliability) to aid problem solving.

generally known to participants. "Correctness" is clear, and "right and wrong" actions known. (3) Method of participation tends to be singular, disciplinary, reductionistic, positivistic, and narrowly ideological (cause and effect, prediction), often with a special interest focus. Thought and actions are thus "bounded." (4) Policy and information are extracted from situations that should be controlled, consequently, authority, control, and dominance are at issue. (5) Problem solving is blueprint-like; a "formula" is known and should be used to address all problems.

In contrast, the policy-oriented, functional professional's work includes views like these: (1) Participants do not know where projects will lead, so work is an open co-learning process; people tend to be flexible. (2) Participants assume multiple realities. Reality is partially socially constructed and must be discovered by participants. "Correctness" and "right and wrong" are to be decided by participants. (3) Method of participation tends to be holistic, interdisciplinary, and broadly ideological, with a common interest focus (empirical, systematic)—integrative. Thought and actions are "unrestricted." (4) Policy understanding and an appropriate focus of attention emerged from interaction with context. Clearly, authority and control are important issues, but focus is on solving common problems fairly. (5) Problem solving is process-like; guidelines are known to address problems, as well as general standards known (e.g., reliability) to aid problem solving.

Active learning

Our systems of conservation need to be upgraded in some cases, but to do so, we must improve decision-making and management policy processes. The problem-oriented working strategy that I support can help us focus systematically and explicitly on our learning capabilities at all levels—individual, organizational, and policy.

Rapid learning

Proactive learning can improve management policy performance by critically and constructively evaluating past performance and learning lessons for application to current problems.[27] We can look for lessons by using a systematic analysis applied to existing situations. Rapid individual learning and co-learning at organizational and policy levels are promising strategies. If we can learn to apply lessons of hindsight and translate them into best practices of foresight, we can expect improved conservation.[28] At the heart of effective conservation problem solving is systematic thinking represented as the problem-oriented approach. Figure 10.1 gives a summary of the problem-oriented approach and its essential tasks focused simultaneously on biophysical, social, and decision-making challenges in any particular case.[29] Problem-solving participants can compare their answers to the questions in that table with one another as a basis for identifying common interests.

The rapid learning and co-learning strategies that I propose are not only about obtaining correct answers but also about understanding the mechanisms of learning itself and how to do better. Rapid learning occurs when we detect and correct errors in performance by simple course changes (i.e., this is called "single-loop" learning) or through assessing our initial underlying assumptions about the work to be done (i.e., this is called "double-loop" learning).[30] Often, the assumptions we make launch us down a path, so it is critical that we are aware of them and their impacts on our thinking and work.

Additionally, underattention to the variables and questions in Figure 10.1 is implicated often in weak, poorly performing programs. Clearly, learning to meet practical conservation goals successfully involves more than technocracy and scientific methods. It requires attending to all the problem-solving tasks and context.[31] It also requires that we attend to organizations and teams.

Organizations and teams

Too often, the structural and cultural design of organizations is overlooked in the learning process.[32] Managing organizations effectively for learning requires attention to the task environment, organizational designs, organizational cultures, and teams.[33] One design—the "congruence model"—details each of these elements and seeks to integrate them in a congruent response to challenges and the work to be done to learn. All conservation work takes place in the context of the organizational work environment. This environment must be understood, if an organizational response is to be tailored to the contextual challenges. We often take organizational structure and function for granted, especially in learning programs. Most task environments show the properties of uncertainty, complexity, diversity, and instability in conservation. These dimensions must be carefully managed and understood, if we hope to learn and be successful.

Organizational designers offer informational processing models about how to operate programs and how learning takes place when confronted with this kind of uncertainty—understanding the operating environment. Every conservation program and effort should have the capacity to process information matched to the demands of the task environment. Of course, the demands of information processing change over time, so organizations need to be flexible. Not all organizational designs or cultures are open and flexible. For example, bureaucracies tend to be inflexible due to their hierarchical structure and reliance on rules, roles, and regulations. But the conservation management policy process often necessitates an organization design and operation that is flexible, quick, and problem-oriented for working teams to be effective.

Predictably, the structure and function of an organization is related to its culture (its cognitive outlook and values). Organizational culture is a system of thought that is a central determinant of any organization's character. The culture is a set of values and cognitive perspectives that are largely shared by members. The culture affects the behavior of members, the organization's ability to effectively meet member needs, and the way the organization copes with the external task environment.

Not all organizations are task-oriented or successful learners, some are more power- or role-oriented, meaning that they are most interested in gaining power over others and becoming bureaucratic rather than solving the real problem at hand. Matching the organizational culture and structural design to the conservation work environment and learning task is key to success. If a weak combination of structural and cultural designs are used, the overall conservation program will likely fail.

Task forces or project teams are often the most effective learning designs for conservation. Small, flexible teams are useful because they can respond quickly to changes in the task environment. To be effective, individual team members must be perceptive, energetic, and willing to work without close supervision or extensive rules and regulations. They must be self-aware of their own values and problem-solving approaches, and able to learn quickly at all levels. Team members should be selected for their ability to contribute to problem solving. A good team can generate needed information rapidly and offer solutions to the conservation task. And, an effective leader is critical. Finally, teams must rely on an explicit approach for conservation problem solving and analyzing organizational problems.

Unfortunately, key organizational elements are often overlooked and too much emphasis is put exclusively on technical matters too often. The working strategy recommended here can make us aware of how we design and manage organizations, attend to decision-making processes, and build and use rapid learning systems as we do our work.

Prototyping

Prototyping is another practical response to the need for innovation, creativity, and flexibility in conservation.[34] Prototypes are small-scale, exploratory interventions for social and environmental management policy systems. They are structured as innovative and interactive processes specifically designed for learning. Prototype projects are the creative, corrigible initiatives that can provide a basis for structuring later pilot projects, although the two are distinct.[35]

Prototyping can be used to upgrade conservation management policy. Successful prototyping requires that participants voluntarily agree to participate, leadership is supportive and cooperative, and the parent organization's design and culture are open and creative. Successful prototyping also requires that participants' primary objective is improving performance, not advancing power dynamics or other personal goals.

Despite their record of positive benefits, prototypes have not been fully utilized in conservation management in the GYE. With an emphasis on learning and the establishment of a self-correcting decision-making process, prototypes have been used elsewhere. For example, it was used in endangered species conservation in Victoria, Australia, and other locations. They have potential in the Greater Yellowstone Ecosystem.

Grounding the Strategy

Grizzly bear #399 is just one case that I could use to demonstrate the advantages of a mixed strategy, in this case "adaptive governance." Adaptive governance is a functional approach to problem solving that can increase the effectiveness over and above conventional, science-based conservation alone. Grounding this working strategy in the field on actual cases might help us become more effective conservationists, both individually and collectively.[36]

Learning projects and networks

Adopting the adaptive governance strategy requires a paradigm shift, one that moves away from the conventional, technical approach dominating today and toward an explicit, problem-oriented, contextual, and pragmatic approach.[37] This working strategy can be used in many situations across the GYE system.[38]

Comparing approaches

One option forward is to shift explicitly and systematically toward adaptive governance.[39] Adaptive governance as used by problem-oriented professionals is outlined in Table 10.2 and contrasted with a traditional, conventional professionalism.[40] In practice, there are significant differences between the two modes of professionalism in practice. These have significant implications and consequences for conservation.[41]

Changing paradigms requires that we shift how we make decisions and, as important, that we change the role in how we understand science in the process.[42] Details and examples of each approach, as well as the rationale behind them, are explored in the books by Professor Ronald Brunner of the University of Colorado and his colleagues.[43] This source offers many examples and applications.

Networking is one way to organize learning projects that enhances using adaptive governance. A network is "an arrangement where two or more autonomous individuals and/or organizations come together to exchange ideas, build relationships, identify common interests, explore options on how to work together, share power, and solve problems of mutual interest."[44] In networks, participants maintain their individual autonomy while interacting with one another on issues of mutual interest (e.g., mammal migrations). Many networks currently exist in the GYE, but few have fully used the adaptive governance strategy. There is great potential for conservation gains awaiting successful learning projects and networks using the adaptive working strategy.

Organizational applications

Increasing pressure on natural areas and wildlife from roads, traffic, tourism, recreation, development, and disease requires us to implement effective approaches, if we want to help humans and wildlife thrive far into the future. One organization that is actively, explicitly, and systematically utilizing the adaptive governance strategy is the 33-year-old Northern Rockies Conservation Cooperative (NRCC), whose goal is conservation in the public interest. There may well be other examples in varying degrees. However, it seems the NRCC is unique in this regard at present.

The NRCC works at multiple scales, from local to regional and international, using the contextual, problem-oriented, adaptive governance strategy. The NRCC's operational paradigm provides a practical theory to model or put differently, a way to map any management policy process revealing how it might be directed to be more effective. Like any good cartographic tool, the framework NRCC uses helps map a route through a management "policy-scape" to a specific destination, highlighting key features, pitfalls, and opportunities (Table 10.2).[45] The chief benefit of the NRCC adaptive and prototypical approach (e.g., High Divide Collaborative, People and Carnivores skill workshops) is the promise of practical efficacy. NRCC employs this working strategy, all the while seeking to cooperate with other people and organizations for the common good. However, the NRCC's operating environment is full of obstacles, as discussed below.[46]

Learning networks

One way forward in the GYE perhaps is to link contextual, pragmatic organizations together into rapid learning networks.[47] There are hundreds of organizations active throughout the GYE, each operating with different beliefs, formulas, targets, and working styles. At present, most of these organizations buy into the scientific management (conventional) paradigm, often combined with environmental advocacy and "politics." Can these organizations be melded into a working, functioning learning network? This could happen, if they can share their experiences interactively in a problem-oriented way, harvesting key lessons through an interorganizational network and quickly turn lessons into more effective action on the ground and in public understanding and support.

However, while many organizations engage with specific targets (e.g., business development of Snow King Mountain in Jackson, Wyoming), few are fully problem-oriented as yet. For example, the YNP's Center for Resources (YCR) specializes in the intelligence (information) function, using biophysical and wildlife science perspectives and methods. In doing so, they do excellent work.

There are ways they could translate this into more efficacious management policy. Both the Greater Yellowstone Coalition (GYC) and Yellowstone to Yukon (Y2Y) specialize in the "debate," really the promotion function in decision-making. There are many other examples of organizations in the GYE beside these few that specialize to different decision process functions. An ideal network would combine organizations that specialize in the different functions so that they can support one another's contributions to the overall, full management policy process. Participating organizations could operate synergistically to complement one another to maximize effectiveness.

It appears that we have the opportunity to upgrade conservation in multiple ways in the GYE. If we understand that the challenges to the GYE's future are localized, ordinary problems (e.g., removing old fences that tangle wildlife, attending to road kills and car collision at specific sites, killing a patch of invasive weeds), then the current mix of organizations might be locally effective. If instead, we understand the challenges to the GYE's future as a mix of systemic and cultural problems, then the current mix of organizations may not be adequate to ensure the conservation future of the GYE. Some organizations, such as Earth Justice, are clearly focused on the legal dimensions of governance challenges. Other organizations vary in conventional approaches targeting various problems piecemeal. We should not abandon work at the ordinary level. However, it is imperative that we begin to incorporate systemic and constitutive work into more programs, GYE-wide. We could use prototyping, networks, and co-learning.

Grounded examples

There are several hundred public and nonprofit organizations working in the Greater Yellowstone Ecosystem currently. These organizations constitute an extraordinary assemblage of human and natural resources. In fact, there are so many organizations working in the region that no one is quite sure how many there are or what each one does. It is equally unclear how effective any one or the overall set of organizations actually is.

Consequently, in 2007, the NRCC initiated an inventory of organizations in the GYE. From this cooperative effort, the Greater Yellowstone Conservation Directory was created. It was widely distributed in print and online throughout the GYE and beyond. The directory provided information about every public and nonprofit organization working in the region, regardless of size or focus. Its intent was to be a complete, comprehensive, and a widely available source of information about conservation organizations working in the GYE. We need an update.

A second purpose of the directory was to encourage and facilitate effective partnerships and networks for rapidly learning. Dr. David Cherney of the University of Colorado estimated that about $150 million is funneled to the GYE nonprofit sector each year, and a like sum funds government agencies in the GYE.[48] Clearly, there is significant economic, intellectual, and functional investment in the region, which could be channeled toward developing a new adaptive governance paradigm, enhanced and rapid co-learning, and more effective conservation. The potential to do this is there!

Consider a few high-profile organizational cases in the GYE. Three of the largest and well-resourced organizations in the GYE are the Greater Yellowstone Coalition (GYC), the Yellowstone Center for Resources (YCR), and the Yellowstone to Yukon Initiative (Y2Y).[49] Each of these entities operates from different beliefs (doctrines) and formulas, but all three primarily rely on the dominant technical paradigm for conservation, but in very different ways.[50]

These organizations' reliance on the science-based paradigm is evident in the way they formulate their missions and in how they talk about the problems facing the GYE. The GYC claims that the single defining component of the Greater Yellowstone Ecosystem is the unique and iconic wildlife that roams throughout the region. To thrive, the wildlife of the GYE needs healthy, connected habitats and large expanses of wild country, so the GYC's wildlife conservation program aims to protect key habitats, maintain and restore migration routes, and advocate for sound policies that protect YNP's iconic animals. While sound, it is clear that these goals are focused on the technical wildlife aspects of conservation, such as population size and growth and corridors, rather than on addressing foundational, human social and decision processes involved.

Second, YCR was created to centralize the park's science and resource management functions. The goals of the YCR are to (1) gather, manage, and analyze data in order to better conserve the park's natural and cultural resources; (2) understand and mitigate the environmental and historic consequences of park management; (3) preserve and curate rare, sensitive, and valuable natural and cultural resources; (4) Work with park partners to meet resource management goals; and (5) promote transfer of knowledge to other park staff, partners, and the public. These goals are largely about what is happening on the ground in the park and how science and indices may be used to address technical problems. YCR is an excellent example of good science in this sense. These goals and work are appropriate for the park, but we need a similar effort ecosystem-wide program. None currently exists.

Finally, the Yellowstone to Yukon Conservation Initiative (Y2Y) is a joint Canada-US nonprofit that promotes connecting and protecting habitat from Yellowstone to the Yukon, so people and nature can thrive. It is the only organization dedicated to securing the long-term ecological health of this entire region. Y2Y takes a scientific and promotional approach to conservation, highlighting and focusing on local issues that affect the region (e.g., connectivity). It has worked with more than three hundred partners, including scientists, conservation groups, landowners, businesses, and government agencies, as well as First Nations (Canada)/Native American (USA) communities to unify this landscape. Y2Y argues that without a unified biophysical vision for this deeply interconnected landscape, local conservation efforts would be isolated and less effective. Therefore, Y2Y seeks to ensure that conservation efforts are aligned in support of large-scale biophysical objectives, thus becoming continentally significant. Today, Y2Y is recognized as one of the planet's leading mountain conservation initiatives. This approach is laudable because it recognizes the significant interconnectivity of biological and human systems.

Other efforts

There are many other noteworthy conservation efforts in the GYE. These include, but are not limited to, the Jackson Hole Conservation Alliance, Teton Conservation District, and Future West. The Charture Institute is another example, under Jonathan Schechter, and a progressive effort.[51] The challenge for all organizations in the GYE is to capitalize on their successes, engage in deep reflectivity, and rapidly learn to apply the contextual paradigm—adaptive governance—introduced in this book for future gains.

There are obstacles to adopting the adaptive governance paradigm, as further noted below. Dr. Cherney's study of environmental NGOs (ENGOs) in the GYE found that many key ENGOs are failing to meet their own formal conservation goals.[52] These organizations recognize their weak performance, he argues, but they see that their limited gains are due to lack of resources. In contrast, Cherney attributed the shortfalls of ENGOs to being trapped by conventional perspectives and methods. He suggested that if ENGOs want to improve their effectiveness, they need to fundamentally rethink themselves. They need to critically examine their deeply held assumptions about what conservation ought to look like and how to go about it. What are the obstacles and challenges to doing this?

Obstacles and challenges

As with all change and the adopting of a new paradigm, there are obstacles. How can problem-oriented participation in programs be improved? How can we be persuaded to make long-term commitments to important common interest concerns, over the short-term gains we too often support today? How can governments be induced to reform their policies and programs to encourage farsighted thinking? All change faces obstacles, especially adaptive governance.

Common elements

Among the many challenges that we face as we try to implement more effective programs is a resistance to adaptive governance.[53] This promising approach runs counter to the conventional, deeply institutionalized, traditional approach. First are basic questions that need attention. For example, if the essence of sustainable conservation is the capacity to take sound, farsighted actions, what strategies and institutional arrangements can be promoted to support those actions? How do we put those in place? And, what do citizens, governments, businesses, and NGOs need to know and be skilled in to successfully implement them? We must address these.

The second challenge is that we have to overcome the perception that long-term actions are considered to have greater political and social risks for decision makers than short-term decisions. Given this, how can initiators of farsighted pursuits using "working ahead" initiatives (adaptive governance) gain acceptance, despite the risk?

These two challenges, and others, are often too big or fuzzy for some individuals, NGOs, and governments to conceive or address realistically. The good news is that we have a sound paradigmatic strategy and problem-solving tools for farsighted thinking and action now. We just need to use them.

In order to increase our capacity for farsighted thinking, we must address our reluctance to do so, become more thoughtful and intentional.[54] Thinking about how trends and conditions may play out in the future is essential. Understanding this makes it more likely that our actions do consider long-term consequences. Of course, even if we think in farsighted ways, we are not guaranteed to avoid problems. Consequently, it is important to have a realistic time scale on which we hope to see changes, and a method for reviewing the outcomes and making course corrections for our actions.

Change targets

For those of us interested in promoting farsighted thinking and action, we need to identify appropriate targets for change. The first target should be the individual, as the individual is crucial for sustainability in all management policy spheres.[55] The second target should be collective action and small group efforts (e.g., teams and organizations). Collective action groups do not arise spontaneously from common objectives. Instead, they require continual effort to mobilize and retain the participation of members. The third target is governments that face both internal and external challenges in pursuing long-term goals. Government officials must resolve to pursue these goals, maintain actions under pressure and against promises of political expediency, and persuade other participants that short-term sacrifices are justified by long-term benefits.

Today, most governments and government agents are unlikely to make short-term sacrifices for long-term gains because they are likely to be scrutinized by their constituents when the benefits do not manifest as quickly as the sacrifices. At the same time, however, shortsighted policy frequently leads to distrust of government stemming from a sense that nothing ever gets done, which undermines those institutions. To get around this challenge, we need an innovative environmental policy that includes local communities and trust building in meeting the goals of long-term conservation targets.

Biggest obstacles

Most people are largely concerned with their own short-term interests. This is an evolutionary reality. People often try to frame their special interests in the language of the common interest as a way of justifying their personalized demands, behaviors, and lifestyles in the context of group interactions. This conventional view assigns ordinary meaning to concrete circumstances and is typically misleading.

The alternative perspective (the problem-oriented, contextual paradigm—adaptive governance) looks for special meaning depending on the situation. The functional perspective that I recommend here can be understood by examining how anthropology classifies cultural behavior, word symbols, rituals and ceremonies, and beliefs or myths. The methods for functional understanding are contained in the working strategy described above: problem orientation, social and decision processes, and standpoint awareness—adaptive governance.

The functional perspective uses a stable frame of reference, unlike the conventional perspective that uses a subjective, variable perspective, often a self-interested one. We need to change our understanding of ourselves and our subjectivity in the context of wildlife-rich places like the GYE. The consequences of our conventional worldview are proving destructive and the evidence of this fact is unfolding all around us daily. Addressing this fundamental problem invites us to a much deeper level of analysis, understanding, and response than convention currently allows.

To gain functional understanding well beyond convention, people must be able to examine and understand themselves and others in the context of any social or decision process. A person with a functional understanding has the ability and responsibility to clarify problematic situations for their more conventional friends, for ordinary participants, and especially for decision makers.

Conclusion

Grizzly bear #399 and her family illustrate the complex challenges we face as we struggle to learn to live with nature.[56] We have yet to adequately address issues of human–wildlife coexistence in the GYE through our organizations and management policy. In fact, conventional approaches all too often only make things worse.

The working paradigm and strategy that I recommend—adaptive governance—offers a pragmatic way to build effective organizations, leadership, and partnerships to address actual problems. It offers a means for more effective problem solving, creating successful networks, and rapid co-learning. In some ways, this paradigm amounts to a new promise and story for the GYE and its future. The task before us is to bring this "new" conservation paradigm and story, with its working strategy, to the forefront in the GYE. As such, if we are successful, we can use it systematically and practically to advance common interests, both for our communities and our relationship to wildlife and the environment that is all around us.

Chapter 11

CREATING A NEW STORY, THE LONG VIEW

The major problems in the world are the result of the differences between how nature works and the way people think.[1]

Society is partway to creating a new story of the Greater Yellowstone Ecosystem. Can we finish our journey to that story? The current record of our behavior in the GYE is decidedly mixed. The new story, our best hope for the future, embodies our growing understanding of what nature is and what an ethic for coexistence should be. The new knowledge and ethic together open a route toward sustainability, if we take that path and the long view. We now know that Yellowstone National Park is only part of a much larger regional open ecosystem connected through ecological processes to the rest of the world. This allows us to view the challenge of conserving the GYE's migrations and carnivores more realistically than a few short decades ago. It also allows us to better see our own actions today, their consequences, and also it details our hoped-for new ethic for the ecosystem and wildlife. We can now see what we need to do as we work ahead toward greater understanding and responsibility.

Briefly, to use the words offered by the physicist and philosopher David Bohm, "What is needed is thus a creative attitude to the whole, allowing for a constantly fresh perception of reality, which requires an unending creation of new meaning."[2] The new meaning and story are about integration, coexistence, and ethics. In closing this book, I offer a general assessment of the whole GYE story to date and provide specific recommendations to move us closer to the needed creative attitude, ethic, and actions toward wildlife and nature. To put it most simply, we need a new system of living with nature and wildlife and a new conservation story to guide us there. What we do in the GYE is of great interest to people worldwide.

The GYE conservation challenge is urgent because of our growing human population and intensifying harmful development, recreation, and other uses, especially on public lands. Data currently shows that humans are impacting the ecosystem in many unintended but detrimental ways. Recognition of these facts is growing in the public and professional sphere. This book is my view on these problems, their cause, and their solutions.

I seek to help bring a new awareness of our growing knowledge about ourselves in this ecosystem and the needed new ethic for human–wildlife–nature relations to full maturity and on-the-ground application. I build on all previous chapters, so what follows is in one sense a review, restatement, and an overview of the book. Here, I first look at the critical juncture we are at now and explore the present as a time of opportunity. Second, I offer an integrated, multifaceted strategy to address diverse challenges before us. If we are to benefit from the promise and reality of a new conservation story, we must overcome many embedded historical, political, social, and psychological challenges. And third, this new story embodies a long-term view, and addresses actual problems, including problems caused by the old story. We need to rethink what nature is and our relationship to wildlife. We need to rethink these in functional, pragmatic, and ethical terms. We need to be successful in doing so.

A New Conservation Story

This book celebrates the achievements made in the creation of Yellowstone National Park and all the work and foresight it has taken to sustain its wildness and wildlife since 1872. It lauds the establishment of Grand Teton National Park (1929) and all public lands in the region. These lands are an irreplaceable public good that we have protected through management and policy over many decades, world wars, depressions, and other crises. Across many human generations, we have continually reaffirmed the value of this large landscape to the people of America and to the world. This accomplishment speaks to our values, commitments, and obligations to the future. It speaks to the kind of people we are and hope to be. However, we are currently in a new time and context, with great change underway and more coming over the horizon for GYE and the world. This change threatens our commitments to this great public good. It threatens the integrity of the region, its lands and wildlife, and the opportunities for future humans to benefit from experiencing the GYE.

Taking stock and foundational matters

Any new story aimed at coexistence and sustainability cannot be based on an overly simple divide between humans and nature. That dualism is built into all that we presently think and do. Given our current and future situation, it's no longer tenable. This distinction no longer makes sense, as maybe it once did. The new conservation story is one that fully recognizes interconnections among human societies and the biophysical environment, whether it's the GYE or anywhere else.

Offering the perspective that I do challenges the status quo—conventional thinking and our overly materialized and technical management—that now dominates our thinking and informs what we, as a society, do to the GYE. The record shows that conventional approaches are not sufficient to address the present, much less the foreseeable challenges we face in the GYE. This book calls for a more thoughtful, grounded, and farsighted management policy and on-the-ground actions. It also calls for an increase in our problem-solving capacity, the building of strategic leadership, rapid co-learning networks, and adaptive governance. The key question is, can we move ourselves toward coexistence, a relatively harmonized relationship between people and nature in the GYE?

An accounting

This book is in part an accounting of human–nature relationships in the GYE. We know that *people* seek *meaning (values)* through our *society (institutions)* using and affecting cultural and natural *resources (environment)*. We also know a lot about the different stories people make up and use in their lives about the GYE today and our presence and role in the ecosystem. This book is about these foundational matters as well.

Let's take stock of the ecosystem first. We now know the GYE is a complex adaptive system that came about over evolutionary time. It is made up of thousands of species that evolved to the environment. The GYE is often simply equated in the public's mind with the visibility of its bears, elk, bison, and other large species. These species are connected through complex food webs, material cycles, and energy flows. This complex adaptive system is under growing threat of disruption. Threats stem from large numbers of people using the GYE in destructive ways and from broader destructive forces like climate change. More change and disruption are coming in the future.

Second, we now know these changes are caused by humans—and our mental level. As our populations and uses grow and become more disruptive, we can expect these harmful changes to accumulate in the GYE. This view is in sharp contrast to the one many people hold today. The conventional view is that GYE is largely pristine and healthy, an example of intact nature functioning smoothly the way that it should. If any problems should arise, it is assumed that we can engineer them away (e.g., with wildlife overpasses, underpasses, and fencing).

Third, this brings up foundational matters: how we think, make meaning, and go about behaving. Problems in the GYE are conventionally viewed as an outward expression onto things (e.g., objects like wildlife on roads). However, reality tells us something different. Problems ultimately are really a function of our personalities, values, beliefs, and culture—our individual and collective behavior. It is easy to ignore this latter inward view that focuses on us, people, which are the real cause of mushrooming threats. It seems it is always easy to look outward and see wildlife as the problem because animals just walk out onto highways into speeding cars. We tend to *externalize* problems.

Fourth, we now know that the external view of the problem could not be further from the truth. We humans build roads in wildlife migration corridors, drive too fast, don't pay attention, and use habitat that is essential for wildlife's survival. We are the ultimate problem. We need to look inward to ourselves as sentient, value-seeking primates and into our societies and cultures to understand how these variables shape our views, beliefs, and meaning-making behavior. To put it another way, our psychology, sociology, philosophy, and policy—our behavior and actions—are out of kilter with the reality of the GYE as a complex and vast living system.

Fifth, GYE is full of conflict. Conflict exists because people at different levels of cognitive and moral development differ on values and foundational matters. We all want different things from GYE. There is tremendous variation among individuals involved in terms of personality, intellect, and experience, as noted earlier. Generally speaking, some of us are contemplative and thoughtful, whereas others are more reactive and vocal.

Finally, what are the foundational matters that are at the base of our nonsustainability, our present inability to live in coexistence with wildlife and the ecosystem? Simply put, it is our beliefs, values, and behavior—the relationship between people and nature—our present story. This relationship between nature and people has always been controversial.[3]

As Wendell Berry, the famous populist agriculturist noted, "We have been for some time in a state of general cultural disorder, and that disorder has now become critical."[4] Since he wrote that statement over forty years ago, the disorder and complexity have exploded.[5] To understand and address the situation in the GYE, we need to go to the very foundation of our human system of thought and our culture.

To elaborate on foundational matters, our dominant cultural story now largely sees the GYE as a collection of material things (e.g., wildlife, open spaces, geysers) that are a spectacle, a backdrop to our entertainment and behavior. For many people, it is a place for recreation, adventure, or business. Such stories allow us to stay nestled inside the status quo of convention, but they are not truth stories, as described by philosopher Richard Rorty earlier in the book. Truth stories are ones that comport with the facts of the matter at hand as broadly understood in society, and especially as presented by traditional science.

Although we do often forget, we are the authors of our stories of meaning (and our facts)—our notions of YNP, the GYE, and public lands. We seldom consider the utility of our own stories, individually or culturally, as we live them out daily. The new story I am calling for is one of holism, a new way to see humans in nature—a "holo-story," as some people call it. The holo-story and the associated movement, as conceived by Dr. David Bohm, encourages an active co-learning that requires we all reach for the creativity needed to address our growing problems.[6] As Professor Noel Castree, the Australian philosopher, pointed out, "When we alter or destroy those things we consider to be 'natural,' it is not the world's intrinsic naturalness that has diminished but rather our capacity to describe it thus."[7] What exists is not "nature" in the conventional everyday sense. Instead it is a complex mix of things that we conventionally describe with labels like "Yellowstone," "wildness," or "coexistence."

Finally, and most fundamentally, our representations of the world in images, words, art, and science are human constructs. These representations are our way of engaging with the material forces we experience. It's clear from all of human experience that we can never step outside of or exist apart from those material forces. Neither can we ever escape their influence on how we live, think, and behave. The GYE is comprised of a set of material forces, and we are now confronted with changing how we represent those forces to ourselves (in stories), if we want a healthy, sustainable future. That is our challenge.

In over our heads

To put it bluntly, we are presently in over our heads in the GYE and throughout much of the world, given the approaching large-scale problems we have made for ourselves (e.g., climate change, extinctions, and land conversion). At its base, the problem is that we are relying on our outdated conventional views for meaning making, a grounding of ourselves and behavior. Put in philosophic terms, this problem is about our struggle— our epistemology (knowing), ontology (reality, truth), and axiology (rightness, ethics)— the underpinnings of our society and culture (see Chapter 1: Stories of People, Nature, Yellowstone, Figure 1.3, and Chapter 6: People and Stories of Meaning). Ignoring our millennia-long struggle to get to the position of meaning and standard of material living that we now hold is to our peril. We often overlook this history and evolutionary psychology for widespread ego-defensive and "self-preservation" reasons, as noted in Chapter 6 and other chapters. The tendency to overlook these fundamental evolutionary, psychological, and social matters is reinforced by our present culture and its institutions. In turn, this omission creates blind spots in our understanding, meaning making, and behavior that have harmful consequences.

Since we are both the agents behind our problems and the presumed solution to those problems, we need to revisit these fundamental matters and determine whether we are up to the task of transforming ourselves. Confronting ourselves requires that we dig down to the bedrock of who we are, what we are doing, and why we are doing it. We need to ask whether our organizational and institutional arrangements—our society and its culture—are capable of effectively dealing with our present problems. Put simply: Are our cognitive and cultural abilities adequate to understand and address the complexity of the problems we are creating for ourselves?

The GYE is a notable place to start addressing the basics and appreciating each other's different capacities and meanings. As I see it, we need to enter into relationships with one another and with nonhuman life as distinct *whole* beings. We need to create a new kind of wholeness, by building bridges to the people less able to deal with today's fragmentation, complexity, and mental demands. We must find ways to help everyone feel as though they are being adequately understood and respected. We need to help each other develop cognitively and morally as people and adults as we collectively struggle to address our real problems. We can do this, if we so choose.

The growing complexity of the world requires a new kind of adult consciousness, a more effective way to deal with complexity.[8] As a society, we presently organize reality too simply. Our presently accepted stories mislead us. Our lack of cognitive complexity and deep self-awareness is the real problem behind the many conventional problems we face. As Chapter 6 explains, we exist at different stages of cognitive and social development: socialized, self-authorizing, and self-transforming minds.[9] Different positions of adult development reflect our ability to comprehend and deal with complexity. We need to recognize that most of the conflict and controversy we are experiencing is because of this difference. If we are to attend to environmental and social problems, we need to be more open, cooperative, and helpful to one another.

Time of opportunity and a new story

Yellowstone, both as a physical, biological place and as a complex idea about coexistence, is a narrative about our ever-evolving relationship to nature and as well about our responsibilities to other life forms and future humans.[10,11] The GYE is a particular place, a particular idea, and a particular hope, to which many of us have a personal attachment. Against the original intent of Yellowstone National Park to protect rare geothermal phenomena and keep the ideas of the frontier alive, now we simply need a workable formula for sustainable coexistence, and in some instances to "leave some places alone as much as we can."[12] This sentiment is in keeping with the highest reading of our goals—past, present, and future—for the entire GYE. Will we be able to adapt management policy effectively?

Critical juncture

Society is at a critical juncture.[13] Presently, we are at a time wherein new understanding and treatment of nature and wildlife is needed. This time can best be understood as constituting a "reflexive moment," a time to reconsider ourselves and the GYE in all senses. This reflexive moment encourages all of us to rethink our historical behavior, present living, and future possibilities.

Our current story is limiting the prospect for a healthy future by perpetuating the overuse of resources that is causing the destruction of ecological processes on which we and all other life depend.[14] Our current system of thought and culture—our conventional story—blinds us to the problems we are creating. It is clear that our ways of meaning making have real, hard physical and social consequences, some clearly harmful, as the GYE case shows.

Our present story allowed us to subdivide the GYE into a matrix of many landscape types (parks, forests, and ranches, public and private), and we use those pieces of the matrix differently (consumption, recreation, conservation). We subdivided jurisdictional authority over the pieces into federal, state, and private lands, causing fragmentation, incoherence, and conflict about uses and management policy. As mushrooming harmful impacts become more visible, the GYE is inviting us to find a new system of thought, grounded in ecological reality, and an ethic of shared responsibility.

The new story ideally calls for integration or coherence in ethics and management policy. There are proven integrative methods that could guide our society into the future for more effective controlled problem solving and social co-learning.[15] This branch of integrative science encourages self-awareness, participatory thought, and actual problem solving. This approach can elevate thought to the level that we need to address the overall challenges in GYE and allow for us to consider time scales longer than at present. There has been some progress to organize the GYE arena socially and policy-wise toward truth stories, greater unity, and coherence to address growing problems. More effort and leadership are needed. Success requires our society to become much more conscious and reflective of what we are doing presently and why, if we are to be successful.[16]

Toward a new, meaningful story

Why is a new story needed? We can now see that the GYE is a special place and the opportunity for sustainable conservation and coexistence is here and now. We must come to understand the deeply rooted philosophic core behind the everyday conventional story that now dominates. We must come to address the realities that we now face, especially those inside ourselves—our current stories of meaning about our right to dominate and use nature as we please. We can no longer afford to recycle that old story and the growing conflict and controversy that comes with it. We can no longer go about dressing up that old story with the simple optimistic view that we humans are in charge and on top of problems. To achieve the promise of the new story, we need to significantly increase our insight, capacity, and skills.

We humans cannot live without our stories of personal and cultural meaning. Meaning making inherently involves the distinction between self and other, or between subject (I, the person) and object (everything else, nature, wildlife). The GYE's story runs deep in our culture today, and in the emotions of individuals. We ceaselessly communicate stories with one another through the social process, both private and public. We do so to seek self-affirmation and meaning. We all want reassurance and security. This book offers strategies and practical applications to do what I see is needed.

Human development involves differentiating and integrating this subject and object relationship in more realistic ways.[17] As our society gains new knowledge and broader understanding about the natural world, we are able to redefine the nature–human relationship more adaptively. The new story should reconfigure our responsibility to all life in the GYE. It should change our individual self-awareness and understanding of our place in the world and in the GYE. The new story requires that each of us be a better companion to one another, to nature, and to all life. We need to build bridges, both socially and practically. And, we need to cross those bridges together, to a more humane and sustainable future. This is the path forward.

Upping our leadership capacity

Upping our capacity to make sense of the world and our place in it is a challenging undertaking. The new GYE story must connect all things—a "holo-connectedness." We must recognize that we are complicated people with different, yet potentially complementary relationships with one another and nature. The future will require much from us. This requires that we develop the capacity to meet this challenge.

Do we have the leadership capacity to transition to a new story of meaning? One of the biggest mistakes leaders make today is how they apply technical means to solve adaptive (systemic and constitutive) challenges, notes Ronald Heifet and Riley Sinder of the Harvard Kennedy School of Government.[18] This technical rendering process is widespread and is largely what officials and experts are doing throughout the GYE.[19] We need strategic leaders who will help us address problems practically, pragmatically, and contextually, especially foundational problems. We also need a system of governance or decision- and policymaking that allows us to integrate management across scales. We need adaptive governance.[20]

Effective leaders provide context, direction, and support for all interested people to come together, create community and meaning, and address real problems. In the GYE today, the established view of the effective leader falsely presumes the completeness of knowledge and skill.[21] Such a view is misleading. We need leaders to reconstruct how we understand problems, to help us work together more effectively, and to find practical ways forward in our shared interest over GYE.

We live in a world with a history of conflict and hostility derived from misunderstanding people's differences. Bridging is not just about telling others the way things ought to be or ought to be done. It is about genuine attempts at understanding and finding shared solutions. The conflict and controversy in GYE is our opportunity to resolve problems about our relationship to each other and to nature.

Forward to a New Story

We need to see the GYE situation as a whole, perhaps for the first time.[22] Any resolution of our problems brings with it a call for changes in our society, stories, and worldviews. Much of this book so far has laid the groundwork for the new story. We now need to look at a few very basic human traits that condition how we think and make meaning now, and what the options are for moving forward to a new story.

Seeing it whole for the first time

The new story is about the work of reconstructing our old story. The new story should give us a way to address the present fragmentation and incoherence and should allow for effective collective action. It should not be designed to coerce, exclude, or marginalize any people.[23] To find a new story, first we need to examine certain features of the human condition.

Addressing meaning

Many of the GYE's challenges exist because of complex social-environmental relations and how we make meaning of them and ourselves now.[24] To best understand this dynamic, let's go to the basics behind the obvious social and environmental problems. To do so, we need to look at the social mechanisms that we now use to create coherence and meaning. These mechanisms are basic to the human existential condition. Reviewing them here again is important.

Put simply, society uses two social mechanisms to seek wholeness, meaning: religion and science. The two mechanisms rely on different starting points about reality and truth, knowing, and ethics. They are used by different individuals to make sense of their lives. This is evident in the stories of Yellowstone's meaning in Chapter 1. Put philosophically, the two mechanisms for meaning making start with different ontologies (views of truth, reality), epistemologies (about knowing), and axiologies (ethics), or ways of understanding what is "real" and "true." The two stand in marked contrast with one another.[25] Notably, neither is inherently "better" or "worse" than the other.

This overview is helpful for understanding how we arrived at the cultural stories in our society. First, religion is a way of understanding the world through a fundamental truth that is prescribed by a god. The doctrine of most religions is to bring salvation, transcendence, and perfection into people's lives. Religion is vitally important to many people. It is easy to see how religion presets a fixed, eternal view of nature. Second, science, by comparison, was invented to find empirical (sensory, repeatable) answers and a wholeness or unification of natural phenomena in the world. Science is traditionally practiced by organizing knowledge in certain ways, testing it by experience and experiment, and providing logical criticism of its internal coherence. It is easy to see how science is an open-ended view of nature, as we learn about it. Is it possible to somehow unite the best of both into a new story?

The two views call for different stories of the GYE, and we all engage with or accept different parts of these two stories. For centuries, science has directly challenged religion, although this is yet another example of a false dichotomy in our conventional thought. Originally, science was the study of the work of god through nature, as Roger Bacon noted in the late Middle Ages.[26] About four hundred years ago, Francis Bacon concluded that science was a way to gain knowledge of autonomous matter (i.e., soil, water, air), and that this knowledge could be used to gain power over nature for the advantage of humanity, thus the old saying, "knowledge is power."[27] This led to the Age of Reason. This story mostly dominates our views of the GYE today. Bacon thought that the scientific method would lead to the perfection of humanity through secular actions as we learned more and more about nature. Today, that story is continued in the notion that science will lead to unending progress, materiality, wholeness, coherence, and, ultimately, meaningfulness.

However, the evidence all around us is contrary to this assumed progress. We are plagued by a litany of social and environmental problems. Are we inadvertently creating problems as a by-product of that old story? The harmful by-products of our technological world based on positivistic, a-contextual science, from chemical and biological weaponry to species extinctions and climate change, discourage us from blindly following a naïve view of science and the presumed "answers" it promises.[28] For some people, faith in science has evaporated. This itself is a harmful by-product. Somehow, we need to transform our old narrow view of science into a new broad sense, wherein the sciences are more problem-oriented, contextual, and take people more into account, including their religion.

Religion and science, in their many diverse and competing forms, have created much of the fragmentation and incoherence in the world today.[29] Each in its own way takes a selective view of humans and our situation, typically giving an incomplete worldview and view of nature.[30] The task of integration, contextuality, and co-learning our way out of our current social and environmental problems is made more difficult because of the antagonism between religion and science.[31] Some futurists and scientists see that this antagonism threatens our very existence as a species.[32]

We are still the human animal that we have always been.[33] Yet, what has changed is that we now have science-based modern technology that gives us great reach to alter the biosphere.[34] We now possess technology that can destroy us. Religion also fragments, divides, and brings incoherence. Yet at the same time, we have the same basic emotions and respond to each other as we did tens of thousands of years ago.[35] Reconciling these disconnects and learning to adapt our most basic human responses to the complexity we have created for ourselves through intellect and technological development is our urgent problem in the Anthropocene, especially for places like GYE.[36]

Addressing complexity

Diverse people have offered ways to address our fragmentation and incoherence problems. One useful set of suggestions to address these problems comes from Dr. David

Bohm.[37] Bohm, realizing that the Cartesian, or dualistic (humans vs. nature), model of reality is too limited, called for a very different kind of thinking. He identified two kinds of substance that exist—mental and physical—that somehow interact to give us our worldview. His main concern has been with understanding the nature of reality in general, and of consciousness in particular, as a way to find a coherent interconnected whole. These two, according to Bohm, are never static or complete but are part of an unending process of movement and enfoldment (to use his word).[38] His position is a complex one, yet it must be given serious consideration in places such as the GYE as we try to move to a new story of meaning.

Bohm draws from Einstein, and their joint work on physics (quantum mechanics and relativity) to come up with a notion about wholeness. He uses Einstein's theory of relativity that claims matter is not made up of fragments, but rather, the basic nature of the universe is unbroken wholeness and flowing movements.[39] Quantum theory (as opposed to everyday Newtonian physics) implies indivisible links of action between objects and their environment. In other words, the whole cannot be analyzed into separate parts with preassigned interactions. Using this, Bohm concluded that these physical theories imply a way to overcome reductionism and the separateness of matter. He says it is through integration. His conclusion has both philosophic and material implications. In the GYE, we tend to separate the humanities and social sciences from the biophysical sciences and technology. But should we see the GYE as a biophysical entity or as a social entity? Since we cannot answer this question without losing a great deal of richness in our understanding, how might we integrate the two worldviews?[40] Can we make our stories more coherent?

Bohm moved well beyond conventional, everyday thinking in a way that might address the problems of fragmentation and incoherence. He argued that an essential feature of the development of the modern, dominant worldview is to ignore or deny the basic philosophic questions from Chapter 1 (see Figures 1.2 and 1.3). Therefore, unfortunately, the integration, exploration of wholeness, and coherence in the GYE have little or no place currently in our personal or collective policy deliberations, so they are generally ignored. This omission drives a wedge between science and its philosophic foundations. Both the scientific and humanistic significance of wholeness is ignored in conventional thought, public dialogue, and everyday discussion, as we talk about GYE, nature, and wildlife, as well as in the technocratic management policy we currently employ.

Bohm argued that neither religion nor science is the main source of fragmentation and incoherence. Instead, he argues underlying both, and the main source of our problems, is ego. Ego, self-interest, and selfishness create a powerful and pervasive hold on us—the identification of the self as separate and distinct from other people and from nature is derived from this. Our egos drive us to place ourselves at the center of our individual and social stories of meaning. In turn, this means that the physical world and other people are defined only in reference to ourselves and our self-interests. This limits our ability to see and comprehend broader connections and systems processes—systems of which we are just a part.

This also holds true for our collective, social ego through political party, family, community, society, nation, and civilization. All human conflict arises in an attempt to protect ego interests and the individual and cultural stories that support them. Inflating our sense of self-worth and ego and interests in our society is generally regarded as supreme notions that should override everything else. The egotistical view offers a secure basis for knowledge for many people about self and everything else and it precludes the awareness that is necessary for constructive, paradigmatic change. Can we move beyond its traps?

Adapting old stories

Constructive change can happen in the GYE, but it requires that we change our thought, behavior, and stories of meaning. Changing our thought is the formula for moving forward. We are a new species in an old landscape in the GYE. Given our harmful impacts on the ecosystem, and the likelihood of more harm to come, it is time we undertake serious introspection and look at ourselves and our impacts realistically. We need to attain a semblance of coherence, harmony, and sustainability in the future. That means adapting our old story.

We are mired in ego-enhancing convention, which is a kind of self-imposed blindness. Psychologists have even identified ego defenses that counter greater awareness, reflection, and insight. Our most natural state is that of "I am," with ourselves at the center of our life story. This is how we identify ourselves and make meaning in the world (see Chapter 1: Stories of People, Nature, Yellowstone and Chapter 6: People and Stories of Meaning). In limiting ourselves this way, we leave out a lot of important knowledge and insights. This is the essence of egoism. Our immediate ego interests override everything, particularly shared and common interests, dignity for all, and a healthy environment. We behave as though our ego is beyond the limits of time, space, and condition. In doing so, we risk the future of the GYE, the health of the rest of the planet, and all its diverse life forms. It is not easy to change our brains that are deeply imprinted with eons of being the human animals that we are. We are still emotional, self-centered, and immediate animals, even as we are capable of much more. Somehow we must come to grips with this fact.

To be more than we currently are requires that we superimpose new ways of seeing ourselves on top of what is intrinsically within us, by gaining new understanding, self-awareness, and inventing viable possibilities. We need more inquiry into ourselves and the causes of ego fragmentation, as well as practical ways to end limited thought and harmful behavior.[41] Some people and programs are already doing so.[42]

Leadership in the GYE is struggling in the midst of this complex situation, as are private citizens, environmental NGOs, and all levels of government.[43] No individual or group is immune to this struggle.[44] A growing number of people are seeking unity and coherence in a wide variety of ways, including art, writing, and activism. They face enormous resistance from convention and the status quo. Some people see the GYE as a great cultural monument reflecting our potential to live in harmony with nature.

Fixing our present system of thought

The GYE is a place to bring family, share experiences, and go home changed, perhaps with stories of new meaning to share with friends back home. Why are some of us revitalized and transformed by experiences in GYE? Can these experiences help us fix our broken old system of thought and ethics?

Our thought and stories

It's first important to understand that much of our thought is a reflex conditioned by the culture and environment in which we live (i.e., the GIVEN of society and culture, see Chapter 6: People and Stories of Meaning). By observing the reflexive dimension of thought in action, we can understand how thought drives us. This is an unexplored avenue that has great potential to address personal and societal challenges in more constructive ways. As Bohm noted, "We become possessed by the 'truth' we think we possess."[45] The new story can change that truth to be more realistic.

Our meaningful stories are derived from our system of thought, which is a kind of cultural systems intelligence that influences all our behavior. People find a great deal of emotional security, identity, and even power from what they believe is true, rational, and right.[46] This plays out in the GYE when we put veterinarians in charge of bison management, assuming that it is a technical problem that can simply be fixed with a vaccine or a capture, test, and slaughter program. It likewise plays out when we put wildlife biologists in charge of grizzly bear management to measure population size, food habits, and movements, assuming this knowledge will result in sustainable, common interest conservation. The evidence suggests that while helpful and informative, these approaches are not sufficient to achieve our stated goals for wildlife conservation in the region.

In using experts as weapons in the contest over stories, we create the appearance that we understand the GYE as nature. We create the appearance that we grasp all the important substantive and contextual matters, including people of the GYE. Rendering bison and bears as technical issues and bringing in disciplinary specialists as cultural weapons against competing ways of seeing the conservation challenge blinds us from seeing our cultural system of thought at play. This book recognizes the strengths of our present approach and does not want to abandon it wholesale, but calls for a more integrated, comprehensive, and contextual approach to conservation and our relationship to nature.

Perhaps to better understand this point is to use a metaphor. Bohm said we are like fish in a tank, in which a glass barrier has been placed.[47] The fish become conditioned reflexively to keep away from the barrier. Even after it is removed, the fish do not cross the line into the once-blocked part of the tank. Our system of thought is that barrier—it is the major barrier to integration and coherence for a better world. We have created a worldview and cultural paradigm that has straightjacketed us in convention. It is now our job to identify the weaknesses in the old story and move beyond the bounds of conventional thinking to a new story and a long view. We can only recognize present limitations and move to more realistic views by paying close attention to what is happening now in our own system of thought. This requires attention, time, analysis, reflection, discussion, and action, interactively.

We can challenge ourselves to better match our mental models with current and projected realities in the GYE. Most people are naïve realists, believing that what they see is reality—the truth.[48] The tenets of our current system of thought tell us that humans are

(1) intelligent, (2) successfully participating in many systems simultaneously, (3) unaware of the complexity of those systems, and (4) mostly unaware of the existence of those systems at all. Consequently, we tend to act in the present, and in doing so, we make decisions that affect the short-term direction of our lives but which fail to live up to what is needed for long-term vision. That view leads people to conclude that some things are just plain true, such as common sense or folk knowledge.[49] Of course, if we take a larger contextual view, we know that is not the case.

A growing number of people see that our present worldview is at the very root of our ecological and social problems.[50] The current story about materialism and progress is deeply embedded in our assumptions and perceptions to such an extent that we cannot recognize the opportunities to escape from those that are before us.[51] Currently, the dominance of conventional story-making pushes holistic, integrated thinking out to the margins, at best. Thus, it is difficult to have an open dialogue that explores a better GYE from a new story and long view.

Changing our thought

Humans need the world to be classified and categorized in order for us to function properly as agents in our environment. As a consequence of this categorization, we see the artificial boundaries we create as *real*, although they are simply human constructs, *stories*. Those boundaries are created by our thought. In the GYE, it is time for us to look into our mental boundaries, our rigidity, and ego-dynamics, as the causes of the fragmentation, incoherence, and complexity that we face. These can be changed.

How we think about things matters greatly in terms of what we do. Thinking about our own reactions in everyday situations, and about key issues in the GYE, shows us that our thought is driving our actions.[52] Thought drives us in a mechanical way more than we would like to admit. We all assume that our thought merely shows us how things are. Hence, we miss seeing how thought participates in our perceptions in very basic ways— our stories. In other words, we see what we want to see and hear what we want to hear. If we can learn to understand the role of our thoughts, it could bring about a deeper and clearer appreciation of what is driving the fragmentation, incoherence, and complexity problems we face. Such a new understanding can open up many new, practical, and creative solutions to our problems, particularly those related to the future of the GYE.

So how do we constructively work across individual and group differences or conflicts to get to that needed new story? First, it would be helpful to realize that individuals are at different positions of development, seek different value demands, and hold different expectations.[53] Recall that adult development occurs along three positions. The first position, which comprises a majority of people, is described as the socialized mind. This position consists of team players and faithful followers of groups who align with their own self-interests. The second position, comprising about one-third of people, is described as the self-authorizing mind. These people drive new agendas, learn to lead, are involved in active problem solving, and show independence. Finally, the third position, comprising only 1 percent of people, is described as the self-transforming mind. These people are meta-leaders. They lead by using multiple frames of reference and holding contradictions in their mind. They are problem finding and problem solving all the time. They are highly independent. Put simply, we need more people in the latter two positions of development and we can work to get ourselves into one of them. We can also aid others in their development. However, we must also accept that some people are unwilling or unable to change. Either way, we should appreciate our differences and work to bridge them by meeting people where they are in the present moment.

Second, we need to employ the very best problem-solving approaches and build the capacity to use them. This requires a self-authorizing mind, at the least. This kind of mind uses a qualitatively different way of constructing experience, conflict, and difference than people with a socialized mind. Self-authorizing people come to set aside the old sense of self, replacing it with a respectful, cooperative, and psychologically whole and distinct self. Getting along with others is not about sharing the same identity and outlook. It's about escaping our self-serving ways and being effective problem solvers and leaders.

Third, we can build bridges through dialogue to progress well beyond conventional thinking. We can benefit both from traveling to other people's cultures of mind as well as being visited by people with other cultures of mind. In all cases, it is crucial that the visitor to someone else's culture of mind comes in with a nonsuperiority stance, to see how reality is being constructed in the other person's culture or personal context. This is easier said than done. We must learn to ask questions and to consider the personal context for anyone's belief system, recognizing this is not easy for some people. We must learn to notice where and how an individual's desire for core values is expressed in the beliefs and goals of another person or group.

And, fourth, the way forward in GYE or elsewhere is to show conflicting parties how to engage in advanced, nonpersonalized problem solving. Advanced problem solving is the art of developing solutions that preserve the most basic elements of each party's position. Contending parties must consider solutions that recognize and reassure the other party that their most deeply felt interests, such as demand for respect, will be addressed. Legitimate conflict management replaces the old conventional dynamics of one winner and one loser. It promotes coherence and progress over stagnant, decisive conflict. We must dialogue constructively across differences to achieve this new, more effective dynamic.

Finding a new order of things

What will it take to find a new order of things—a new culture of mind, a new system of thought, a new story to best address our present and growing problems?[54] Generally speaking, we humans have adapted well on this planet. Still it seems highly likely that in our local, regional, national, and global communities, many aspects of our living are maladaptive.

Tuning our systems intelligence

We are now faced with such weighty problems that the prevalent story driving our thinking, dialogue, and action will not help us very much. As our environment changes, we hold on to old thoughts and feelings as reflexive responses. These are defensive reflexes that function to resist change. Examples include climate change deniers. People who hold rigidly to old notions and reactions may not have open personalities or intellects of the kind needed to bring about constructive change. As a result, they resist change and struggle to keep their thought system intact, while the world changes and evidence builds up against them. Their thought reflexes are not adaptive enough to save them from the confusion and problems caused by fragmentation and incoherent behavior. To the extent this is true, we need much more than past thought to help us address future problems in the GYE and well beyond.

Our problems in the GYE do not have to be fatal.[55] We need more conscious thinking and public dialogue to address the GYE's future. In fact, the GYE could become a global model for such efforts, if we can gain the needed self-awareness and organization to first change how we think and dialogue about the GYE. Such thinking and dialogue can lead to flashes of intuition and insight, thus changing our thought reflexes. It can promote co-learning and exposure to different ways of thinking. Our current social and environmental dilemmas are sinking into some peoples' consciousnesses, but this is not enough. Are we smart enough to adapt before the obvious need to do so becomes inescapable?

Creating a holo-story and movement

How can we build the needed holo-story and movement? Again, recall that the holo-movement (a large-scale holistic movement) is a movement about thought, systems, and social learning that takes us well beyond convention and our old dominant story. Presently, the holo-story and movement is small, underorganized, and little effective, given the magnitude of the problems we face. Fortunately, it is growing. To move to a full-fledged holo-story and movement, we must progress well beyond the present system of thought, personal experiences, and mechanistic assumptions in our old story. Conventional thought is a superficial kind of engagement with the world that we must transcend.

In addition to convention dominating in society, academia is a major part of our problem today.[56] Academics must liberate themselves from the habituated and limited cognitive and psychological bounds of academic disciplines.[57] As Bohm said, "If we don't do anything about thought, we won't get anywhere. We may momentarily relieve the population problem, the economic problem, and so on, but they will come back in another way."[58] We need a kind of social co-learning that is rapid and contextual, that creates new situations and climates for dialogue and problem solving, and that helps us find new practical ways of seeing, understanding, and behaving.[59] Colleges and universities can lead the way.

People across all human domains of expression offer the same sentiment about the needed new story and holo-movement. Creativity is needed. Take this example from an artist friend:

> The creative process for me is an act of contemplation inspired and guided by my love for Nature and my deep concern for the environment. Painting is one way that I honor and

give thanks to the earth. I may break up my compositions to give a greater emphasis to the harmonious rhythmic patterns connecting parts to the whole—diversity with unity. My aim is to feel and express the interconnectedness within Nature—how all the bits communicate with each other to create their own inner music.[60]

This sentiment speaks to the openness and groundedness needed in GYE at all levels. In short, the way toward a new holo-story and movement as called for in this book seems clear. However, it is harder to bring about a new holo-story and movement than to describe it.

Taking the Long View

To be practical in transitioning to a new story and living, we must turn to pragmatism, realistic hope, and collective action. The role of the practical, pragmatic genius or leadership is not to overly complicate things but to simplify the complicated so that we can get to the root of the complex. We could certainly use that kind of genius in the GYE but cannot count on it showing up. So, how can we accelerate the social co-learning and integrated problem solving that we need? How can we grow the holo-story and movement in practical ways in the GYE?

Pragmatic hope—the bedrock

It should be obvious by now that our human social systems—culture, institutions, and constitutive processes—all affect our understanding of ourselves and nature. The new story provides an opportunity to bring our culture, thoughts, and actions to fuller self-awareness and into better harmony with nature. A growing number of people, individually and collectively, are actively co-learning at higher levels to adjust the current story to one that is more fitting for present conditions and the oncoming, foreseeable challenges. I could list dozens presently.

Reclaiming pragmatic hope

Pragmatic hope should become the bedrock of our holo-movement transition. We live in a time of growing anxiety and declining morale.[61] Much of the news these days, combined with the unsuccessful struggles of too many people to cope using too few resources, leads to hopelessness, even nihilism, and a loss of faith in our institutions and ourselves. To turn this around constructively, education is the best tool for hope. Scholars, educators, community leaders, and thought leaders—all of us, for that matter—have a duty to cultivate pragmatic hope. Pragmatic hope is not just unfounded optimism. Rather it is meaningful hope that truly responds to actual problems. Pragmatic hope can go a long way in helping us overcome our present mess. Pragmatic hope is something we can create and foster within ourselves and our communities, not simply something that we feel individually.

Pragmatism has a long history. It is a philosophical tradition that began in the United States around 1870 from philosophers William James, John Dewey, and Charles Sanders

Peirce. Peirce described pragmatism in saying, "Consider what effects, that might conceivably have practical bearings, we conceive the object of our conception to have. Then, our conception of these effects is the whole of our conception of the object."[62] Pragmatism considers thought to be an instrument for prediction, problem solving, and action, rejecting the idea that the function of thought is to represent reality. Pragmatists contend that most philosophical topics, such as the nature of knowledge, language, meaning, belief, and science, are best viewed in terms of their practical uses.

Pragmatic hope is about having practical perspectives, knowledge, and skills. It is an optimistic state of mind based on an expectation of positive circumstances in one's life and the world at large. When using pragmatic hope in cooperative work, it is about an awareness of the powerful influence we have over the challenges at hand (in this case, our existing GYE story).

Realizing particular goals (e.g., saving grizzly bears or preserving migrations in GYE) requires that we take on the habits of pragmatic hope. It requires that we live in a place of hopefulness in our daily lives, work, and community. This is more easily done in a community that is inclusive and welcoming to open inquiry. It is easier in a community held together by a shared sense of unity and respect. It is certainly easier in a community focused on dealing realistically with the challenges of the present and future, while understanding the past. Pragmatic hope must be connected to all that we do. Such hope can sustain us in dealing with our challenges, overcoming low morale, and managing high anxiety. However, it is crucial to note that pragmatic hope must be realistic.

Building pragmatic hope

What are the actions needed to build and sustain pragmatic hope? Ecologist Andrew Knight said, "We need to reframe hope to a pragmatic practice in our conservation, education, and science."[63] Hope is a belief that our goals can be achieved by our own actions, or that luck or universal intelligence will intervene on our behalf. The latter is wishful thinking, a comforting fantasy. Knight believed pragmatically in the former, requiring us to build hope based on our own actions.

There are several things we must do to build pragmatic hope. First, we must dare to deal with our actual situation. This makes pragmatic hope both necessary and, at the same time, difficult. We need to gain a realistic picture of the challenges we face. Most importantly, we must distinguish the characteristics of pragmatic hope from the naïve notions common, including "fake" news, in society these days. To engage in pragmatic hope, we must recognize that the hope lies within ourselves and our ability to affect change where we wish to see it.

Second, to sustain ourselves in this time, we must dare to believe in the utility of what we are doing as citizens, parents, officials, or activists. It can be challenging to sustain that belief. Many of the challenges we now face in the GYE are large-scale, complex, and outside our sphere of influence. It may often feel as though we are too small or inconsequential to deal with problems of such magnitude, but raising awareness of our collective challenges is the first step. Simply calling for small acts of good will (e.g., recycling, not idling your car, carpooling, avoiding recreating in wildlife habitat, leashing your dogs) is worth doing but is not enough to address the actual challenges we face. These challenges are multifaceted and as such, they require more than technical and individual-level solutions. They require us to operate at all levels (ordinary, systemic, and constitutive problems).

Third, much more knowledge, skill, and organizing are needed. Daring to do more can be arduous. But if we are to affect change, we must upgrade ourselves as thoughtful and self-aware citizens. We must engage with other people in helpful ways through dialogue, cooperative work, and writing for open publications. This will increase public awareness and engagement and will help to bring these conversations to the forefront of thought in the GYE. There is significant change underway in the GYE, and if we are to stave off the irreversible changes on the horizon, we must dare to do more. We must dare to live and work with pragmatic hope.

In the end, all of our personal actions should be grounded in pragmatism. Actions should flow from a commitment to making things better.[64] Blind optimism does not allow us to realistically orient ourselves to the problem or to develop effective, informed solutions to the actual problems we face.[65] Pragmatic hope addresses the naïveté of blind optimism, while retaining the goal of mitigating despair.

Using pragmatic hope

Imagine that we want to build a metaphorical bridge from the past to a sustainable future, working across our differing personal and political views. If we want to help people take hopeful action toward sustainability, we must bridge at least some of our differences. There is little benefit in preaching to people about the need for self-direction or self-development. Telling other people that they are, in some way, wrong for being who they are is unproductive and polarizing. Rather, we should embrace differences, be respectful, and work to bridge different understandings practically. This requires empathy.

Empathy can be a new kind of consciousness. It can help us build bridges to the other side of the philosophical chasm among people through two steps. First, we must firmly anchor the bridge on our side by attending to basic and foundational matters through self-awareness (Figure 1.3). In other words, we must consider why we are on our side of the chasm. Second, we must understand that the bridge cannot be built without the consent and support of those on the other side. We can gain that support only through cooperative problem solving and by welcoming others to the adult development process of thought and action, if we can. This process cannot involve disdain for people who disagree with us. We are only able to work together effectively if we approach the conflict with mutual respect and contextual understanding of the basic values of each side.

In the GYE and beyond, many people assume the cooperative bridge already exists and blame those on the other side, those who disagree with them, for not walking over. In reality, however, the bridge may not exist at all, may be anchored to only one side, or may be too weak for some people to cross by themselves. It is also important to consider the possibility that the bridge does not carry either side to common interest outcomes. All these flaws in the bridge building system reduce the capacity for sustained cooperation. Instead, we should be open and inviting to everyone, while also remaining aware of when our core values or goals are being challenged. This will help us construct a secure bridge among individuals and participant groups. Only when others recognize the bridge's existence and function can they choose to cross it or meet us in the middle. There are many instances in the GYE where we can try out the bridge-building idea practically (e.g., grizzly bear conservation, large mammal migrations, wildness area conservation). But empathy is only part of the remedy. Realistic problem solving is another. As is contextuality and self-awareness of one's standpoint, or window on the world.

It is a mistake for us to assume that most members of society are equipped with the intellectual and emotional skills necessary to build a cooperative bridge. Instead, we must quickly learn together how to build bridges across divisive lines where we can. This requires that we move to a new order of consciousness and self-awareness directed at developing self-authorizing minds, thereby rejecting our socialized minds of convention holding us in place.

In the coming years, it is the role of our social and educational institutions, such as colleges and universities, to engage this issue and work to nurture self-authorizing minds. As an educator at a professional school for the past 35 years, it is clear to me that many students are hungry for a higher order of consciousness and the knowledge and skills that come with it. Hopefully, people in the GYE region will come to seek that deeper consciousness, knowledge, and skill, too.

Collective transformative impact

Large-scale, constructive change in the GYE can bring about the transformations needed to bring us to a new story, living, and ethic. This requires broad, cross-sectorial goal clarification and commitment, coordination, and cooperation—a collective action. It requires better conflict management than we are using now, more skilled leadership, and better knowledge creation founded in practice-based work. We need an overall co-learning system to best understand the challenges we face and to help bring about change in our thinking and behavior in the GYE for sustainable conservation. Storytelling as an interactive process can be used to bring about sound public dialogue, new openness to ideas and other people, and co-learning.

Anchoring new modes of cooperation

We need to anchor our dialogue in truthful, pragmatic, and hopeful ways. We need to ensure that our hope is founded on sound assumptions and experiences, not abstract, inadequate, or faulty notions. Failure to rigorously question what we know and how we know it invites disaster. Building bridges in our communities invites constructive dialogue and can help us avoid such disaster.

To build a bridge we need to do several things. First, bridging helps us individually develop as adults to levels of consciousness and capacity sufficient to meet actual challenges.[66] We need to move beyond identifying with externally generated values, referents of nature, political correctness formulas, or slogan thinking. We must move on from convention and construct ourselves as authentic, grounded adults who can position and propel ourselves to higher orders of mental complexity. This is perhaps the most important of all bridging requirements.

Second, bridging can help us understand that we are not complete, independent individuals, despite what we might think. We are, in fact, flawed organisms functioning within a complex fabric of social interactions. Conflict among us is a by-product of our ego-centric ways of understanding and our pretension of completeness. Yet, few people are at a high enough order of development to see conflict as a signal of our overidentification with conventional systems of thought. We long for a higher order of consciousness, more effective problem-solving skills, and a new sense of our relationship and connection to each other and nature.

Third, we all strive to be successful and respected in our private and public lives. It is easy to fall into the many conventional traps that seem to confer success and respect, at least superficially. To address the GYE's actual problems requires much more of us than only using what conventional assumptions and understanding now permit. We need to be alert to conventional traps and recognize how to escape them.[67]

Creating new modes of cooperation

It might become easier to develop and cooperate as problems grow and become more visible as an immediate *real* crisis. In the GYE, we are currently struggling to adequately or realistically diagnose the challenges before us. Perhaps this is because many of us are operating at levels of consciousness below the levels that the challenges demand of us. Many people are working at cross-purposes with their own goals and those of other people. Fortunately, we are gradually evolving higher levels of consciousness as our problems become more fragmented and complex—visible. We are beginning to more fully understand the complexity we face.

The GYE is a crucible for our struggle for personal and social development in a rapidly changing social and environmental world of our own creation. To address the challenges we face, we need to greatly accelerate our movement toward effective collective action. Pragmatism, with its controlled problem solving, is by far the best, most practical understanding of hope that can guide individual and collective action. It can ground us constructively and integrate our actions for the common interest. How do we get there? Professor Keely Maxwell, an anthropologist at the US Environmental Protection Agency, called for better integration of the social sciences to change research and then to transform that research into action on the ground.[68] Integrating the biophysical sciences with the social sciences and humanities requires a high-order integrative framework.[69] Fortunately for us, such a practical meta-framework already exists and is being put to use, as introduced, described, and utilized in this book (the policy sciences).[70]

In the GYE, for instance, the Northern Rockies Conservation Cooperative (NRCC) has led work using this hopeful, collective, pragmatic approach for the past 30 years. The NRCC has sought to provide leadership and teach problem-solving skills across diverse cases. The NRCC has worked to find shared goals and to set in place cooperative actions that make a positive difference. In the coming years, we need to intensify contact with one another and accelerate movement toward effective problem solving. Doing this takes more than good intentions and good will to achieve effective conservation management policy.

Working forward institutionally

Many of the GYE's problems are embedded in the interplay of culture, especially the governmental and commercial sectors, with nature. This is an institutional problem. For many of the problems in the GYE today, including the persistent institutional policy problems, there are strategic responses worth trying. Some approaches have a proven track record of helping. What are they?

First, to make needed gains, key funders should support strategically pragmatic interventions. The growing interest in the philanthropic community toward venture and social entrepreneurship could be harnessed to aid the GYE. This support could help identify and grow the needed high-performing nonprofits in GYE, such as the NRCC and other worthwhile initiatives. In doing so, this kind of philanthropy could lend support to practice-based work, rapid co-learning, and transformation.

Second, bridging can help us learn about the true fragmentation, incoherence, and complexity that we now face. Building social bridges around areas of conflict can increase dialogue and help us identify realistic demands and expectations about what needs to be done and how to do it. Currently, we are trying to find expression for our dim understanding of the dilemma that we are creating for ourselves in the GYE, while using only convention and old paradigmatic formulas. That is like looking in the rear-view mirror as we accelerate forward.

At the ground level, my colleagues and I have used these recommendations for collective impact with positive effects, especially in education. As President Theodore Roosevelt said, it is "far better [...] to dare mighty things, to win glorious triumphs, even though checkered by failure [...] than to rank with those poor spirits who neither enjoy nor suffer much, because they live in a gray twilight that knows not victory nor defeat."[71] In the end, it takes all of us working collectively to make an impact.

Making the new Greater Yellowstone story

We now know what to do strategically and tactically about the problems we face in the GYE. We need to change our relationship to nature, recalibrate our society, and reframe its story of meaning, practically, philosophically, and psychologically. This is a tall order, and it will take time, but we can use the tools of pragmatic hope, self-authorizing cognition, and cooperative bridging to get there.

Adjusting our relationships

The world's scientists recently gave a second warning to humanity. A 2017 manifesto signed by 15,364 scientists from 184 countries proclaimed, "Fundamental changes [are] urgently needed to avoid the consequences our present course would bring."[72] The first notice came in 1993 from more than 1,700 independent scientists, including the majority of the then-living Nobel laureates in the sciences. This first manifesto called on humankind to stop environmental destruction and cautioned that "a great change in our stewardship of the Earth and life on it is required, if vast human misery is to be avoided."[73]

As stated by Nobel Peace Prize winner Vaclav Havel, former President of Czechoslovakia, in 1985,

> I think there are good reasons for suggesting that the modern age has ended. Today, many things indicate that we are going through a transitional period, when it seems that something is on its way out and something else is painfully being born. It is as if something were crumbling, decaying, and exhausting itself, while something else, still indistinct, were arising from the rubble.[74]

This speaks directly to the dynamics we are all experiencing in the GYE today and well beyond. We are leaving our old paradigms behind, without any idea of where we are going. Will we rise to the possibility of future change for the better?

There is no *out there* without an *in here* and without us, the subject of our own stories, paradigms, and self-constructed worldviews. The growing problems confronting us are internal, reflected in the stories we create about our relationships to nature, though they manifest in the external or natural world. These stories reside inside our own psychodynamics, philosophic assumptions, beliefs, and actions. We are discovering the hard way that some of those stories are destructive. Living out conventional beliefs, as we do today, has brought about unprecedented problems for life on our planet. These problems are only likely to grow in the foreseeable future, unless we do something foundationally different.

Confronting our future

Recall that to learn how to adjust our society and its culture requires looking at our beliefs and stories about ourselves and nature. The relationship can be understood using a simple story line: *People* seek *Meaning (Values)* through *Society (Institutions)* using *Resources (Environment)* (see Chapter 1 for explanation). These variables and their relationship can be changed to help us adapt.[75] We know quite a lot about these four variables, their interaction, and consequences in the GYE. We also know about our basic and foundational philosophic assumptions that undergird our present beliefs and actions. These assumptions need to be adjusted, as well as our daily practices (our seeking values through our institutions). All four elements—people, meaning, society, and resources—are open for change through our stories of meaning about the GYE and about our place in and responsibility for nature.[76] We can do it.

To make needed adjustments, we must look beyond our current capitalist, economics-based framework. As Professor Robert Schiller, the Nobel economist at Yale, said in his remarkably candid discussion, "Real progress will come from outside of economics."[77] He called for a broader and integrated framework to understand our cultural and environmental limitations. We can use the four variables of the heuristic and the integrative problem-solving framework at the center of this book for the social learning and transformation needed. We can use them to mobilize all the wisdom we can muster for the conservation of the GYE.[78] At the same time, we need a more encompassing dialectic and public dialogue that can bring us to a new truth for the GYE. Our present challenge is to find a new truthful story of meaning that is more realistic, and one that will create a sustainable relationship with humans and nature.

The future of the Greater Yellowstone Ecosystem, as both a large-scale landscape challenge and a complex social and policy challenge, requires our very best knowledge and skills. It requires a truthful public dialogue and, ultimately, a new paradigm and story. We have the resources, knowledge, and organization to do so. Now is time to go to work.

Our new Greater Yellowstone Ecosystem story

Our new story of meaning should be one of healthy communities in healthy environments, including all plants and animals. It should be one full of pragmatic hope and constructive action. We need to dedicate ourselves to our unfinished work to meet our promises—past, present, and future. In the end, the GYE story that we need is one that will help us individually and culturally recommit ourselves to the high ideals of our civilization. The time to write and live our new story is now.

I end with a few final recommendations. First, we need to build on all the good work of the people who went before us, to perpetuate the gift that we inherited in the GYE. This is about practice-based learning, harvesting lessons of hindsight, and translating them into lessons of foresight. We need to put those needed learning systems in place immediately. We need to pass that gift on to future generations.

Second, we need to see the GYE holistically as a large open system, as opposed to discrete objects in our constructed reality. This recommendation turns our dominant paradigm on its head. In the end, we urgently need thoughtful, practical, action grounded in attentive participation. We need environmental concern and responsible living to ensure a healthy future for the GYE.

Third, there is no simple prescription to address the GYE's problems. Yet there are simple things each of us can do to be part of the solution. The Greater Yellowstone Ecosystem's challenges are complex, and the sooner we move to a more adaptive mode of living, the better. Too often through our own indifference, apathy, and cynicism, we justify our lack of engagement as citizens concerned with the GYE. We need to stop making excuses. We must start showing up at meetings, making it clear to those in power that this place matters to us and that we are not going away.

Fourth, we need to help each other become realistically and pragmatically aware of what is at stake. Certainly, we need better thinking and understanding, if we are to secure a healthy future for the GYE. However, it will be what resides in our human hearts and our emotional attachments that will make all the difference. I encourage us all to contemplate how we should use the GYE today and in the future and who gets to decide that usage. We have the right as citizens to engage decision makers on this important matter. If we are not exercising our democratic rights of direct participation, then we are abdicating the power we hold in our hands.[79]

Fifth, I recommend we ponder the words of Dr. Seuss's *The Lorax*, a parable for our time, again. This wonderful children's tale concludes, "Unless someone like you cares a whole awful lot, nothing is going to get better. It's not."[80] We can project our future based on past experiences and current conditions. Or we can alter those trends and conditions to bring about a better outcome. The time to act is now with pragmatism, hope, and effective action for the future of the GYE.

Sixth, we cannot continue postponing a deep and wide-ranging public discussion about the GYE's future. Our discussion must be a full adult discussion. It must be honest, serious, nuanced, mature, and realistic. I call for education in the broadest sense and rapid co-learning that can be taught and used widely throughout the GYE.[81] We can educate ourselves about our precarious future and how to address it constructively.

Seventh, the needed discussions cannot be bounded in punditry or narrow advocacy, or in faith-based neoliberal capitalistic economics or overreliance on traditional science. We cannot assume that technology will save us. Citizens must forcefully demand that politicians, officials, journalists, and other change agents lead these discussions in an open commitment of genuine concern for the GYE and our common future. We can and must build on existing opportunities and open new ones.

Conclusion

The Greater Yellowstone Ecosystem story is really a story of our deep, unresolved personal and cultural issues, including our basic and fundamental concerns about our relationship with each other and the natural world. In our modern world, society is creating many more ways to tell stories—social media is one of many. If we use these platforms wisely, they can help us create the needed transformation toward genuine sustainability. We should employ all social vehicles to build a shared new conservation paradigm. The future that we create for ourselves, and all life on the planet, is ours to make. What kind of future will we have?

The time of pristine forests, abundant wildlife, and nature writ large will someday end—*if we do nothing*. I don't know when this will happen. Scientists suggest it may not be too far off. It will happen well before any of us want—*if we do nothing*. As the famous ecologist and nature writer Dr. Robert Pyle wrote in his 1993 *The Thunder Tree Lessons from an Urban Wildland*,

> I believe that one of the greatest causes of the ecological crises is the state of personal alienation from nature in which many people live. We lack a widespread sense of intimacy with the living world. Natural History has never been more popular in some ways, yet few people organize their lives around nature, or even allow it to affect them profoundly. Our depth of contact is too often wanting.[82]

We must take time to advocate for America's wild backyard, acknowledging that the GYE cannot advocate for itself. The Greater Yellowstone Ecosystem's future is in our hands.

I leave you, the reader, with a call for action—all hands on deck!

NOTES

Chapter 1 Stories of People, Nature, Yellowstone

1 Douglas W. Smith, personal communication, 2017.
2 Dan Wenk, "Wenk: Tell People What They Need to Hear, Not What They Want to Hear," *Mountain Journal*, December 18, 2017.
3 Dennis Glick in Todd Wilkinson, "Are We Loving Yellowstone to Death?" *National Geographic Magazine*, May 2016. Also see: David Quammen, "Yellowstone: The Battle for the American West," *National Geographic Magazine*, May 2016; Associated Press, "Misbehavior Abounds as Park Agency Turns 100," *Providence Journal*, August 29, 2016, 12.
4 Marguerite Holloway, "Your Children's Yellowstone Will Be Radically Different," *New York Times*, November 15, 2018.
5 Dave Hallac in Quammen, "Yellowstone," 131.
6 David Brooks, "The Problem with Meaning," *New York Times*, January 5, 2015; Steve Horan and Ruth W. Crocker, *People of Yellowstone* (Old Mystic, CT: Elm Groove Press, 2017).
7 Timothy Terway, "Sustained, in Significance with(out) Context and Ourselves: Expert Environmental Knowledge and 'Social-Ecological-Systems'" (PhD diss., Yale University, 2018).
8 This was the last diary entry of Helen Mettler (1910–1926), who died in a fall at Taggart Canyon, Wyoming, while on vacation with her parents Helen Fleischmann and John Wyckoff Mettler.
9 Horan and Crocker, *People of Yellowstone.*
10 Personal communication originally at Murie Center weekend retreat and reconfirmed, October 18, 2017.
11 Personal communication, October 18, 2017.
12 Susan Marsh, *A Hunger for High County: One Woman's Journey to the Wild in Yellowstone Country* (Corvallis: Oregon State University Press, 2014).
13 Mary Beth Baptiste, *Altitude Adjustment: A Quest for Love, Home, and Meaning in the Tetons* (Washington, DC: Rowman & Littlefield, 2014), 251.
14 Earle F. Layser, *Darkness Follows Light: A Memoir of Love, Place, and Bereavement* (Alta, WY: Dancing Pine, 2016).
15 Richard Slotkin, *Regeneration through Violence: The Mythology of the American Frontier: 1600–1860* (New York: University of Oklahoma Press, 1973); Richard Slotkin, *The Fatal Environment: The Myth of the Frontier in the Age of Industrialization* (New York: University of Oklahoma Press, 1985); Richard Slotkin, *Gunfighter Nation: The Myth of the Frontier in Twentieth-Century America* (New York: University of Oklahoma Press, 1992).
16 Personal communication, August 18, 2016.
17 Denise Casey, an artist based in Jackson, Wyoming, says, "What I love about painting is the total engagement in the process, the heightened awareness, complete focus, and the sense of purpose and fulfillment that it gives. It's a challenge to embrace the uncertainty of not knowing how a painting will develop or how I'll get to the end, but I'm always excited to see what reveals itself as I proceed. Many artists call it 'dancing' with the work—making a mark, being open to what it says or means in the overall work, and then responding to it with another mark." To learn more or view Casey's artwork, visit her website: http://www.denisecaseyart.com/.
18 R. J. Turner says, "Through my photography I hope to capture the beauty and raw power of the natural world that I have experienced and that is forever unfolding before us."

19 Rick Bass, *For a Little While: New and Selected, Stories* (New York: Little, Brown, 2016), 8–10.

20 Figure derived from: Noel Castree, *Making Sense of Nature* (New York: Routledge, 2014), 16–20.

21 For more on Jack Turner, see: Rachel Becker, "The World's Wilderness Is Disappearing," *The Verg*, September 8, 2016; Timothy Farnham, *Saving Nature's Legacy: Origins of the Idea of Biological Diversity* (New Haven, CT: Yale University Press, 2007); Gavin Van Horn and John Hausdoerffer, eds., *Wildness: Relations of People and Place* (Chicago: University of Chicago Press); Leath Tonion, "Not on Any Map: Jack Turner on Our Lost Intimacy with the Natural World," *Chicago Sun*, August 2014, 4–18; Jack Turner, *The Abstract Wild* (Tucson: University of Arizona Press, 1996), xiv.; Ed Zahniser, "The Muries, Saving Wilderness, and Kinship with Untamed Land" (Commentary, Wilderness Rendezvous, Moose, WY, October 12, 2014).

22 Turner, *The Abstract Wild*.

23 There are many historic and conflicting stories of meaning about Greater Yellowstone. Today, many stories are highly personal about adventure, finding one's self in the Greater Yellowstone experience. However, our personal stories often leave out the controversies that operate over large spatial or temporal scales such as wildlife conservation, recreation management, and collision between exploitative and conservative values and policy. For a more thorough exploration of the myriad narratives and voices coming out of the GYE today, from local recreationalists to Native American tribes to academic experts in ecology and policy management, explore the following sources. The best overview is perhaps Liza J. Nicholas, *Becoming Western: Stories of Culture and Identity in the Cowboy State* (Lawrence: University of Nebraska Press, 2006). Additional titles include, Mark Barringer, *Selling of Yellowstone: Capitalism and the Construction of Nature* (Lawrence: University of Kansas Press, 2002); Susan G. Clark and Marion Vernon, "Governance Challenges in Joint Inter-Jurisdictional Management: Grand Teton National Park, Wyoming, Elk Case," *Environmental Management*, no. 56 (2015): 286–99; Susan G. Clark and Marion Vernon, "Elk Management and Policy in Greater Yellowstone: Assessing the Constitutive Process," *Policy Sciences*, no. 49 (2016): 1–22; Christina Cromley, "Beyond Boundaries: Learning from Bison Management in Greater Yellowstone Greater Yellowstone," in *Finding Common Ground: Governance and Natural Resources in the American West*, ed. Ronald D. Brunner et al. (New Haven, CT: Yale University Press, 2002), 126–58; Justin Farrell, *The Battle for Yellowstone: Morality and the Sacred, Roots of Environmental Conflict* (Princeton, NJ: Princeton University Press, 2015); Justin Farrell, *Billionaire Wilderness* (Princeton, NJ: Princeton University Press, In Press); Geoffrey O'Gara, *What You See in Clear Water: Indians, Whites, and a Battle over Water in the American West* (New York: Knopf, 2000); Paul Robbins, "The Politics of Barstool Biology: Environmental Knowledge and Power in Greater Northern Yellowstone," *Geoforum*, no. 37 (2006): 185–99; Hannah Jaicks, "The Conflicts of Coexistence: Rethinking Humans' Placements and Connections to Predators in the Greater Yellowstone Greater Yellowstone Ecosystem" (PhD diss., New York: New York University, 2015); Marian E. Vernon, Zachary Bischoff-Mattson, and Susan G. Clark, "Discourses of Elk Hunting and Grizzly Bear Incidents in Grand Teton National Park, Wyoming," *Human Dimensions of Wildlife*, no. 21 (2015): 65–85; Marian E. Vernon and Susan G. Clark, "Addressing a Persistent Policy Problem: The Elk Hunt in Grand Teton National Park, Wyoming," *Society and Natural Resources*, no. 29 (2016): 836–51; Michael Yochim, *Protecting Yellowstone: Science and Politics in National Park Management* (Albuquerque: University of New Mexico Press, 2013).

24 Baum Buscher and Robert Fletcher, "Accumulation by Conservation," *New Political Economy*, no. 20 (2015): 273–98; George Monbiot, *How Did We Get into This Mess?* (Brooklyn, NY: Verso Books, 2016).

25 Martin Murie, "What Would the Muries Say about Recreation, Conservation and Wildness," *Mountain Journal*, April 10, 2018; Emilene Ostlind, "The Muries: Wilderness leaders in Wyoming," *WyoFile*, March 24, 2014.

26 *Jackson Hole Daily*, March 31, 2016.

27 Kathleen Parker, "What to Do When Voting Isn't Enough?" *Herald Net*, March 29, 2016.

28 Scott Turner, "Cauldron of Democracy: American Pluralism and the Fight over Yellowstone Bison," National Center for Case Study Teaching in Science.

29 Cromley, "Beyond Boundaries," 126–58.

30 P. J. White, *Can't Chew the Leather Anymore: Musing on Wildlife Conservation in Yellowstone from a Broken-Down Biologist* (Mammoth, WY: Yellowstone Center for Resources, 2016).

31 Steven Sloman and Hilip Ferbach, *The Knowledge Illusion: Why We Never Think Alone* (New York: Riverhead Books, 2017); Gukkuab Tett, *The Silo Effect: The Peril of Expertise and the Promise of Breaking Down Barriers* (New York: Simon & Schuster, 2017). On this issue, Wilfrid Sellars says, "The aim of philosophy, abstractly formulated is to understand how things in the broadest possible sense of the term hang together in the broadest possible sense of the term." Thus, "philosophy is a reflectively conducted, higher-order inquiry that is continuous with but distinguishable from any of the special disciplines, and the understanding it aims at must have *practical force*, guiding our activities, both theoretical and practical." Sellars' works can be found in Kevin Scharp and Robert B. Brandom, eds., *In the Space of Reasons: Selected, Essays of Wilfrid Sellars* (Cambridge, MA: Harvard University Press, 2007); Wilfrid Sellars, *Science, Perception and Reality* (Atascadero, CA: Ridgeview, 1991); Wilfrid Sellars, *Empiricism and the Philosophy of Mind* (Cambridge, MA: Harvard University Press, 1997); Jeffery F. Sicha, ed., *Pure Pragmatics and Possible Worlds: The Early Essays of Wilfrid Sellars* (Atascadero, CA: Ridgeview, 1980).

32 Many authors, including Nobel Prize winners, have written about the problem of "bounded rationality." Bounded rationality is the notion that our capacity for thought is restricted and limited by the finite amount of information that we have available to us, the cognitive limitations of our minds (e.g., personal histories of decision-making, values and concepts that we accept as GIVEN, social context), and the conditions under which we are asked to make logical decisions.

33 Shawn Achor, "The Happiness Advantage: Linking Positive Brains to Performance," YouTube video, 12:29, posted by "TEDx Talks," June 30, 2011, https://www.youtube.com/watch?v=GXy__kBVq1M.

34 Mike Koshmri, "Park: 399's Cub Dead," *Jackson Hole Daily*, June 21, 2016, 1–2.

35 Glenn Albreht, as cited in Robert Mcfarlane, "Generation Anthropocene: How Humans Have Altered, the Planet Forever," *Guardian News Media*, April 1, 2016. Additional readings include Noel Castree, "Neoliberalism and the Biophysical Environment," *Environment and Society: Advances in Research*, no. 1 (2016): 5–45.

36 Glenn Albreht in Mcfarlane, "Generation Anthropocene."

37 Gabriela Lichtenstein, "Book Review," *Conservation Biology*, 30, no. 5 (2016): 1135–37.

38 Susan G. Clark et al., eds., *Large-Scale Conservation in the Common Interest* (New York: Springer, 2014).

39 Terway, "Sustained in Significance with(out) Context and Ourselves."

40 C. S. Ridenour, *Mixed Methods Research* (Carbondale: Southern Illinois University Press, 2008).

41 Noel Castree, "Contract Research, Universities and the 'Knowledge Society': Back to the Future," in *The Knowledge Business: The Commodification of Urban and Housing Research*, ed. C. Allen and R. Imrie (Farnham: Ashgate, 2010), 221–45.

42 James G. Speth, *Red Sky at Morning America and the Crises of the Global Environment* (New Haven, CT: Yale University Press, 2004); James G. Speth, *The Bridge at the Edge of the World: Capitalism, the Environment and Crossing from Crisis to Sustainability* (New Haven, CT: Yale University Press, 2008); James G. Speth, *Angels by the River: A Memoir* (White River Junction, VT: Chelsea Green, 2014).

43 The Pogo story and cartoon come from Walter C. Kelly Jr. (1913–1973), a cartoonist, animator, and puppeteer. He started the syndicated series Pogo in 1949. He used satire and politics to criticize and offer perspective on current events.

44 Paul Schullery, *Searching for Yellowstone: Ecology and Wonder in the Last Wilderness* (Boston, MA: Houghton Mifflin, 1997), 248–65.

45 Rudolf Carnap, *Meaning and Necessity: A Study of Semantics and Modal Logic* (Chicago: Chicago University Press, 1956); C. I. Lewis, "Experience and Meaning," *Philosophic Review*, no. 43

(1934): 125–46; Hilary Putnam, "The Meaning of Meaning," in *Language, Mind and Knowledge*, ed. Keith Gundeson (Minneapolis: University of Minnesota Press), 131–93.

46 Richard Rorty, *Philosophy and the Mirror of Nature* (Princeton, NJ: Princeton University Press, 1979), 167.

47 Robert Righter, *Parks, Politics & Passion* (Moose, WY: Grand Teton National History Association, 2014), 273, 277.

48 To learn more, see, David Bohm, "Fragmentation and Wholeness in Religion and Science," *Zygon*, no. 20 (1985): 125–33; David Selby, "Reaching into the Holomovement: A Bohmian Perspective on Social Learning for Sustainability," in *Social Learning toward a Sustainable World*, ed. A. E. J. Wals (Netherlands: Wagnering, 2007), 165–80.

49 Wendell Bell, *Foundations of Future Studies* (New Brunswick, NJ: Transaction, 2003), xxiii, 20.

50 Bell, *Foundations of Future Studies*, xxiii, 20.

Chapter 2 Yellowstone and Significance

1 National Park Service Organic Act, 1916.

2 US Department of the Interior, "Greater Yellowstone Ecosystem," 2015, https://www.nps.gov/yell/planyourvisit/upload/RI_2016_FINAL_Ecosystem_web.pdf.

3 Timothy R. New, *An Introduction of Invertebrate Conservation Biology* (Oxford: Oxford University Press, 1995).

4 Steven Poole, *Butterflies of Grand Teton and Yellowstone National Parks* (Moose, WY: Grand Teton Natural History Association, 2009).

5 Kevin J. Gaston, *Rarity* (New York: Chapman & Hall, 1994).

6 John A. Shivik, *The Predator Paradox: Ending the War with Wolves, Bears, Cougars, and Coyotes* (Boston, MA: Beacon, 2014).

7 Tim W. Clark et al., eds., *Carnivores in Ecosystems: The Yellowstone Experience* (New Haven, CT: Yale University Press, 1999); Susan G. Clark and Murray B. Rutherford, eds., *Large Carnivore Conservation Integrating Science and Policy in the North American West* (New Haven, CT: Yale University Press, 2014).

8 Tim W. Clark, Paul C. Paquet, and Petyon Curlee, "Large Carnivore Conservation in the Rocky Mountains of the United States and Canada," *Conservation Biology*, 10, no. 4 (1996): 936–1058.

9 Rick Burroughs and Tim W. Clark, "Ecosystem Management: A Comparison of Greater Yellowstone and Georges Bank," *Environmental Management*, 19 (1995): 649–63; Steve Primm and Tim W. Clark, "The Greater Yellowstone Policy Debate: What Is the Policy Problem?" *Policy Sciences*, 29 (1996): 137–66.

10 Tim W. Clark, Murray B. Rutherford, and Denise Casey, eds., *Coexisting with Large Carnivores: Lessons from Wyoming* (Washington, DC: Island Press, 2005).

11 Jason Sunder, "Biologist Will Explore Living with Carnivores," *Jackson Hole News & Guide*, July 29, 2015.

12 To learn more, see: D. D. Bjornlie et al., "Methods to Estimate Distribution and Range Extent of Grizzly Bears in the Greater Yellowstone Ecosystem," *Wildlife Society Bulletin*, 38 (2014): 182–87; David J. Mattson, "Divvying Up the Dead: Grizzly Bears in a Post-ESA World," *Counterpunch*, February 12, 2016; David J. Mattson, "A Recipe for Killing: The 'Trust Us' Argument of State Grizzly Bear Managers," *Counterpunch*, October 25, 2016; Angus M. Theuermer Jr., "Yellowstone Superintendent Fears Fewer Visitors Will See Grizzlies," *WyoFile*, December 6, 2016; F. T. van Manen et al., "Re-evaluation of Yellowstone Grizzly Bear Population Dynamics Not Supported by Empirical Data: Response to Doak & Cutler," *Conservation Letters*, 7 (2014): 323–31; F. T. van Manen et al., "Density Dependence, Whitebark Pine, and Vital Rates of Grizzly Bears," *Journal of Wildlife Management*, 80 (2016): 300–13; P. J. White, Kerry A. Gunther, and F.T. van Manen, eds., *Yellowstone Grizzly Bears. Ecology and Conservation of an Icon of Wildness* (Yellowstone National Park, WY: Yellowstone Forever, 2017).

13 There is much reporting on grizzly bears. Examples include: S. Cherry et al., "Evaluating Estimators of the Numbers of Females with Cubs-of-the-Year in the Yellowstone Grizzly Bear Population," *Journal of Agricultural Biological and Environmental Statistics*, 12 (2007): 195–215; Mike Koshmrl, "Upper Green Stockmen Navigate Constant Griz Threat," *Jackson Hole News & Guide*, November 1, 2015; Mike Koshmrl, "2015 Grizzly Death Toll Tops All Previous Years," *Jackson Hole News & Guide*, December 23, 2015; Kent Nelson, "Voodoo Science: Yellowstone Grizzly Count Is a 'Flexible Fiction,'" *Jackson Hole News & Guide*, December 9, 2015; Louisa Willcox, "Cattle in Grizzly Country," *Counterpunch*, May 26, 2016; Yellowstone National Park, US Geological Survey, *Yellowstone Grizzly Bears: Ecology and Conservation of an Icon of Wildness* (Yellowstone National Park, WY, 2017).

14 Cherry et al., "Evaluating Estimators of the Numbers of Females," 195–215.

15 Here is a small sample of news articles on grizzly bears: Karin Bruillard, "The True Story of Two Fatal Grizzly Bear Attacks That Changed Our Relationship with Wildlife," *Washington Post*, August 3, 2017, www.washingtonpost.com/news/animalia/wp/2017/08/03; Franz Camenzind, "Why Grizzlies Should Stay Endangered," *Jackson Hole News & Guide*, April 8, 2015; Brian Kevin, "Everybody Hates Chuck Schwartz," *Sierra Club Magazine*, January/February 2011: 26–31, 102; Mike Koshmrl, "Mead Urges Action on Grizzly Bear Status," *Jackson Hole News & Guide*, May 28, 2014; Mike Koshmrl, "Bears Get into Trouble in Teton, Upper Green," *Jackson Hole News & Guide*, August 2–3, 2014; Mike Koshmrl, "Wolf Pack Is Biggest in West," *Jackson Hole News & Guide*, April 15, 2015; Mike Koshmrl, "People Killing Grizzly Bears at Alarming Rate," *Jackson Hole News & Guide*, October 21, 2015; Mike Koshmrl, "Teton Park's Wildlife Brigade Keeps People, Griz Safely Apart," *Jackson Hole News & Guide*, May 25, 2016; Renee Montagne, "Scientist Battle over Fate of Yellowstone Grizzlies," *NPR*, December 13, 2013; Louisa Willcox, "Yellowstone Grizzlies and the Betrayal of the Public Trust," *NRDC Switchboard*, December 31, 2010.

16 Mellissa Thomasa, "Open Season: An American Icon Is Caught in the Crosshairs of an Environmental, Economic, and Spiritual Battle after Its Removal from the Endangered Species List," *Planet Jackson Hole*, August 23, 2017: 12–15.

17 The Interagency Grizzly Bear Committee (IGBC) and Study Team webpage, www. gbconline.org, is tasked with enacting the species recovery plan for the grizzly bear that is mandated by the Endangered Species Act. After the grizzly was delisted in 2007, and relisted in 2009, the IGBC has recommended that the grizzly once again be delisted, which would strip the species of federal protection. The IGBC cites a near doubling in population and steady growth rate as evidence for recovery, although population numbers are not the only metric for recovery. Other factors, such as genetic diversity, population resilience, home range distribution, and rates of human-bear conflict are also necessary considerations. In contrast, the *Grizzly Times*, offers a different view of the human relationship with grizzlies. The *Grizzly Times* seeks to educate and inform citizens, journalists, and policy makers about grizzly bear ecology and to engage the public and advocate for the continued protection of the species.

18 Franz Camenzind, "Wolf Hunt Is Not Just about Numbers," *Jackson Hole News & Guide*, January 23, 2013; Mike Koshmrl, "Wolf-Driven Trophic Cascade Is Overblown," *Jackson Hole News & Guide*, May 14, 2014; Mike Koshmrl, "Fed Judge Kills Wolf Hunt," *Jackson Hole News & Guide*, September 24, 2014; Thomas D. Mangelsen, "Dead Wolf Display Was an Act of Hate," *Jackson Hole News & Guide*, October 23, 2013; Lynn Stutter, "The Truth about Wolves," *The All American*, 1, no. 6 (2010): 1–3, 16–20; Todd Wilkinson, "Wolf Trapping Exacts Unintended Causalities," *Jackson Hole News & Guide*, January 9, 2013; Todd Wilkinson, "Make Wolf Poachers Pay for Their Crimes," *Jackson Hole News & Guide*, January 16, 2013; Kelly Wood, "Gay Wolf/Timber Wolf Threatens Human Life," *The All American*, 1, no. 6 (2010): 1, 3–4, 12.

19 Christina M. Cromley, "The Killing of Grizzly Bear 209: Identifying Norms for Grizzly Bear Management," in *Foundations of Natural Resources Policy and Management*, ed. Susan G. Clark,

Andrew Willard, and Christina M. Cromley (New Haven, CT: Yale University Press, 2000), 173–220.

20 Lydia A. Lawhon, "Is There Harmony in the Howling? An Analysis of the Wolf Policy Subsystem in Wyoming" (PhD diss., University of Colorado, 2016); Doug W. Smith, Daniel R. Stahler, and Daniel R. MacNulty, *Yellowstone Wolves* (Chicago: University of Chicago Press), 2020.

21 Denise Casey and Tim W. Clark, *Tales of the Wolf: Fifty-One Stories of Wolf Encounters in the Wild* (Moose, WY: Homestead Press, 1992).

22 Douglas W. Smith, "Celebrating 20 Years of Wolves," *Yellowstone Science*, 24, no. 1 (2016): 1–81.

23 Nate Blankeslee, *American Wolf: A True Story of Survival and Obsession* (New York: Crown, 2017).

24 Casey and Clark, *Tales of the Wolf.*

25 D. J. Mattson and S. G. Clark, "People, Politics, and Cougar Management," in *Cougar: Ecology & Conservation*, ed. M. Hornocker and S. Neigri (Chicago: University of Chicago Press, 2010), 206–20.

26 See Penny Maldonado, "Cougar Could Cope with Nature, Not Man," *Jackson Hole News & Guide*, December 18, 2013; Thomas D. Mangelsen, "Stop Killing Mountain Lions for Sport," *Jackson Hole News & Guide*, May 15, 2013.

27 Mountain lions are the subject of much public attention, a selection: Rebecca Huntington, "Commission Locks in Higher Cougar Quotas," *Jackson Hole News & Guide*, August 6, 2003; Mike Koshmrl, "Lives of Mountain Lions Unfold in Nighttime Videos," *Jackson Hole News & Guide*, April 17, 2013; Mike Koshmrl, "Cougar Spooks Gregory Lane," *Jackson Hole Daily*, August 29, 2013: 1A, 24A; Mike Koshmrl, "Lions Still Lurk in Buffalo Valley," *Jackson Hole News & Guide*, December 16, 2015: 1A, 16 A; Timothy C. Mayo, "Cougar Call Arrogant," *Jackson Hole News & Guide*, August 6, 2003.

28 Take these sample headlines that capture the drama: "Wyoming's lions escape trapping plan," "Commission locks in higher cougar quotas," and "Stop killing mountain lions for sport." All are from the *Jackson Hole News & Guide* and *Jackson Hole Daily*. There are many more similar headlines, almost weekly.

29 Kerry M. Murphy, "The Ecology of the Cougar (*Puma concolor*) in Northern Yellowstone Ecosystem: Interactions with Prey, Bears, and Humans" (PhD diss., University of Idaho, 1998); Tony K. Ruth and Mark Elbroch, "The Carcass Chronicles: Carnivory, Nutrient Flow, and Biodiversity," *Wild Felid Monitor*, 7, no. 1 (2014): 13–17; Tony Ruth, P. C. Buotte, and Marurce G. Hornocker, *Yellowstone Cougars: Ecology before and during Wolf Reestablishment* (Boulder: University of Colorado Press, In Press); Dan Stahler and Colby Anton, *Yellowstone Cougar Project Annual Report* (Mammoth, WY: Yellowstone National Park, 2015).

30 Mark Elbroch, "Wyoming's Lions Escape Trapping Plan," *Jackson Hole News & Guide*, February 10, 2016: 5A.

31 To learn more about the effects of hunting and wolf competition on cougar survivorship, see L. M. Elbroch et al., "Rapid Human-Facilitated Shift in Top Predator Regime Following the Recovery of a Large Carnivore," *Ecology* (in review). To learn about prey selection and kill rates for cougars (e.g., cougar dietary ecology), see L. M. Elbroch et al., "Individual- and Population-Level Prey Selection by a Solitary Predator, as Determined with Two Estimates of Prey Availability," *Canadian Journal of Zoology*, 94 (2016): 275–82. To learn about behavior and social organization, see L. M. Elbroch, H. Quigley, and A. Caragiulo, "Spatial Associations in a Solitary Predator: Using Genetic Tools and GPS Technology to Assess Cougar Social Behaviors in the Southern Yellowstone Ecosystem," *Acta Ethologica*, 18 (2015): 127–36. L. M. Elbroch et al., "Spatial Overlap in a Solitary Carnivore: Support for the Land-Tenure, Kinship, or Resource Dispersion Hypotheses?" *Journal of Animal Ecology*, 85 (2016): 487–96. L. Mark Elbroch and Howard Quigley, "Social Interactions in a Solitary Carnivore," *Current Zoology*, 63, no. 4 (2017): 357–62. For information on rates of cougar competition with wolves, see L. M. Elbroch et al., "Recolonizing Wolves Impact the Realized Niche of Resident Cougars," *Zoological Studies*, no. 54 (2015).

32 To follow updates from Rebecca Waters, visit the Wolverine Foundation website: http://www.wolverinefoundation.org/. Waters updates a blog regularly with information about new research, policy, and management of wolverines.

33 Keith B. Aubry, Kevin S. McKelvey, and Jeffery P. Copeland, "Distribution and Broadscale Habitat Relations of the Wolverine in the Contiguous United States," *Journal of Wildlife Management*, 71, no. 7 (2006): 1–10; Jeff. P. Copeland et al., "The Bioclimatic Envelope of the Wolverine (*Gulo gulo*): Do Climate Constraints Limit Its Geographic Distribution?" *Canadian Journal of Zoology*, 88 (2009): 233–46.

34 Robert M. Inman et al., eds., *Greater Yellowstone Wolverine Program* (Bozeman, MT: Wildlife Conservation Society, 2008); Kevin S. Mckelvey et al., "Climate Change Predicted to Shift Wolverine Distributions, Connectivity, and Dispersal Corridors," *Ecological Applications*, 21, no. 8 (2011): 2882–97; Kevin S. Mckelvey et al., "Recovery of Wolverines in the Western United States: Recent Extirpation and Recolonization or Range Retraction and Expansion?" *Journal of Wildlife Management*, 78, no. 2 (2013): 325–34; Michael K. Schwartz et al., "Wolverine Gene Flow across a Narrow Climate Niche," *Ecology*, 90, no. 11 (2000): 3222–32.

35 Mckelvey et al., "Recovery of wolverines in the Western United States."

36 Inman, *Greater Yellowstone Wolverine Program*.

37 Craig Benjamin, "A Big Win for Wolverines," *Planet Jackson Hole*, April 27, 2016, 4; B. Jodi Petson, "Backstory: Wolverines," *High Country News*, May 2, 2016, 12.

38 Mike Koshmrl, "Teton Wolverine Study Key in Move to Protect," *Jackson Hole News & Guide*, February 6, 2013; Mike Koshmrl, "No More Wolverines, Gov. Mead Tells Feds," *Jackson Hole News & Guide*, June 11, 2014; Associated Press, "USFWS Biologists Flip-Flops on Animal," *Jackson Hole Daily*, July 8, 2014, 1–8.

39 S. M. Webb et al., "Distribution of Female Wolverines Relative to Snow Cover, Alberta, Canada," *Journal of Wildlife Management*, 80 (2016): 1461–70.

40 Jim A. Estes et al., "Trophic Downgrading of Planet Earth," *Science*, 333 (2011): 301–6; W. J. Ripple et al., "Status and Ecological Effects of the World's Largest Carnivores," *Science*, 343 (2014): 1241–48.

41 Matt J. Kaufman, J. F. Brodie, and E. S. Jules, "Are Wolves Saving Yellowstone's Aspen? A Landscape-Level Test of a Behaviorally Mediated Trophic Cascade," *Ecology*, 91 (2010): 2742–55; Arthur Middleton et al., "Grizzly Bear Predation Links the Loss of Native Trout to the Demography of Migratory Elk in Yellowstone," *Proceeding of the Royal Society* (2013): 280.

42 Joe Riis, *Yellowstone Migrations* (Seattle, WA: Braided River, 2017).

43 David Quammen, "Mysteries of Great Migrations: What Guides Them into the Unknown?" *National Geographic*, 218, no. 5 (2010): 28–89.

44 Matt J. Kauffman et al., *Wild Migrations: Atlas of Wyoming's Ungulates* (Corvallis: Oregon State University Press, 2018).

45 Regan Lyons, "Conservation Priorities for Maintaining Large Mammal Migrations in Greater Yellowstone" (master's thesis, Duke University, 2006).

46 Joel Berger, "The Last Mile: How to Sustain Long-Distance Migrations of Mammals," *Conservation Biology*, 18 (2004): 320–31.

47 There is significant conflict over elk management, which is reflected in news articles from towns surrounding YNP. For example, Mike Koshmrl, "Closing Feedgrounds for CWD a Long Shot," *Jackson Hole News & Guide*, December 9, 2015; Mike Koshmrl, "New Feedground Plan Bans Anti-Wolf Efforts," *Jackson Hole News & Guide*, December 9, 2015; Timothy May, "Park's 'Elk Reduction' a Travesty of Hunting," *Jackson Hole News & Guide*, 2016. Marion Vernon has also thoroughly investigated the clashing stories over elk management in Grand Teton National Park, on the adjacent National Elk Refuge, and throughout western Wyoming. Her papers include Susan G. Clark and Marion Vernon, "Governance Challenges in Joint Inter-Jurisdictional Management: Grand Teton National Park, Wyoming, Elk Case," *Environmental Management*, 56 (2015): 286–99; Susan G. Clark and Marion Vernon, "Elk Management and Policy in Greater Yellowstone: Assessing the Constitutive Process," *Policy Sciences*, 49 (2016): 1–22; Marian

E. Vernon, Zachary Bischoff-Mattson, and Susan G. Clark, "Discourses of Elk Hunting and Grizzly Bear Incidents in Grand Teton National Park, Wyoming," *Human Dimensions of Wildlife*, 21, no. 1 (2015): 65–85; Marian E. Vernon and Susan G. Clark, "Addressing a Persistent Policy Problem: The Elk Hunt in Grand Teton National Park, Wyoming," *Society and Natural Resources*, 29, no. 7 (2016): 836–51.

48 Joel Berger, "The Last Mile: How to Sustain Long-Distance Migrations of Mammals," *Conservation Biology*, 18 (2004): 320–31.

49 David N. Cherney and Susan G. Clark, "The American West's Longest Large Mammal Migration: Clarifying and Securing the Common Interest, *Policy Sciences*, 42 (2009): 95–111.

50 See Hall Sawyer, *Seasonal Distribution Patterns and Migration Routes of Mule Deer in the Red Desert and Jack Morrow Hills Planning Area* (Laramie, WY: Western Ecosystems Technology, 2014); Hall Sawyer et al., *The Red Desert to Hoback Mule Deer Migration Assessment* (Laramie, WY: Wyoming Migration Initiative, University of Wyoming, 2014); US Department of the Interior, *Yellowstone Resources and Issues Handbook 2016* (Mammoth, WY, 2016).

51 Northern Rockies Conservation Cooperative, "Integrating Science into Policy: A Case Study of Mule Deer in Jackson Hole," in *Joint Information Meeting Agenda Documentation—November 2, 2009* (Jackson Hole, Wyoming, 2009).

52 Tim W. Clark and Thomas M. Campbell, 1998. "Mule Deer Studies in Jackson Hole, Wyoming," unpublished as of 2021.

53 Sawyer et al., *Red Desert to Hoback Mule Deer Migration Assessment*; Angus M. Thuermer Jr., "Researcher Seeks Reasons for Demise of Mule Deer," *WyoFile*, January 20, 2015.

54 Robert A. Garrott, P. J. White, and Fred G. R. Watson, *The Ecology of Large Mammals in Central Yellowstone: Sixteen Years of Integrated Field Studies* (New York: Elsevier, 2009); P. J. White, Robert A. Garrott, and Glenn E. Plumb, *Yellowstone's Wildlife in Transition* (Cambridge, MA: Harvard University Press, 2013).

55 Rebecca Huntington, "You Will Be Missed, Boris the Bison," *Jackson Hole News & Guide*, February 22, 2006; Little Thunder (Pte Oyate), Rosalie Geist, and Darrell Geist, "Decades in, Bison Fate Still an American Shame," *Jackson Hole News & Guide*, May 7, 2014; Associated Press, "Y'stone Proposes Culling Roughly 900 Park Bison," *Jackson Hole Daily*, August 12, 2014, 7; "Yellowstone Chief: Bison Slaughter to Continue for Now," *Associated Press*, January 12, 2016.

56 Examples of headlines include "Decades in, bison fate still an American shame," "Yellowstone chief: Bison slaughter to continue for now," "You will be missed, Boris the Bison," "Montana defies logic in killing bull bison," and "Y'stone proposes culling roughly 900 park bison." All are from the all from the *Jackson Hole News & Guide* and *Jackson Hole Daily*. There are many more similar headlines, almost weekly.

57 See US Department of the Interior, *Yellowstone Resources and Issues Handbook 2015* (Mammoth, WY, 2015).

58 Aly B. Courtemanch, "Seasonal Habitat Selection and Impacts of Backcountry Recreation on a Formerly Migratory Bighorn Sheep Population in Northwest Wyoming, USA" (master's thesis, University of Wyoming, 2014); Aly B. Courtemanch et al., "Alternative Foraging Strategies Enable a Mountain Ungulate to Persist after Migration Loss," *Ecosphere*, 8 (2017).

59 Mike Koshmrl, "Skiers Back Sheep Closures," *Jackson Hole News & Guide*, July 9, 2014; Mike Koshmrl, "Bighorns Get Their Checkup," *Jackson Hole News & Guide*, July 9, 2015; Angus M. Thuermer Jr., "House Would Seed Clouds Move Sheep, Seize Federal Land," *WyoFile*, February 4, 2015; Todd Wilkinson, "Dead Bighorns Blamed on Livestock Disease," *Jackson Hole News & Guide*, March 25, 2015.

60 Francis Moody, "Tracking the Travels of Goats," *Jackson Hole News & Guide*, May 8, 2016.

61 Mark Huffman, "1 More Moose Killed, but Overall Toll Falls," *Jackson Hole News & Guide*, February 13, 2013; Mike Koshmrl, "Bilers Crossed Moose Habitat," *Jackson Hole News & Guide*, March 4, 2015; Michael Polhamus, "Planners OK New Park, Want Moose Protected," *Jackson Hole News & Guide*, June 25, 2014.

62 J. S. Johnson et al., "Migratory and Winter Activity of Bats in Yellowstone National Park," *Journal of Mammalogy*, 98, no. 1 (2017): 211–21; National Audubon Society, *Audubon's Birds and Climate Change Report* (New York: National Audubon Society, 2014).

63 Mike Koshmrl, "Tracking the Travels of Goats," *Jackson Hole News & Guide*, May 8, 2016.

64 A major threat is disease. To learn more, explore these new articles: Mike Koshmrl, "Supervisor Says CWD Worse in Feedgrounds," *Jackson Hole News & Guide*, January 28, 2014; Mike Koshmrl, "CWD Study Feeds Debate," *Jackson Hole News & Guide*, July 30, 2014; Mike Koshmrl, "CWD Study Predicts Hunting Will Suffer," *Jackson Hole News & Guide*, August 6, 2014; Mike Koshmrl, "Wasting Disease Nears Park," *Jackson Hole News & Guide*, November 18, 2014; Mike Koshmrl, "Wasting Disease Crept Further West This Year," *Jackson Hole News & Guide*, November 26, 2014; Angus M. Thuermer Jr., "19% of Deer Herd a Year Dies of Chronic Wasting Disease," *Jackson Hole Daily*, December 16, 2015.

65 The principle strategy to combat reduction of migrations is to protect or make corridors of connectivity as explained in Jodi A. Hilty et al., *Corridor Ecology: The Science and Practice of Linking Landscapes for Biodiversity Conservation* (Washington, DC: Island Press, 2006).

66 Robert L. Fischman and J. B. Hyman, "The Legal Challenge of Protecting Animal Migrations as Phenomena of Abundance," *Virginia Environmental Law Journal*, 28, no. 2 (2010): 173–238.

Chapter 3 Greater Yellowstone as a System

1 Wendall Berry, *Memories of the Future* (New York: Transaction, 2012).

2 Molly Loomis, "Public Lands in Public Hands." *Jackson Hole Magazine*, March, 2, 2018.

3 Mike Koshmrl, "Commissioners Tell Public Lands Group to Reconvene," *Jackson Hole News & Guide*, September 5, 2018, 1.

4 Mike Koshmrl, "Country Gives Up on Land Initiative: No Recommendation Coming from Commissioners on New Wilderness," *Jackson Hole News & Guide*, October 17, 2018.

  The WSA story was reported on heavily, including the following articles: Mike Koshmrl, "Fractured Lands Committee Walks Away," *Jackson Hole News & Guide*, August 9, 2018, https://www.jhnewsandguide.com/jackson_hole_daily/local/fractured-lands-committee-walks-away/article_5ec143fd-53e7-5243-850d-1d8b4a230770.html; Mike Koshmrl, "Split Decision Leaves Wilderness Areas in Limbo," *Jackson Hole News & Guide*, August 15, 2018, https://www.jhnewsandguide.com/news/environmental/split-decision-leaves-wilderness-study-areas-in-limbo/article_8493091d-4c21-58bb-8ed7-2c02d195d28c.html; Erik Molvar, "Wyoming Wilderness Initiative Goes Off the Rails," *WyoFile*, May 29, 2018, https://www.wyofile.com/wyoming-wilderness-initiative-goes-off-the-rails/; Joy Ufford, "Interests Collide over Shoal Creek Wilderness Study Area," Wyoming News Exchange, February 16, 2018, https://trib.com/news/state-and-regional/interests-collide-over-shoal-creek-wilderness-study-area/article_0d5dae61-bdee-539e-9bc9-798b33fc4977.html; George Wuerthnner, "Wyoming Public Lands Initiative Undemocratic," *The Wildlife News*, February 21, 2018, https://www.thewildlifenews.com/2018/02/21/wyoming-public-lands-initiative-undemocratic/.

5 Tim W. Clark and Dusty Zaunbrecher, "The Greater Yellowstone Ecosystem: The Ecosystem Concept in Natural Resources Policy and Management," *Renewable Resources Journal* (spring 1987): 8–11.

6 Frank Benjamin Golley, *A History of the Ecosystem Concept in Ecology* (New Haven, CT: Yale University Press, 1993).

7 Golley, *History of the Ecosystem Concept in Ecology*.

8 Andrew Abbott, "Things of Boundaries," *Social Research*, 62, no. 4 (1995).

9 To learn more about effective cross-boundary management, explore the following titles: James K. Agee and Darryl R. Johnson, *Ecosystem Management for Parks and Wilderness* (Seattle: University of Washington Press, 1988); F. Stuart Chapin, III, Gary P. Kofinas, and Carl Folke, eds., *Principles of Ecosystem Stewardship* (New York: Springer, 2009); Piermaria Corona and Boris Zeide, eds.,

Contested Issues of Ecosystem Management (New York: Food Products Press, 1999); John C. Freemuth, *Islands under Siege: National Parks and the Politics of External Threats* (Lawrence: University of Kansas Press, 1991); Herman A. Karl et al., eds., *Restoring Lands-Coordinating Science, Politics, and Action* (New York: Springer, 2012); Richard L. Knight and Peter B. Landres, eds., *Stewardship across Boundaries* (Washington, DC: Island Press, 1998); Andrew Hansen and Linda Phillips, "Trends in Vital Signs for Greater Yellowstone: Application of a Wildland Health Index," *Ecosphere*, 9, no. 8 (2018): 1–28; James M. Sweeney, ed., *Management of Dynamic Ecosystems* (Washington, DC: Island Press, 1990).

10 Theodosius Dobzhansky, "Nothing in Biology Makes Sense Except in the Light of Evolution," *American Biology Teacher*, no. 3 (1973): 125–29.

11 G. Evelyn Hutchinson, *Ecological Theater and Evolutionary Play* (New Haven, CT: Yale University Press, 1965); G. Evelyn Hutchinson, *The Kindly Frits of the Earth* (New Haven, CT: Yale University Press, 1979).

12 Lawrence S. Dillion, *Evolution: Concepts and Consequences* (Saint Louis, MO: C.V. Mosby, 1973), 9; I. Michael Lerner, *Heredity, Evolution, and Society* (San Francisco, CA: W.H. Freeman, 1968); L. Krauss, *A Universe from Nothing: Why There Is Something Rather Than Nothing* (New York: Free Press, 2012); M. Scheiner, "Toward a Conceptual Framework for Biology," *Quarterly Review of Biology*, no. 3 (2010): 293–318.

13 Aulay Machenzie, Andy S. Ball, and Sonia. R. Virdee, *Ecology: Instant Notes* (New York: Springer, 1998); Carl R. Pratt, *Ecology* (Springhouse, PA: Springhouse Corporation, 1995).

14 Hansen and Phillips, "Trends in Vital Signs for Greater Yellowstone."

15 James A. Pritchard, Preserving Yellowstone's Neural Conditions: Science and the Perception of Nature (Lincoln: University of Nebraska, 1999), 312–13.

16 Robert L. Flood, "The Relationship of 'Systems Thinking' to Action Research," *Systems Practice and Action Research*, no. 23 (2010): 269–84.

17 Donella H. Meadows, *Thinking in System: A Primer* (London: Earthscan, 2008).

18 As an example, see Pauline L. Kamath et al., "Genomics Reveals Historic and Contemporary Transmission Dynamics of a Bacterial Disease among Wildlife and Livestock," *Nature Communications*, no. 7 (2016): 1–10.

19 Kamath et al., "Genomics Reveals Historic and Contemporary Transmission Dynamics."

20 Donella Meadows, "Lines in the Mind, Not in the World," *Donella Meadows Archives*, December 24, 1987, http://donellameadows.org/archives/lines-in-the-mind-not-in-the-world/.

21 Douglas P. Wheeler, "Ecosystem Management: An Organizing Principle for Land Use," in *Land Use in America*, ed. Henry L. Diamond and Patrick Noonan (Washington, DC: Island Press, 1996), 155–72.

22 To learn more about the spatial, temporal, and complexity dimensions of the GYE in an easily readable format, with examples, see Tim W. Clark, *The Natural World of Jackson Hole: An Ecological Primer* (Moose, WY: Grand Teton Natural History Association, 1999).

23 David Schneider, "The Rise of the Concept of Scale in Ecology: The Concept of Scale Is Evolving from Verbal Expression to Quantitative Expression," *BioScience*, 51, no. 7 (2001): 545–53.

24 Charles C. Chester, Conservation across Borders: Biodiversity in an Independent World (Washington, DC: Island Press, 2006).

25 David Bohm, *Thought as a System* (London: Routledge, 1992).

26 For more reading, see Russell L. Ackoff and Sheldon Rovin, *Redesigning Society* (Redwood City, CA: Stanford Business Books, 2003); Meadows, "Lines in the Mind, Not in the World."

27 Flood, "Relationship of 'Systems Thinking' to Action Research," 269–84.

28 Susan G. Clark and Richard L. Wallace, "Integration and Interdisciplinarity: Concepts, Frameworks, and Education," *Policy Sciences*, no. 2 (2015): 233–55.

29 John H. Holland, "Complex Adaptive Systems," *Daedalus*, no. 1 (1992): 17–30.

30 Jianguo Liu et al., "Complexity of Coupled Human and Natural Systems," *Science*, no. 317 (2007): 1513–16.

31 National Science Foundation, *Complex Environmental Systems: Synthesis for Earth, Life, and Society in the 21st Century*, by Stephanie Pfirman and the Advisory Committee for Environmental Research and Education (2003).

32 Tory Taylor, *On the Trail of the Mountain Shoshone Sheep Eaters* (Dubois, WY: Wind River, 2017).

33 Charlie Craighead, *History of Grand Teton National Park* (Moose, WY: Grand Teton Association, 2006); Aubrey Haines, *The Yellowstone Story: A History of Our First National Park* (Boulder: University of Colorado Press, 1977); Joel C. Janetski, *Indians of Yellowstone Park* (Salt Lake City, UT: Bonneville Books, 1987); T. Perdue and M. D. Green, *North American Indians: A Very Short Introduction* (Oxford: Oxford University Press, 2010); Paul Schullery and Sarah Stevenson, eds., *People and Place: The Human Experience in Greater Yellowstone*, Proceedings of the Fourth Biennial Scientific Conference on the Greater Yellowstone Ecosystem (Yellowstone National Park, WY: Yellowstone Center for Resources, 2004); Mark David Spence, *Dispossessing the Wilderness: Indian Removal and the National Parks* (Oxford: Oxford University Press, 1999).

34 George Black, *The Epic Story of Yellowstone: Empire of Shadows* (New York: St. Martin's Press, 2012).

35 Wall text, Grand Teton National Park Visitor's Center, Moose, WY.

36 Wall text, Grand Teton National Park Visitor's Center, Moose, WY.

37 Hansen and Phillips, "Trends in Vital Signs for Greater Yellowstone," 1–28.

38 Wall text, Grand Teton National Park Visitor's Center, Moose, WY.

39 Wall text, Grand Teton National Park Visitor's Center, Moose, WY.

40 Wall text, Grand Teton National Park Visitor's Center, Moose, WY.

41 Wall text, Grand Teton National Park Visitor's Center, Moose, WY.

42 Wall text, Grand Teton National Park Visitor's Center, Moose, WY.

43 Wayne N. Johnson, Shine Not in Reflected Glory: The Untold Story of Grand Teton National Park (Morgan Hill, CA: Bookstand, 2017).

44 US Department of the Interior, National Parks Service, Quotes—National Park Service 50th Anniversary (1916–1966), by Elizabeth H. Coiner (1966), 24–25, http://www.nps.gov/history/history/online_books/npsg/quotes/index.htm.

45 Alan Taylor, American Colonies (New York: Viking, 2001); and Alan Taylor, American Revolutions (New York: W. W. Norton, 2001).

46 Wendell Bell, Foundations of Future Studies: Human Science for a New Era (New Brunswick, NJ: Transaction, 1997), 297–311.

47 Bell, Foundations of Future Studies, 297–311.

Chapter 4 Boundaries and Context

1 Donella Meadows, Global Citizens Columns, 1978. Between 1986 and her death in 2001, Donella Meadows wrote a weekly column, The Global Citizen. These columns—well over seven hundred of them!—were syndicated by 20 newspapers nationwide, published in other independent journals, and even eventually collected into a book.

2 Charles R. Preston, "The Greater Yellowstone Ecosystem: Where Do We Draw the Lines?" in *Invisible Boundaries: Exploring Yellowstone's Great Animal Migrations*, ed. H. Clifford (Cody, WY: Buffalo Bill Center of the West, 2016), 22–35.

3 Tim W. Clark and Dusty Zaunbrecher, "The Greater Yellowstone Ecosystem: The Ecosystem Concept in Natural Resources Policy and Management," *Renewable Resources Journal Summer* (Spring 1987): 8–15.

4 Josh Morse and Susan G. Clark, "Corridors of Conflict: Learning to Coexist with Long-Distance Mule Deer Migrations, Wyoming," in *Human-Wildlife Interactions: Turning Conflict in*

Coexistence, ed. Beatrice Frank, Jenny A. Gilkman, and Silvio Marchini (Cambridge: Cambridge University Press, 2019), 150–76.

5 Harold D. Lasswell, *A Pre-View of Policy Sciences* (New York: American Elsevier, 1971).

6 To explore this topic, see Richard N. L. Andrews, *Managing the Environment, Managing Ourselves: A History of American Environmental Policy* (New Haven, CT: Yale University Press, 1999); Susan G. Clark, A. R. Willard, and Christian M. Cromley, *Foundations of Natural Resources Policy and Management* (New Haven, CT: Yale University Press, 2000); Clarence J. Glacken, *Traces on the Rhodian Shore: Nature and Culture in Western Thought from Ancient Times to the End of the Eighteenth Century* (Berkeley: University of California Press, 1967).

7 George Perkins Marsh, *Man and Nature: Or, Physical Geography as Modified by Human Action* (New York: Scribner, 1964).

8 Reed Noss, "Context Matters: Considerations for Large-Scale Conservation," *Conservation Biology*, no. 3 (2002): 10–19.

9 Susan G. Clark, *The Policy Process: A Practical Guide for Natural Resource Professionals* (New Haven, CT: Yale University Press, 2002).

10 Bill Ascher and Robert Healey, *Natural Resource Policymaking in Developing Countries* (Durham, NC: Duke University Press, 1990).

11 For examples of ecological models, see Andrew J. Hansen et al., "Exposure of U.S. National Parks to Land Use and Climate Change 1900–2100," *Ecological Applications*, no. 3 (2014): 484–502; Nathan B. Piekielek and Andrew J. Hansen, "Extent of Fragmentation of Coarse-Scale Habitats in and Around US National Parks," *Biological Conservation*, no. 155 (2012): 13–22; Monica G. Turner, "Landscape Ecology: The Effect of Pattern on Process," *Annual Review of Ecology and Systematics*, no. 20 (1989): 171–97; Monica G. Turner, "Disturbance and Landscape Dynamics in a changing world," *Ecology*, no. 91 (2010): 2833–49.

12 Richard P. Reading, Susan G. Clark, and Stephen R. Kellert, "Attitudes and Knowledge of People Living in the Greater Yellowstone Ecosystem," *Society and Natural Resources*, no. 7 (1994): 349–65.

13 Wilfrid Sellars, *Empiricism and the Philosophy of Mind* (Cambridge, MA: Harvard University Press, 2000).

14 Todd Wilkinson, "Shift Asks Fun Hogs to Reflect on Impacts," *Jackson Hole News & Guide*, October 1, 2014.

15 See the column by Maury Jones in the *Jackson Hole News & Guide*—Cowboy Common Sense. Some examples of pieces she has written: Maury Jones, "Why Wyoming Should Manage Public Lands," *Jackson Hole News and Guide*, June 29, 2016; Maury Jones, "Feds' Policies Fueling Sagebrush Rebellion," *Jackson Hole News and Guide*, June 15, 2016.

16 Arnold Arluke and Clinton Sanders, *Regarding Animals* (Philadelphia, PA: Temple University Press, 1996), 6.

17 Willem Devries and Trimm Triplett, *Knowledge, Mind and the Given* (Cambridge, MA: Hackett, 2000); Ernst Hegel, *Phenomenology of Mind*, trans. A. V. Miller (Oxford: Clarendon Press, 1977); Richard Rorty, *Philosophy and the Mirror of Nature* (Princeton, NJ: Princeton University Press, 1979); Wilfrid Sellars, *Empiricism and the Philosophy of Mind* (Cambridge, MA: Harvard University Press, 2000).

18 Devries and Triplett, *Knowledge, Mind and the Given*; Rorty, *Philosophy and the Mirror of Nature*; Sellars, *Empiricism and the Philosophy of Mind*.

19 Martin Heidegger, *What Is a Thing?*, trans. W. B. Barton Jr. and Vera Deutsch (Washington, DC: Gateway Editions, 1970).

20 Sellars, *Empiricism and the Philosophy of Mind*.

21 Susan G. Clark et al., "College and University Environmental Programs as a Policy Problem (Part 1): Integrating Knowledge, Education, and Action for a Better World?" *Environmental Management*, no. 47 (2011): 701–15; Susan G. Clark et al., "College and University Environmental Programs as a Policy Problem (Part 2): Strategies for Improvement," *Environmental Management*, no. 47 (2011): 716–26.

22 Ralph Waldo Emerson, "Self-Reliance," in *Essays: First Series* (New Delhi: Astral International, 2019).

23 Upton Sinclair, *I, Candidate for Governor: And How I Got Licked* (Berkeley: University of California Press, 1935).

24 Steve Primm and Tim W. Clark, "The Greater Yellowstone Policy Debate: What Is the Policy Problem?" *Policy Sciences*, no. 29 (1996): 137–66.

25 Susan G. Clark and Murray B. Rutherford, "The Institutional System of Wildlife Management: Making It More Effective," in *Coexisting with Large Carnivores: Lessons from Greater Yellowstone*, ed. Susan G. Clark, Murray B. Rutherford, and Denise Casey (Washington, DC: Island Press, 2005), 211–25; Dylan Taylor and Susan. G. Clark, "Management Context: People, Animals, and Institutions," in *Coexisting with Large Carnivores: Lessons from Greater Yellowstone*, ed. Susan G. Clark, Murray B. Rutherford, and Denise Casey (Washington, DC: Island Press, 2005); Samuel Western, *Pushed Off the Mountain, Sold Down the River: Wyoming's Search for Its Soul* (Moose, WY: Homestead Press, 2002).

26 John R. Moses, "Feds Sued over Teton Park Grizzly Kill Rules," *Jackson Hole News and Guide*, April 7, 2015; Scott Streater, "Mead Pushing Jewell to Quickly Delist Yellowstone Grizzlies," *Planet Jackson Hole*, May 27, 2014.

27 Mickey Bellman, "Gray Wolf Population Management Seeks to Balance Nature," *Statesman Journal*, March 31, 2015; R. Pumphrey, "Public Federal Lands Make Montana Special" [letter to the editor], *Billings Gazette*, 2015; Todd Wilkinson, "GOP Faces Backlash in Selling Public Land," *Jackson Hole News and Guide*, January 7, 2015.

28 Phil Taylor, "Quietly Philanthropic Tycoon Makes His Mark in the West," *Environment and Energy News*, March 24, 2015; Margaret Webster, "Let's Stop Exploiting Public Lands" [letter to the editor], *Billings Gazette*, March 29, 2015.

29 To learn more about the Greater Yellowstone Coordinating Committee, visit their website at https://www.fedgycc.org/.

30 Bellman, "Gray Wolf Population Management Seeks to Balance Nature"; Warren Cornwall, "Have Returning Wolves Really Saved Yellowstone?" *High Country News*, December 8, 2014; Regan Lyons, "WCS Greater Yellowstone Ungulate Migration" (master's thesis, Duke University, 2006).

31 J. Gibson, "Big Money Trying to Buy Up Wildlife, Access" [letter to the editor], *Billings Gazette*, 2015; Ben Graham, "Forest Service Land Loses Public Zoning," *Jackson Hole New and Guide*, March 6, 2015.

32 Arluke and Sanders, *Regarding Animals*, 191.

33 Zachary A. Smith and John Freemuth, eds., *Environmental Politics and Policy in the West* (Boulder: University of Colorado Press, 2016).

34 Hannah Jaicks, *The Conflicts of Coexistence: Rethinking Humans' Placements and Connections to Predators in the Greater Yellowstone Ecosystem* (PhD diss., New York University, 2015).

35 Justin Farrell, *The Battle for Yellowstone: Morality and the Sacred Roots of Environmental Conflict* (Princeton, NJ: Princeton University Press, 2016).

36 John Gunther, *Inside U.S.A.* (New York: New Press, 1947); T. A. Larson, *History of Wyoming* (Lincoln: University of Nebraska Press, 1978).

37 Phil Roberts, David Roberts, and Steven L. Roberts, *Wyoming Almanac* (Cheyenne, WY: Skyline West Press, 2013)

38 Larson, *History of Wyoming*; Don Pitcher, *Wyoming Handbook: Including Yellowstone and Grand Teton National Parks* (Emeryville, CA: Avalon Travel, 2000).

39 Wyoming Humanities Council, *A Wyoming Civility Reader: Heal Up and Hair Over* (Cheyenne, WY: Wyoming Humanities Council, 2013), 1–2; Wyoming Humanities Council, *Welcome to Wyoming: A Guide to Newcomers* (Cheyenne, WY: Wyoming Humanities Council), 54–55.

40 To explore the changing context and culture in Wyoming, read: Western, *Pushed Off the Mountain*.

41 John Daugherty, *A Place Called Jackson Hole: The Historic Resource Study of Grand Teton National Park* (Moose, WY: Grand Teton National Park, 1999); Fern K. Nelson, *This Was Jackson's Hole* (Glendo, WY: High Plains Press, 1994).

42 *Jackson/Teton County Comprehensive Plan* (Jackson, WY, 2012), http://www.tetoncountywy.gov/DocumentCenter/View/1837/JacksonTeton-County-Comprehensive-Plan-April-6-2012-PDF.

43 Jonathan Schechter, "Jackson Hole Compass," *Jackson Hole News & Guide*, 2017, 6–56.

44 Ray Rasker and Andrew J. Hansen, "Natural Amenities and Population Growth in the Greater Yellowstone region," *Human Ecology Review*, no. 7 (2000): 30–40; Ray Rasker, "An Exploration into the Economic Impact of Industrial Development versus Conservation on Western Public Lands," *Society and Natural Resources*, no. 19 (2006): 191–207.

45 U.S. Census Bureau, *Population Trends and Projected Growth* (2010).

46 U.S. Department of Agriculture, *Understanding Wyoming's Land Resources: Land-Use Patterns and Development Trends* by Jeffrey D. Hamerlinck, Scott Lieske, and William L. Gribb (Laramie: University of Wyoming Press, 2013).

47 Jaicks, *Conflicts of Coexistence*; P. C. Jobes, "The Greater Yellowstone Social System," *Conservation Biology*, no. 3 (1991): 387–94.

48 Rasker and Hansen, "Natural Amenities and Population Growth in the Greater Yellowstone region," 30–40.

49 Jackson Hole Chamber of Commerce, *Jackson Hole Economic Indicators* (Jackson Hole, WY: Chamber of Commerce, 2015), http://files.ctctcdn.com/17daeeb6001/2f4ecc4d-2c29-4cd9-b230-0982e71c9169.pdf; *Jackson/Teton County Comprehensive Plan* (Jackson, WY, 2012), http://www.tetoncountywy.gov/DocumentCenter/View/1837/JacksonTeton-County-Comprehensive-Plan-April-6-2012-PDF.

50 Derek Thompson, "Where Did All the Workers Go? 60 Years of Economic Change in 1 Graph," *The Atlantic*, January 26, 2012; Reid Wilson, "Which of the 11 American Nations Do You Live In?" *Washington Post*, November 8, 2013.

51 William Asher and Barbara Hirschfeder-Ascher, *Revitalizing Political Psychology: The Legacy of Harold D. Lasswell* (Mahwah, NJ: Lawrence Erlbaum Associates, 2005), 116–38.

52 Ralf Seppelt et al., "Synchronized Peak-Rate Years of Global Resource Use," *Ecology and Society*, 19, no. 4 (2014): 50–58.

53 See National Intelligence Council and related organizations reports. For example, Wilhelm Agrell and Gregory F. Treverton, *National Intelligence and Science: Beyond the Great Divide in Analysis and Policy* (Oxford: Oxford University Press, 2015); Tatu Vanhanen, *The Limits of Democratization: Climate, Intelligence, and Resource Distribution* (Whitefish, MT: Washington Summit, 2009); National Intelligence Council, *Global Trends Paradox of Progress: 2017 Report of the National Intelligence Council, Promise or Peril of the Future, War, Population, Energy, Climate, Terrorism, Populist Anti-Establishment Politics* (Washington, DC: National Intelligence Council, 2017).

54 Parag Khanna, "A New Map for America," *New York Times*, April 15, 2016.

55 "President Map, Election 2012," *New York Times*, November 6, 2012, https://www.nytimes.com/elections/2012/results/president.html; "Live Coverage of the Midterm Election," *New York Times*, November 5, 2014, https://www.nytimes.com/elections/2014/liveblog?utm_source=top_nav&utm_medium=web&utm_campaign=election-2014.

56 Ronald D. Brunner, "Myth and American politics," *Policy Sciences*, no. 27 (1994): 1–18.

57 A case that illustrates these points is in Tim W. Clark, *Averting Extinction: Reconstructing Endangered Species Recovery* (New Haven, CT: Yale University Press, 1997).

58 Robert J. Bruelle, "Institutionalizing Delay: Foundation Funding and the Creation of U.S. Climate Change Counter-Movement Organizations," *Climate Change*, no. 122 (2014): 681–94; Riley E. Dunlap and Peter J. Jacques, "Climate Change Denial Books and Conservative Think Tanks: Exploring the Connection," *American Behavioral Scientist*, no. 6 (2013): 699–731; Aaron M. McCright and Riley E. Dunlap, "Anti-Reflexivity: The American Conservative Movement's Success in Undermining Climate Science and Policy," *Theory, Culture & Society*, no. 27 (2010): 100–33.

59 Schuyler Null, "Ten Billion: UN Updates Population Projections, Assumptions on Peak Growth Shattered," *New Security Beat*, May 12, 2011; United Nations, Department of Economic and

Social Affairs, *World Population Prospects: 2012 Revision* (New York, 2013), https://population. un.org/wpp/.

60 Richard G. Newell and Stuart Iler, "The Global Energy Outlook," in *Energy & Security: Towards a New Foreign Policy Strategy*, ed. J. Kalicky and D. Goldwyn (Washington, DC: Johns Hopkins University Press, 2013).

61 See figures 4.3 and 4.4 in U.S. Environmental Protection Agency, *Climate Change Indicators in the United States* (Washington, DC: U.S. Environmental Protection Agency, 2014).

62 Walter E. Westman, *Ecology, Impact Assessment, and Environmental Planning* (New York: John Wiley & Sons, 1985).

63 Timothy Terway, "Sustained in Significance with(out) Context and Ourselves: Expert Environmental Knowledge and 'Social-Ecological-Systems'" (PhD diss., Yale University, 2018).

64 Harold D. Lasswell and Abraham Kaplan, *Power and Society: A Framework for Political Inquiry* (New Haven, CT: Yale University Press, 1950).

Chapter 5 Controversy and Society

1 John A. Livingston, *The Fallacy of Wildlife Conservation* (Toronto: McClelland and Stewart, 1981).

2 Rob Chaney, "Review of Grizzly Attack Raises Warning for Mountain Bikers," *Missoulian*, March 6, 2017; Madison Park, "Montana Officials Search for Bear That Killed Cyclist," *CNN*, July 1, 2016; Wesley Yin, "Bear Kills Biker in Montana, in Seventh Fatal Grizzly Attack since 2010 in the Northern Rockies," *Washington Post*, June 30, 2016.

3 Steve Horan and Ruth W. Crocker, *People in Yellowstone* (Old Mystic, CT: Elm Grove Press, 2017).

4 Cody Cottier, "Future of King Packs Hearing," *Jackson Hole News & Guide*, November 14, 2018.

5 Harold D. Lasswell, "Future Systems of Identity in the World Community," in *The Future of the International Legal Order, Volume 4*, ed. Cyril E. Black and Richard A. Falk (Princeton, NJ: Princeton University Press, 1972), 20–48.

6 Susan G. Clark and Richard L. Wallace, "Integration and Interdisciplinarity: Concepts, Frameworks and Education," *Policy Sciences*, 48, no. 2 (2015): 233–55.

7 Examples include: Gloria Dickie, "Grizzly Face-Off," *High Country News*, May 16, 2016; Isa Jones, "The Life and Times of Grizzly 399," *Jackson Hole News & Guide*, July 6, 2016; Mike Koshmrl, "Fight Heats Up to Ban State-Run Griz Hunts," *Jackson Hole News & Guide*, May 18, 2016; Todd Wilkinson, "The New West: Democracy Falters in Delisting of Greater Yellowstone Bears," *Explore Big Sky*, April 15, 2016.

8 For a more complete description, see Peter Berger, *Invitation to Sociology: A Humanistic Perspective* (New York: Anchor Books, 1963).

9 A classic case of this social process is Jordan F. Smith, *Engineering Eden: The True Story of a Violent Death, A Trial, and the Fight over Controlling Nature* (New York: Crown, 2016). Another salient reading is Florence R. Shepard and Susan Marsh, *Saving Wyoming's Hoback: The Grassroots Movement That Stopped Natural Gas Development* (Salt Lake City: University of Utah Press, 2017).

10 Peter Berger, *Invitation to Sociology* (New York: Double Day, 1963); Peter Berger and Thomas Luckmann, *The Social Construction of Reality: A Treatise in the Sociology of Knowledge* (Harmondsworth: Penguin, 1971).

11 Emerson et al., unpublished. Data available from the National Elk Refuge archives, Jackson, Wyoming.

12 Christina M. Cromley, "The Killing of Grizzly Bear 209: Identifying Norms for Grizzly Bear Management," in *Foundations of Natural Resources Policy and Management*, ed. Susan G. Clark, Andrew Willard, and Christina M. Cromley (New Haven, CT: Yale University Press, 2000), 173–220.

13 Christina M. Cromley, "Bison Management in Greater Yellowstone," in *Finding Common Ground: Governance and Natural Resources in the American West*, ed. Ronald D. Brunner et al. (New Haven, CT: Yale University Press, 2002), 126–58.

14 Todd Wilkinson, "The Killing Fields Await Yellowstone Bison Once Again in Montana," *Mountain Journal*, December 15, 2017.

15 Allan Mazur, *The Dynamics of Technical Controversy* (Washington DC: Communication Press, 1981).

16 Susan G. Clark, "Political Myths and the Management of Large Carnivores," unpublished.

17 Alexander Wilson, *The Culture of Nature: North American Landscape from Disney to the Exxon Valdez* (Toronto: University of Toronto Press, 1991).

18 Harold D. Lasswell and Myres S. McDougal, *Jurisprudence for a Free Society: Studies in Law, Science, and Policy* (Leiden: Martinus Nijohoff, 1992).

19 Magali Larson, *The Rise of Professionalism: A Sociological Analysis* (Berkeley: University of California Press, 1977).

20 Alenjero Flores and Tim W. Clark, "Finding Common Ground in Biological Conservation: Beyond the Anthropocentric vs. Biocentric Controversy," *Yale School of Forestry and Environmental Studies, Bulletin Series*, 105 (2001): 241–52; Louisa Willcox, "Why Wyoming's Thugs Should Not Be Trusted with Our Grizzly Bears," *Counterpunch*, April 15, 2016.

21 Hannah Jaicks, "The Conflicts of Coexistence: Rethinking Humans' Placements and Connections with Predators in the Greater Yellowstone Ecosystem" (PhD diss., University of New York, 2016); Todd Wilkinson, "Long Journeys Home," *Jackson Hole Magazine*, December 23, 2016.

22 Harold D. Lasswell and Abraham Kaplan, *Power and Society: A Framework for Political Inquiry* (New Haven, CT: Yale University Press, 1950), 103–41; Jack Solomon, *The Symbols of Our Time: The Secret Meanings of Everyday Life* (New York: HarperCollins, 1990).

23 Justin Farrell, *The Battle for Yellowstone: Morality and the Sacred Roots of Environmental Conflict* (Princeton, NJ: Princeton University Press, 2015); Justin Farrell, *Billionaire Wilderness* (Princeton, NJ: Princeton University Press, 2019).

24 Elif Batuman, "Bison Bison Bison," *The New Yorker*, May 13, 2016; P. J. White, Rick L. Wallen, and David E. Hallac, *Yellowstone Bison: Conserving an American Icon in Modern Society* (Mammoth, WY: Yellowstone Association, 2015).

25 Dan Wenk, "Wenk: Tell People What They Need to Hear, Not What They Want to Hear," *Mountain Journal*, December 18, 2017.

26 "People and Carnivores" is a project of the Northern Rockies Conservation Cooperative, Jackson, Wyoming. The NRCC connects people, ideas, and resources to advance rangeland stewardship and carnivore conservation in the American West. NRCC works with ranchers, hunters and outfitters, rural residents, land managers, and scientists to keep grizzlies, wolves, and other carnivores in the wild and out of trouble. The People and Carnivores project is now carried out by Keystone Conservation, Bozeman, Montana. To learn more, visit www.peopleandcarnivores.org.

27 Mike Koshmrle, "Environmental Groups Walk Away from Upper Green Talks," *Jackson Hole News & Guide*, March 21, 2018.

28 Mike Koshmrle, "Hunters Kill 12 Wolves in First 40 Hours of Wyoming Hunting Season," *Jackson Hole News & Guide*, October 3, 2017.

29 Molly Absolon, "Fighting about Saving the Things We All Love," *Jackson Hole News & Guide*, July 22, 2015; Kelsey Dayton, "Storer to Talk about Conservation Challenges," *Jackson Hole News & Guide*, September 2, 2015; Denny Emory, "Have We Lost Our Moral Compass?" *Jackson Hole News & Guide*, October 3, 2012; Laura Linn Meadows, "Wilderness Helps Preserve Ways of Life," *Jackson Hole News & Guide*, October 12, 2011; Jack Nichols, "Think Small, Preserve What We Have," *Jackson Hole News & Guide*, September 14, 2016; Todd Wilkinson, "Handing Off the Griz," *Jackson Hole Magazine*, May 27, 2015; Todd Wilkinson, "Growth of Recreation Affects Bridger-Teton," *Jackson Hole News & Guide*, July 1, 2015.

30 Susan G. Clark, *Ensuring Greater Yellowstone's Future: Choices for Leaders and Citizens* (New Haven, CT: Yale University Press, 2008).

31 Tim W. Clark and Denise Casey, *Tales of the Grizzly: Thirty-Nine Stories of Grizzly Bear Encounters in the Wilderness* (Moose, WY: Homestead Press, 1992).

32 Lasswell and Kaplan, *Power and Society*.

33 Robert W. Righter, *Peaks, Politics & Passion: Grand Teton Comes of Age* (Moose, WY: Grand Teton Association, 2014).

34 Grand Teton National Park, *Grand Teton National Park: Moose-Wilson Corridor Road Safety Audit*, ed. Victoria Brinkly and Craig Allred (Moose, WY: Grand Teton National Park, 2014).

35 Grand Teton National Park, *Grand Teton National Park*.

36 Grand Teton National Park, *Grand Teton National Park*.

37 See Ian Morris, *Why the West Rules—for Now: The Patterns of History, and What They Reveal about the Future* (New York: Farrar, Straus, and Giroux, 2010); Richard Tarnas, *The Passion of the Western Mind: Understanding the Ideas That Have Shaped Our World View* (New York: Ballantine Books, 1991).

38 Michael E. Soule and John Terbourgh, *Continental Conservation: Scientific Foundations of Regional Reserve Networks* (Washington, DC: Island Press, 1999).

39 J. Weston Phippen, "Kill Every Buffalo You Can! Every Buffalo Dead Is an Indian Gone," *The Atlantic*, May 12, 2016, https://www.theatlantic.com/national/archive/2016/05/the-buffalo-killers/482349/.

40 Clyde V. Arnspiger, *Personality in Social Process: Values and Strategies of Individuals in a Free Society* (Chicago: Follett, 1959).

Chapter 6 People and Stories of Meaning

1 Wendell Berry, *Fidelity: Five Stories* (New York: Pantheon Books, 1992).

2 Oliver Sacks, *The River of Consciousness* (New York: Knopf, 2017). This collection of essays contains reflections on the evolution of life and the evolution of ideas, the workings of memory, the process of consciousness, and the nature of creativity. It also considers Sacks's own misrememberings and his experience of illness.

3 Nicole Krause, "A Last Glimpse into the Mind of Oliver Sacks," *New York Times*, December 4, 2017; Sacks, *River of Consciousness*.

4 Liz Kearney, "An Interview with Yellowstone Social Scientist Ryan Atwell," *Yellowstone Insider*, June 22, 2016, https://yellowstoneinsider.com/2016/06/22/an-interview-with-yellowstone-social-scientist-ryan-atwell/; US Department of the Interior, National Parks Service, *Yellowstone National Park Visitor Use Study* (Vermont, 2017), https://www.nps.gov/yell/getinvolved/upload/R-YELL_VUS_FINAL-Report.pdf; US Department of the Interior, National Parks Service, *Transportation and Vehicle Mobility Study* (2017), https://www.nps.gov/yell/getinvolved/upload/Yellowstone-Transportation-Mobility-Study_lo-res.pdf.

5 US Department of the Interior, *Yellowstone National Park Visitor Use Study*; US Department of the Interior, *Transportation and Vehicle Mobility Study*.

6 US Department of the Interior, *Yellowstone National Park Visitor Use Study*.

7 Edward O. Wilson, *Biophilia* (Cambridge, MA: Harvard University Press, 1984).

8 Sonoran Institute, *Getting Ahead in Greater Yellowstone: Making the Most of Our Competitive Advantage*, by Ray Rasker and Ben Alexander (2003), https://sonoraninstitute.org/files/pdf/getting-ahead-in-greater-yellowstone-making-the-most-of-our-competitive-advantage-06012003.pdf.

9 Robert Kegan and Lisa L. Lahey, *Immunity to Change: How to Overcome It and Unlock Potential in Yourself and Your Organization* (Boston, MA: Harvard Business Press, 2009).

10 Tony Wagner et al., *Change Leadership: A Practical Guide to Transforming Our Schools* (New York: Jossey-Bass, 2009).

11 Clyde V. Arnspiger, *Personality in Social Process: Values and Strategies of Individuals in a Free Society* (Chicago: Follett, 1959).

12 Gary Cox, *How to Be an Existentialist* (New York: Bloomsbury, 2009).

13 Daniel Kahneman, *Thinking, Fast and Slow* (New York: Farrar, Straus and Giroux, 2011).

14 Kahneman, *Thinking, Fast and Slow*.

15 Associated Press, "Environmentalists Take Action on Energy Plans," *Jackson Hole News & Guide*, July 28, 2011.

16 Cory Hatch, "Wildlife Foundation Gets New Director," *Jackson Hole News & Guide*, October 12, 2011.

17 Richard Anderson, "Body of Missing Climber Recovered in Tetons," *Jackson Hole News & Guide*, July 16, 2012, https://www.jhnewsandguide.com/news/top_stories/article_6b43b0f0-dd9c-524f-9660-9ace3218ccf0.html.

18 "The basic human motive" is discussed in *The Individual Psychology of Alfred Adler*, ed. Heinz L. Ansbacher and R. Rowena Ansbacher (New York: Basic Books, 1956).

19 The striving for superiority is functionally equivalent to Abraham Maslow's "self-actualization," by which people desire self-fulfillment in ways unique to themselves. Abraham Maslow, *Motivation and Personality* (New York: Van Nostrand Reinhold, 1970); Abraham Maslow, *The Farther Reaches of Human Nature* (New York: Viking Press, 1971); James McMartin, *Personality Psychology* (Thousand Oaks, CA: Sage, 1995); Henry A. Murray, *Explorations in Personality* (New York: Oxford University Press, 1938).

20 Carl R. Rogers, *On Becoming a Person* (Boston, MA: Houghton Mifflin, 1961).

21 Rogers, *On Becoming a Person.*

22 Hans Vaihinger, *The Philosophy of "As If,"* trans. C.K. Ogden (New York: Harcourt Brace, 1925).

23 To learn more about "group think," see Irving L. Janis, *Victims of Groupthink: A Psychological Study of Foreign-Policy Decisions and Fiascoes* (Boston, MA: Houghton Mifflin, 1972); Irving L. Janis, *Groupthink: Psychological Studies of Policy Decisions and Fiascoes* (Boston, MA: Houghton Mifflin, 1982); Irving L. Janis, *Crucial Decisions: Leadership in Policymaking and Crisis Management* (New York: Free Press, 1989).

24 Christopher Wanjekj, "How Your Brain Wiring Drives Social Interactions," *LiveScience*, November 14, 2017, https://www.livescience.com/60937-social-brain-wiring.html.

25 John Allman, Nicole Tetreault, Atiya Hakeem, Kebreten Manaye, Katerina Semendeferi, Joseph Erwin, Soyoung Park, Virgine Goubert, and Patrick Hof, "The von Economo Neurons in Fronto-Insular and Anterior Cingulate Cortex, *Annals of the New York Academy of Sciences*, 1225, no. 1 (2011): 59–71.

26 David C. Geary, *The Origins of Mind: Evolution of Brain, Cognition, and General Intelligence* (Washington, DC: American Psychological Association, 2005), 136–55.

27 Leath Tonino, "Not on Any Map: Jack Turner on Our Lost Intimacy with the Natural World," *The Sun*, August 2014, https://www.thesunmagazine.org/issues/464/not-on-any-map.

28 Jacob Bronowski, "The Reach of Imagination," *American Scholar*, no. 2 (1967): 193–201; Susan Petrilli, *The Self as a Sign, the World, and the Other* (New Brunswick, NJ: Transaction, 2013).

29 Frederich Nietzsche, "On Truth and Lie in an Extra-Moral Sense," in *The Portable Nietzsche*, trans. Walter Kaufmann (London: Penguin Classics, 1994).

30 Noel Castree, *Making Sense of Nature* (New York: Routledge, 2014).

31 Megan Bang, Douglas L. Medin, and Scott Atran, "Cultural Mosaics and Mental Models of Nature," *Proceeding of the National Academy of Sciences*, no. 35 (2007): 13868–74.

32 Walter Lippman, *Public Opinion* (New York: Harcourt Brace, 1922). This book is a critical assessment of functional democratic government, especially of the irrational and often self-serving social perceptions that influence individual behavior and prevent optimal societal cohesion.

33 Donald Schon, *The Reflective Practitioner: How Professionals Think in Action* (New York: Basic Books, 1983); Donald Schon, *Educating the Reflective Practitioner* (San Francisco, CA: Jose-Bass, 1987).

34 Susan Wolf, *Meaning in Life and Why It Matters* (Princeton, NJ: Princeton University Press, 2010).

35 Mike Koshmrl, "Grizzly Delisting Is Complete," *Jackson Hole News & Guide*, June 23, 2017, https://www.jhnewsandguide.com/news/environmental/article_92879907-12d1-5d4c-aa4c-206bd697f66c.html.

36 Mike Koshmrl, "Wolf Numbers? Harder to Say," *Jackson Hole News & Guide*, June 21, 2017, https://www.jhnewsandguide.com/news/environmental/article_b8213f21-66bb-54d0-9312-5600dd480081.html.

37 Will Buckingham et al., *The Philosophy Book: Big Ideas Simply Explained* (New York: DK, 2011), 12.

38 A. C. Grayling. *The History of Philosophy* (New York: Penguin, 2019); Richard Tarnas, *The Passion of the Western Mind: Understanding the Ideas That Have Shaped Our World View* (New York: Ballantine Books, 1991).

39 Carlos M. N. Eire, *Reformations: The Early Modern World, 1450–1650* (New Haven, CT: Yale University Press, 2016).

40 Ivan Illich, *Deschooling Society* (New York: Harper & Row, 1971).

41 Nietzsche, "On Truth and Lie in an Extra-Moral Sense."

42 Darrel P. Rowbottom and Sarah J. Aiston, "The Myth of 'Scientific Method' in Contemporary Educational Research," *Journal of Philosophy of Education*, no. 2 (2006): 137–56.

43 Roy Scranton, *Learning How to Die in the Anthropocene: Reflections on the End of Civilization* (San Francisco, CA: City Lights, 2015).

44 Richard Rorty, *Philosophy and the Mirror of Nature* (Princeton, NJ: Princeton University Press, 1979), 170.

45 David Bohm, "Meaning and Information," in *The Search for Meaning: The New Spirit in Science and Philosophy*, ed. Paavo Pylkkanen (Wellingborough: Crucible, 1988): 1–25; David Bohm, *Changing Consciousness* (New York: Harper Collins, 1991); David Bohm, *Thought as a System* (London: Routledge, 1994).

Chapter 7 Coherence and Policy

1 Myers S. McDougal, "Legal Basis for Securing the Integrity of the Earth-Space Environment," *Journal of Natural Resources and Environmental Law*, 8 (1992–93): 177–207.

2 Susan G. Clark and Murray B. Rutherford, eds., *Large Carnivore Conservation: Integrating Science and Policy in the North American West* (Chicago: University of Chicago Press, 2014).

3 This definition follows Harold D. Lasswell and Myres S. McDougal, *Jurisprudence for a Free Society: Studies in Law, Science, and Policy* (Leiden: Martinus Nijohoff, 1992).

4 Susan G. Clark, Catherine H. Picard, and Aaron M. Hohl, "A Problem-Oriented View of Large-Scale Conservation," in *Large-Scale Conservation in the Common Interest*, ed. Susan G. Clark et al. (New York: Springer, 2014), 1–14; Susan G. Clark, Catherine H. Picard, and Aaron M. Hohl, "The Importance of People, Institutions, and Resources in Large-Scale Conservation," in *Large-Scale Conservation in the Common Interest*, ed. Susan G. Clark et al. (New York: Springer, 2014), 15–28.

5 Myres S. McDougal, Harold D. Lasswell, and L. Chen, *Human Rights and World Public Order: Basic Policies of an International Law of Human Dignity* (New Haven, CT: Yale University Press, 1980), 205.

6 Lasswell and McDougal, *Jurisprudence for a Free Society*.

7 This classification comes from Susan G. Clark, *Ensuring Greater Yellowstone's Future: Choices of Leaders and Citizens* (New Haven, CT: Yale University Press, 2008).

8 There has been a lot of conflict over elk management. For a popular review, see Mike Koshmrl, "Closing Feedgrounds for CWD a Long Shot," *Jackson Hole News & Guide*, December 9, 2015, https://www.jhnewsandguide.com/news/environmental/article_5b946232-47bf-5cf1-ac3c-9b58347956e6.html; Mike Koshmrl, "New Feedground Plan Bans Anti-Wolf Efforts," *Jackson Hole News & Guide*, December 9, 2015, https://www.jhnewsandguide.com/news/environmental/article_c5598d86-7464-53ea-a5d2-3b4c63dcafc9.html; Todd Wilkinson, "Park's 'Elk Reduction' a Travesty of Hunting," *Jackson Hole News & Guide*, October 28, 2015, https://www.jhnewsandguide.com/opinion/columnists/the_new_west_todd_wilkinson/article_01f00652-254f-55f2-aafb-3f66e223122d.html. For an academic review of the conflict, see Susan G. Clark and Marion Vernon, "Governance Challenges in Joint Inter-Jurisdictional

Management: Grand Teton National Park, Wyoming, Elk Case," *Environmental Management*, 2 (2015): 286–99; Susan G. Clark and Marion Vernon, "Elk Management and Policy in Greater Yellowstone: Assessing the Constitutive Process," *Policy Sciences*, 2 (2017): 295–316; Marian E. Vernon, Zachary Bischoff-Mattson, and Susan G. Clark, "Discourses of Elk Hunting and Grizzly Bear Incidents in Grand Teton National Park, Wyoming," *Human Dimensions of Wildlife*, 1 (2015): 65–85; Marian E. Vernon and Susan G. Clark, "Addressing a Persistent Policy Problem: The Elk Hunt in Grand Teton National Park, Wyoming," *Society and Natural Resources*, 7 (2016): 836–51; Paul Schullery, ed., *The Yellowstone Wolf: A Guide and Sourcebook* (Norman: University of Oklahoma Press, 2003).

9 Rocky Barker, *Scorched Earth: How the Fires of Yellowstone Changed America* (Washington, DC: Island Press, 2005); Robert W. Righter, *Peaks, Politics & Passion: Grand Teton Comes of Age* (Moose, WY: Grand Teton Association, 2014), 149.

10 Jason Vogel, David N. Cherney, and Elizabeth A. Lowham, "The Policy Sciences as a Transdisciplinary Approach for Policy Studies," in *The Oxford Handbook of Interdisciplinarity*, ed. Robert Frodeman, Julie Thompson, and Carl Mitcham (Oxford: Oxford University Press, 2017), 358–69.

11 Vernon and Clark, "Addressing a Persistent Policy Problem," 836–51.

12 Clark and Vernon, "Elk Management and Policy in Greater Yellowstone," 295–316.

13 Tim W. Clark, *Averting Extinction: Reconstructing Endangered Species Recovery* (New Haven, CT: Yale University Press, 1997); Brian Miller, Richard Reading, and Steve Forrest, *Prairie Night: Black-Footed Ferrets and the Recovery of Endangered Species* (Washington, DC: Smithsonian Institution Press, 1996).

14 Marian E. Vernon, Zachary Bischoff-Mattson, and Susan G. Clark, "Discourses of Elk Hunting and Grizzly Bear Incidents in Grand Teton National Park, Wyoming," *Human Dimensions of Wildlife*, 1 (2015): 65–85.

15 Steven L. Yaffee, *Prohibitive Policy: Implementing the Federal Endangered Species Act* (Cambridge, MA: MIT Press, 1982), 90, 99.

16 Archie Carr, "Letter to Diversity Section," *Conservation Biology*, 1 (1987): 81–86.

17 Noel Snyder, "California Condor Recovery Program," in *Raptor Research Reports No. 5: Raptor Conservation in the Next 50 Years*, ed. Stanley E. Senner, Clayton M. White, and Jimmie R. Parrish (Provo, UT: Press Publishing Limited, 1985), 56–81.

18 Susan G. Clark, *The Policy Process: A Practical Guide for Natural Resource Professionals* (New Haven, CT: Yale University Press, 2002), 111–26.

19 Susan G. Clark and Richard L. Wallace, "Integration via Interdisciplinarity: Concepts, Frameworks, and Education," *Policy Sciences*, 2 (2015): 233–55.

20 Ronald D. Brunner et al., *Finding Common Ground: Governance and Natural Resources in the American West* (New Haven, CT: Yale University Press, 2002); Ronald D. Brunner et al., *Adaptive Governance: Integrating Science, Policy, and Decision Making* (New York: Columbia University Press, 2005).

21 Ronald D. Brunner, "A Paradigm of Practice," *Policy Sciences* (2006): 135–67.

22 Clark, *Policy Process*, 74–76.

23 Johanna Love, Richard Anderson, and Kevin Olsen, "Grizzly Battle Has Just Begun," *Jackson Hole News & Guide*, June 28, 2017, https://www.jhnewsandguide.com/opinion/editorial/article_dab429ef-e4c6-5677-b57c-5704a25ad5c2.html.

24 Erik Molvar, "Save the Elk: Shut Feedlots, add Wolves," *The Wildlife News*, June 29, 2017, http://www.thewildlifenews.com/2017/06/29/save-the-elk-shut-feedlots-add-wolves/.

25 Adrian Treves et al., "Predators and the Public Trust," *Biological Reviews* 92 (2017): 248–70.

26 Associated Press, "Zinke: Unbind Public Land, *Jackson Hole Daily*, June 28, 2017.

27 Clark and Wallace, "Integration via Interdisciplinarity," 233–55; Timothy Terway, "Sustained in Significance with(out) Context and Ourselves: Expert Environmental Knowledge and 'Social-Ecological-Systems'" (PhD diss., Yale University, 2018).

28 Gerardo Ceballos, Paul R. Ehrlich, and Rodolfo Dirzo, "Biological Annihilation via the Ongoing Sixth Mass Extinction Signaled by Vertebrate Population Losses and Declines," *Proceedings of the National Academy of Science*, 30 (2017): E6089-E6096.

29 Michael Bonnett, "Normalizing Catastrophe: Sustainability and Scientism," *Environmental Education Research*, 2 (2012): 187–97.

30 P. J. White, Rich L. Wallen, and David E. Hallac, eds., *Yellowstone Bison: Conserving an American Icon in Modern Society* (Mammoth, WY: Yellowstone Association and National Park Service, 2015).

31 Roberto S. Foa and Yascha Mounk, "The Democratic Disconnect," *Journal of Democracy*, 3 (2016): 5–17.

32 Neal Deavers, "The Overton Bubble and Political Control," *Entelekheia*, July 20, 2017, http:// www.entelekheia.fr/2017/07/20/the-overton-bubble/.

33 Richard Wallace, *Yellowstone's Uncertain Future: Is Education the Answer?* (In preparation); Richard Wallace, *Yellowstone's Educational Landscape: An Assessment of Content, Methods, and Outcomes* (in preparation).

34 Bram Buscher and Robert Fletcher, "Accumulation by Conservation," *New Political Economy*, 2 (2015): 273–98.

35 Yuval N. Harari, *Homo Deus: A Brief History of Tomorrow* (New York: HarperCollins, 2017). To learn more about the work that is being done on this issue in the GYE, consider the organization "Wyoming Untrapped," which is based in Jackson, Wyoming, and led by Lisa Robertson. This organization promotes an overall ethic of compassionate coexistence for our wildlife and other natural resources in our rapidly changing society.

36 Murray Bookchin, *Toward an Ecological Society* (Buffalo, NY: Black Rose Books, 1996).

37 Lasswell and McDougal, *Jurisprudence for a Free Society*.

38 Myres S. McDougal, Harold D. Lasswell, and W. Michel Reisman, "The World Constitutive Process of Authoritative Decision," in *International Law Essays: A Supplement to International Law in Contemporary Practice*, ed. Myres S. McDougal and W. Michael Reisman (New York: Foundation Press, 1981), 191–282.

39 Consider the following:

I think the difficulty is this fragmentation. [...] All thought is broken up into bits. Like this nation, this country, this industry, this profession and so on [...] And they can't meet. That comes about because thought has developed traditionally in a way such that it claims not to be effecting anything but just telling you the way things are. Therefore, people cannot see that they are creating a problem and then apparently trying to solve it. [...] Wholeness is a kind of attitude or approach to the whole of life. If we can have a coherent approach to reality then reality will respond coherently to us. (David Bohm, "Wholeness and Fragmentation" (speech, Amsterdam, 1990), https://creativesystemsthinking.wordpress.com/2014/10/01/wholeness-a-coherent-approach-to-reality-david-bohm/)

40 Peter M. Haas, "Introduction: Epistemic Communities and International Policy Coordination," *International Organization*, 1 (1992): 1–35.

41 Gary W. Trompf, "The Classification of the Sciences and the Quest for Interdisciplinarity: A Brief History of Ideas from Ancient Philosophy to Contemporary Science," *Environmental Conservation*, 2 (2011): 113–26.

42 Disciplines are especially important as tools of cognition and communication. See Pierre Bourdieu, *Outline of a Theory of Practice*, trans. Richard Nice (Cambridge: Cambridge University Press, 1977), 78–97.

43 Terway, "Sustained in Significance with(out) Context and Ourselves."

44 Noel Castree, *Making Sense of Nature* (New York: Routledge, 2014), 213–81.

45 *Yellowstone Science* is a publication devoted to Yellowstone's natural and cultural resources. *Yellowstone Science* features articles about research, conferences, or other special events in the Greater Yellowstone Ecosystem, provides scientists with an opportunity to communicate and exchange ideas, and keeps the public informed about scientific endeavors in and around the park. *Yellowstone Science* was first published in 1992.

46 Ronald D. Brunner and Tim W. Clark, "A Practice-Based Approach to Ecosystem Management," *Conservation Biology*, 1 (2002): 48–58.

47 Brunner et al., *Finding Common* Ground; Brunner et al., *Adaptive* Governance; Ronald D. Brunner
    and Amanda H. Lynch, *Adaptive Governance and Climate Change* (Boston, MA: American
    Meteorological Society, 2010); Toddi Steelman, "Adaptive Governance," in *Handbook on
    Theories of Governance*, ed. Christopher Ansell and Jacob Torfing (Northampton, MA: Edward
    Elgar, 2016), 538–50.

48 Brunner and Clark, "Practice-Based Approach to Ecosystem Management," 48–58.

49 Brunner et al., *Finding Common* Ground; Brunner et al., *Adaptive* Governance; Brunner and
    Lynch, *Adaptive Governance and Climate Change*; Ronald D. Brunner, "Adaptive Governance
    as a Reform Strategy," *Policy Sciences*, 4 (2010): 301–41; Steelman, "Adaptive Governance,"
    538–50.

50 The Northern Rockies Conservation Cooperative (NRCC), founded in 1987, is a 501(c)3
    nonprofit organization headquartered in Jackson Hole, Wyoming. NRCC advances
    conservation for the common good in three important ways: (1) generating reliable knowledge
    and scientific data, (2) fostering effective leadership, and (3) enabling innovative policies and
    practices. Although NRCC works globally, they place a special focus on the Greater Yellowstone
    Ecosystem (GYE). One of the last intact ecosystems in America, the GYE presents a unique
    opportunity to develop innovative conservation solutions that can be applied wherever
    problems arise.

51 Buscher and Fletcher, "Accumulation by Conservation," 273–98; Bill Joy, "Why the Future
    Doesn't Need Us," *Wired*, April 1, 2000, https://www.wired.com/2000/04/joy-2/; George
    Mombiot, "Neoliberalism—the Ideology at the Root of All Our Problems," *The Guardian*,
    April 15, 2016, https://www.theguardian.com/books/2016/apr/15/neoliberalism-ideology-
    problem-george-monbiot.

52 Ian Morris, *Why the West Rules—for Now: The Patterns of History, and What They Reveal about the
    Future* (New York: Farrar, Straus and Giroux, 2010).

53 Richard P. Reading, Tim W. Clark, and Stephen R. Kellert, "Attitudes and Knowledge of People
    Living in the Greater Yellowstone Ecosystem," *Society & Natural Resources*, 4 (1994): 349–65.

54 Jonathan Schechter, "Leading the Region: Jackson Hole's Role in the Greater Teton
    Community," *Jackson Hole Compass 2017 Edition*, July 3, 2017.

55 Richard Sennett, *Respect in a World of Inequality* (New York: W.W. Norton, 2003); Richard
    Sennett, *Together: The Rituals, Pleasures and Politics of Cooperation* (New Haven, CT: Yale University
    Press, 2012).

Chapter 8 Challenges and Future

1 Keith Ginger, *Jackson Hole News & Guide* (2016).

2 Oscar Venter et al., "Sixteen Years of Change in the Global Terrestrial Human Footprint
   and Implications for Biodiversity Conservation," *Nature Communications*, 7 (2016): 1–11;
   Peter M. Vitousek et al., "Human Alteration of the Global Nitrogen Cycle: Sources and
   Consequences," *Ecological Applications*, 3 (1997): 737–50.

3 Clive Hamilton, "Define the Anthropocene in Terms of the Whole Earth," *Nature*, 536
   (2016): 251; Amanda H. Lynch and Siri Veland, *Urgency in the Anthropocene* (Cambridge,
   MA: MIT Press, 2018); Roy Scranton, *Learning How to Die in the Anthropocene: Reflections on the
   End of a Civilization* (San Francisco, CA: City Lights, 2015); Elisabeth Young-Bruehl, *Why Arendt
   Matters* (New Haven, CT: Yale University Press, 2006).

4 Tom Butler, *Overdevelopment, Overpopulation, Overshoot* (New York: Goff Books, 2015), 1; Jurriaan
   M. De Vos et al., "Estimating the Normal Background Rate of Species Extinction," *Conservation
   Biology*, 2 (2014): 452–62. Tom Butler is the editorial projects director for the Foundation for
   Deep Ecology and president of the Northeast Wilderness Trust. His other books include
   Wildlands Philanthropy and *ENERGY: Overdevelopment and the Delusion of Endless Growth*.

5 Susan G. Clark, *The Policy Process: A Practical Guide for Natural Resource Professionals* (New Haven, CT: Yale University Press, 2002), 111–26.

6 Clark, *Policy Process*, 17–55.

7 Paul Schullery, *Searching for Yellowstone: Ecology and Wonder in the Last Wilderness* (Helena: Montana Historical Society Press, 2004).

8 John W. Peters and John Varley, "Forward," in *Knowing Yellowstone: Science in America's First National Park*, ed. Jerry Johnson (Lanham, MD: Taylor Trade, 2010); Elizabeth Shanahan and Mark McBeth, "The Science of Storytelling: Policy Marketing and Wicked Problems in the Greater Yellowstone Ecosystem," in *Knowing Yellowstone: Science in America's First National Park*, ed. Jerry Johnson (Laham, MD: Taylor Trade, 2010), 141–59; Cathy Whitlock, "Using Yellowstone's Past to Understand the Future," in *Knowing Yellowstone: Science in America's First National Park*, ed. Jerry Johnson (Laham, MD: Taylor Trade, 2010), 33–48.

9 Historians Paul Schullery and Lee Whittlesy, for example, have written on many facets of Yellowstone that provide an overview of the human social process in the region over the last hundred and more years.

10 Ray Rasker and Andrew Hansen, "Natural Amenities and Population Growth in the Greater Yellowstone Region," *Human Ecology Review*, 7, no. 2 (2000): 30–40.

11 Journalist Todd Wilkinson has written about the overall growth in Greater Yellowstone. See Todd Wilkinson, "Are We Loving Yellowstone to Death?" *National Geographic*, May 2016, https://www.nationalgeographic.com/magazine/2016/05/yellowstone-national-parks-land-use/; Todd Wilkinson, "The New West: Will Human Growth Sink Yellowstone's Golden Age?" *Explore Big Sky*, June 10, 2016, http://www.explorebigsky.com/the-new-west-will-human-growth-sink-yellowstones-golden-age; Todd Wilkinson, "The Big Picture: Thinking about Greater Yellowstone's Elephant in the Room," *Mountain Journal*, October 5, 2017, https://mountainjournal.org/blind-spots-killing-greater-yellowstone; Todd Wilkinson, "The New West: Growth Is Bringing Huge Ecological, Economic Costs to Greater Yellowstone," *Explore Big Sky*, January 19, 2018, http://www.explorebigsky.com/the-new-west-growth-is-bringing-huge-ecological-economic-costs-to-greater-yellowstone.

12 US Department of the Interior, National Parks Service, *Yellowstone National Park Visitor Use Study* (Vermont, 2017), https://www.nps.gov/yell/getinvolved/upload/R-YELL_VUS_FINAL-Report.pdf; US Department of the Interior, National Parks Service, *Transportation and Vehicle Mobility Study* (2017), https://www.nps.gov/yell/getinvolved/upload/Yellowstone-Transportation-Mobility-Study_lo-res.pdf.

13 Shawn Otto, *The War on Science: Who's Waging It, Why It Matters. What Can We Do about It?* (Minneapolis, MN: Milkweed, 2016); Todd Wilkinson, *Science under Siege: The Politician's War on Nature and Truth* (Boulder, CO: Johnson Books, 1998).

14 James Gleik, *Chaos: Making a New Science* (New York: Penguin Books, 1987).

15 Roger Lewin, *Complexity: Life at the Edge of Chaos* (Chicago: University of Chicago Press, 1999).

16 Sidney Dekker, *Drift into Failure: From Hunting Broken Components to Understanding Complex Systems* (Boca Raton, FL: CRC Press, 2011).

17 For a description of the "governmentality" problem, explore James Ferguson, *The Anti-Politics Machine: "Development," Depoliticization, and Bureaucratic Power in Lesotho* (Minneapolis: University of Minnesota Press, 1994); Tania Murray Li, *The Will to Improve: Governmentality, Development, and the Practice of Politics* (Durham, NC: Duke University Press, 2007); Judith A. Merkle, *Management and Ideology: The Legacy of the International Scientific Management Movement* (Berkeley: University of California Press, 1980).

18 Susan G. Clark and Marion Vernon, "Governance Challenges in Joint Inter-Jurisdictional Management: Grand Teton National Park, Wyoming, Elk Case," *Environmental Management*, 2 (2015): 286–99; Susan G. Clark and Marion Vernon, "Elk Management and Policy in Greater Yellowstone: Assessing the Constitutive Process," *Policy Sciences*, 2 (2016): 295–316; Marian E. Vernon, Zachary Bischoff-Mattson, and Susan G. Clark, "Discourses of Elk Hunting and

Grizzly Bear Incidents in Grand Teton National Park, Wyoming," *Human Dimensions of Wildlife*, 1 (2015): 65–85; Marian E. Vernon and Susan G. Clark, "Addressing a Persistent Policy Problem: The Elk Hunt in Grand Teton National Park, Wyoming," *Society and Natural Resources*, 7 (2016): 836–51.

19 Examples and descriptions in Susan G. Clark, Murray B. Rutherford, and David J. Mattson, "Large Carnivores, People, and Governance," in *Large Carnivore Conservation: Integrating Science and Policy in the North American West*, ed. Susan G. Clark and Murray B. Rutherford (Chicago: Chicago University Press, 2014), 1–28; David J. Mattson and Susan G. Clark, "People, Politics, and Cougar Management," in *Cougar Ecology and Conservation*, ed. Maurice Hornocker and Sharon Negri (Chicago: Chicago University Press, 2010), 206–20.

20 Robert Putnam, "Bowling Alone: America's Declining Social Capital," *Journal of Democracy* 1 (1995): 65–78.

21 For example, see Will Steffen et al., "Planetary Boundaries: Guiding Human Development on a Changing Planet," *Science*, 6223 (2015): 736–52.

22 Garry Jacobs and Winston Nagan, "The Global Values Discourse," *Eruditio*, 1 (2012): 136–49; Myres S. McDougal, Harold D. Lasswell, and Lung-Chu Chen, *Human Rights and World Public Order: The Basic Policies of an International Law of Human Dignity* (New Haven, CT: Yale University Press, 1980).

23 Harrison Brown, *The Challenge of Man's Future* (New York: Viking Press, 1954); Jared Diamond, *Collapse: How Societies Choose to Fail or Succeed* (New York: Penguin Books, 2005); Buckminster Fuller, *Utopia or Oblivion: The Prospects for Humanity* (Zurich, CH: Lars Müller, 2008); Robert Heilbroner, *An Inquiry into the Human Prospect: Looked at Again for the 1990s* (New York: W.W. Norton, 1991); Jane Jacobs, *Dark Age Ahead* (New York: Vintage Books, 2005); James Lovelock, *A Rough Ride to the Future* (New York: Penguin Books, 2014); Ian Morris, *Why the West Rules—for Now: The Patterns of History, and What They Reveal about the Future* (New York: Farrar, Straus, and Giroux, 2010); Dmitry Orlov, *The Five Stages of Collapse: Laying the Groundwork for Social, Political, and Economic Revolution* (New York: New Society, 2013); Paul Robbins and Sarah A. Moore, "Ecological Anxiety Disorder: Diagnosis the Politics of the Anthropocene," *Cultural Geographies*, 1 (2012): 3–19; Joseph A. Tainter, "Archaeology of Overshoot and Collapse," *Annual Review of Anthropology*, 35 (2006): 59–74.

24 F. H. Bormann, "Ecology: A Personal History," *Annual Review of Energy and Environment*, 21 (1996): 1–29.

25 Bormann, "Ecology," 1–29.

26 Bormann, "Ecology," 27.

27 Ian Morris, *Why the West Rules—for Now*.

28 Janet A. Weiss, "The Powers of Problem Definition: The Case of Government Paperwork," *Policy Sciences*, 2 (1989): 97–121.

29 Susan G. Clark, "Making Sense of Wolves," in *The Wolves of Yellowstone*, ed. Doug Smith, Daniel Stahler, and Daniel MacNulty (Chicago: University of Chicago Press, in press).

30 Clark and Vernon, "Elk Management and Policy in Southern Greater Yellowstone," 295–316.

31 See, for example, Claudio Campagna and Daniel Guevara, "Conservation in No-Man's Land," in *Keeping the Wild: Against the Domestication of Earth*, ed. George Wuerthner, Eileen Crist and Tom Butler (Washington, DC: Island Press, 2014), 55–65.

32 See Noel Castree, *Making Sense of Nature* (New York: Routledge, 2014); Philippe Descola, *Beyond Nature and Culture*, trans. Janet Lloyd (Chicago: University of Chicago Press, 2013); Michael Dove, "Human Ecology," *Oxford Bibliographies*, June 25, 2013, http://www.oxfordbibliographies. com/view/document/obo-9780199830060/obo-9780199830060-0050.xml; Tim Ingold, *The Perception of the Environment: Essays in Livelihood, Dwelling and Skill* (London: Routledge, 2000); Eduardo Kohn, *How Forests Think: Towards an Anthropology beyond the Human* (Berkeley: University of California Press, 2013); Bruno Latour, *We Have Never Been Modern*, trans. Catherine Porter

(Cambridge, MA: Harvard University Press, 1993); James D. Proctor, "Geography, Paradox and Environmental Ethics," *Progress in Human Geography*, 2 (1998): 234–55; James D. Proctor, "The Social Construction of Nature: Relativist Accusations, Pragmatists and Critical Realist Responses," *Annals of the Association of American Geographers*, 3 (1998): 352–76; James D. Proctor, *Envisioning Nature, Science, and Religion* (Conshohocken, PA: Templeton Press, 2009), 293–312; James D. Proctor, "Replacing Nature in Environmental Studies and Sciences," *Journal of Environmental Studies and Sciences*, 4 (2016): 748–52; Sarah Whatmore, *Hybrid Geographies: Natures Cultures Spaces* (London: Sage, 2002).

33 For a partial listing of ordinary challenges, see David Quammen, "National Parks: Nature's Dead End," *New York Times*, July 28, 1996, https://www.nytimes.com/1996/07/28/opinion/national-parks-nature-s-dead-end.html; A. M. Thuermer Jr., "Yellowstone Declared 'in Danger,'" *Jackson Hole News*, December 6, 1995; US Department of the Interior, *Yellowstone Resources & Issues: An Annual Compendium of Information about Yellowstone National Park* (Wyoming, 2017); World Resources Institute, *People and Ecosystems: The Fraying Web of Life* (Washington, DC: 2000), https://wriorg.s3.amazonaws.com/s3fs-public/pdf/world_resources_2000-2001_people_and_ecosystems.pdf?_ga=2.212213842.778451446.154742918 4-777763522.1547429184.

34 Judith R. Gordon, *A Diagnostic Approach to Organizational Behavior* (Upper Saddle River, NJ: Prentice Hall, 1983). Irving L. Janis, *Victims of Groupthink: A Psychological Study of Foreign-Policy Decisions and Fiascoes* (Boston, MA: Houghton Mifflin, 1972); Irving L. Janis, *Groupthink: Psychological Studies of Policy Decisions and Fiascoes* (Boston, MA: Houghton Mifflin, 1982); Irving L. Janis, *Crucial Decisions: Leadership in Policymaking and Crisis Management* (New York: Free Press, 1989).

35 Daniel Katz and Robert L. Kahn, *The Social Psychology of Organizations* (Hoboken, NJ: John Wiley & Sons, 1966).

36 L. Naughton-Treves, "Whose Animals? A History of Property Rights to Wildlife in Toro, Western Uganda," *Land Degradation and Development*, 4 (1999): 311–28; Martin Nie et al., "Fish and Wildlife Management on Federal Lands: Debunking State Supremacy," *Environmental Law*, 4 (2017): 797–932.

37 Robert G. Healy and William Ascher, "Knowledge in the Policy Process: Incorporating New Environmental Knowledge in Natural Resource Policy Making," *Policy Sciences*, 1 (1995): 1–19.

38 Mahnoush Arsanjani, *International Regulation of Internal Resources: A Study of Law and Policy* (Charlottesville: University of Virginia Press, 1981); Douglas M. Johnston, *The International Law of Fisheries: A Framework for Policy-Oriented Inquires* (New Haven, CT: Yale University Press, 1987); Myres S. McDougal, *Studies in World Public Order* (New Haven, CT: Yale University Press, 1960); Myers S. McDougal and W. T. Burke, *The Public Order of the Oceans: A Contemporary International Law of the Sea* (New Haven, CT: Yale University Press, 1962); Myers S. McDougal, Harold D. Lasswell, and Ivan A. Vlasic, *Law and Public Order in Space* (New Haven, CT: Yale University Press, 1963); Emilio J. Sahurie, *The International Law of Antarctica* (New Haven, CT: Yale University Press, 1992).

39 Aldo Leopold, *A Sand County Almanac: With Other Essays on Conservation from Round River* (Oxford: Oxford University Press, 1971).

40 Aldo Leopold, *A Sand County Almanac: With Other Essays on Conservation from Round River* (Oxford: Oxford University Press, 1971); Myres S. McDougal and W. Michael Reisman, eds., *International Law Essays: A Supplement to International Law in Contemporary Perspective* (Mineola, NY: Foundation Press, 1981), 237–64.

41 To learn more about the cultural or constitutive process, see Harold D. Lasswell, *Pre-view of Policy Sciences* (New York: Elsevier, 1971), 98–111; McDougal, Lasswell, and Reisman, "The World Constitutive Process of Authoritative Decision," 237–64; Myres S. McDougal, Harold D. Lasswell, and W. Michael Reisman, "Theories about International Law: Prologue to a Configurative Jurisprudence," in *International Law Essays: A Supplement to International Law in Contemporary Perspective*, ed. Myres S. McDougal and W. Michael Reisman (Mineola,

NY: Foundation Press, 1981), 43–141; W. Michael Reisman, "Law from a Policy Perspective," in *International Law Essays: A Supplement to International Law in Contemporary Perspective*, ed. Myres S. McDougal and W. Michael Reisman (Mineola, NY: Foundation Press, 1981), 9–14.

42 Jason M. Vogel, *Persistent Policy Problems* (PhD diss., University of Colorado, 2006).

43 Vogel, *Persistent Policy Problems*, iii.

44 Susan G. Clark, *Ensuring Greater Yellowstone's Future: Choices for Leaders and Citizens* (New Haven, CT: Yale University Press, 2008).

45 Horst W. J. Rittel and Melvin M. Webber, "Dilemmas in a General Theory of Planning," *Policy Sciences*, 2 (1973): 155–69.

46 Elizabeth Kolbert, *The Sixth Extinction: An Unnatural History* (New York: Henry Holt, 2014).

47 Kolbert, *Sixth Extinction*.

48 Kolbert, *Sixth Extinction*.

49 Michael McCarthy, *The Moth Snowstorm: Nature and Joy* (New York: New York Review of Books, 2016).

50 Gabriela Lichtenstein, "Book Review," *Conservation Biology*, 30, no. 5 (2016): 1135–37.

51 Michael Fotos, personal communication, April 8, 2016; James C. Scott, *Seeing Like a State: How Certain Schemes to Improve the Human Condition Have Failed* (New Haven, CT: Yale University Press, 1998).

52 Fotos, personal communication, April 8, 2016.

53 Fotos, personal communication, April 8, 2016.

54 Lawrence D. Berg, Edward H. Huijbens, and Henrik G. Larsen, "Producing Anxiety in the Neoliberal University," *Canadian Geographer*, 2 (2016): 168–80; Wendy Brown, "Neoliberalized knowledge," *History of the Present*, 1 (2011): 113–29; Noel Castree, "Contract Research, Universities, and the 'Knowledge Society': Back to the Future," in *The Knowledge Business: The Commodification of Urban and Housing Research*, ed. Chris Allen and Rob Imrie (Farnham: Ashgate, 2010), 221–40; David Hursh, Joseph Henderson, and David Greenwood, "Environmental Education in a Neoliberal Climate," *Environmental Education Research*, 3 (2015): 299–318; Yancey Orr and Raymond Orr, "The Death of Socrates," *Australian Universities Review*, 2 (2016): 15–25.

55 Richard W. Wallace and Susan G. Clark, "The Role of Education in Conservation of the Greater Yellowstone Ecosystem: An Assessment and Recommendations for the Future" (in preparation).

56 This innate desire is reflected in the popularity of books, which discuss ways of connecting with nature in meaningful and transformative ways. Some examples: Robert Llewellyn and Joan Maloof, *The Living Forest: A Visual Journey into the Heart of the Woods* (Portland, OR: Timber Press, 2017); Matt Sewell, *A Charm of Goldfinches: And Other Wild Gatherings* (New York: Ten Speed Press, 2017); Sophie Walker, *The Japanese Garden* (New York: Phaidon Press, 2017).

57 One example from the business world comes from Robert Kegan and Lisa Laskow Lahey, *An Everyone Culture: Becoming a Deliberately Developmental Organization* (Cambridge, MA: Harvard Business School, 2016).

58 Fotos, personal communication, April 8, 2016.

Chapter 9 Learning and Transforming

1 Theodore Roosevelt, speech at the Lincoln Club Dinner, 1899, http://www.notable-quotes. com/r/roosevelt_theodore.html.

2 John Quay and Jayson Seaman, *John Dewey and Education Outdoors: Making Sense of the 'Educational Situation' through More than a Century of Progressive Reforms* (Rotterdam: Sense, 2013).

3 Susan G. Clark, *The Policy Process: A Practical Guide for Natural Resource Professionals* (New Haven, CT: Yale University Press, 2002), 153–72.

4 John Dewey, "My Pedagogic Creed," *School Journal*, 54 (1897): 77–80.

5 Richard W. Wallace and Susan G. Clark, *The Role of Education in Conservation of the Greater Yellowstone Ecosystem: An Assessment and Recommendations for the future* (in preparation).
6 John Dewey, *Experience and Education* (Indianapolis, IN: Kappa Delta Pi, 1938), 5.
7 For an introduction to policy sciences, see Clark, *Policy Process*.
8 Dewey, *Experience and Education*, 68–69.
9 John Dewey, *Democracy and Education* (New York: Free Press, 1916), 169.
10 John Dewey, *How We Think* (Boston, MA: D.C. Health, 1910).
11 John Dewey, *Studies in Logical Theory* (Chicago: Chicago University Press, 1909), 3–18.
12 Dewey, *How We Think*, 106–7.
13 Dewey, *Democracy and Education*, 176.
14 Some excellent introductions and overviews of transformative learning are Marilyn M. Taylor, *Emergent Learning for Wisdom* (New York: Palgrave Macmillan, 2011); Edward W. Taylor and Patricia Cranton, eds., *The Handbook of Transformative Learning: Theory, Research, and Practice* (San Francisco, CA: Jossey-Bass, 2012).
15 Susan G. Clark and Richard L. Wallace, "Integration and Interdisciplinarity: Concepts, Frameworks and Education," *Policy Sciences*, 2 (2015): 233–55.
16 Clark, *Policy Process*.
17 Emily C. Chamberlain, "Perspectives on Grizzly Bear Management in Banff National Park and the Bow River Watershed, Alberta: A Q Methodology Study" (master's thesis, Simon Fraser University, 2006); Emily C. Chamberlain, Murray B. Rutherford, and Michael L. Gibeau, "Human Perspectives and Conservation of Grizzly Bears in Banff National Park, Canada," *Conservation Practice and Policy*, 3 (2012): 420–31; Jutta K. Kölhi, "Stakeholder Views on Grizzly Bear Management in the Banff-Bow Valley: A Before-After Q-Methodology Study" (master's thesis, Simon Fraser University, 2010); J. Daniel Oppenheimer and Lauren Richie, "Collaborative Grizzly Bear Management in Banff National Park: Learning from a Prototype," in *Large Carnivore Conservation: Integrating Science and Policy in the North American West*, ed. Susan G. Clark and Murray B. Rutherford (Chicago: University of Chicago Press, 2014), 215–50; Lauren J. Richie, Daniel Oppenheimer, and Susan G. Clark, "Social Process in Grizzly Bear Management: Lessons for Collaborative Governance and Natural Resource Policy," *Policy Sciences*, 3 (2012): 265–91; Murray B. Rutherford et al., "Interdisciplinary Problem Solving Workshops for Grizzly Bear Conservation in Banff National Park, Canada," *Policy Sciences*, 2 (2009): 163 87.
18 A few examples: William Ascher, "Strategies of Influence for the Conservation of the Sangha River Basin: Insights from the Policy Sciences," *Yale FES Bulletin*, 102 (1998): 259–71; Matthew Auer, "Contexts, Multiple Methods, and Values in the Study of Common-Pool Resources," *Journal of Policy Analysis and Management*, 1 (2006): 215–27; Radford Byerly and Roger A. Pielke Jr., "The Changing Ecology of United States Science," *Science*, 269 (1995): 1531–32.
19 Michael Gibeau, "Of Bears, Chess, and Checkers: Moving Away from Pure Science to Solve Problems," *The Wildlife Professional*, 6 (spring 2012): 62–64.
20 Robert Kegan, "What 'Form' Transforms? A Constructive-Developmental Approach to Transformative Learning," in *Contemporary Theories of Learning: Learning Theorists ... in Their Own Words*, ed. Knud Illeris (New York: Routledge, 2009), 32–52.
21 To learn more, explore the following titles: Bonne J. B. Leonard, "Integrative Learning as a Developmental Process: A Grounded Theory of College Student's Experiences in Integrative Studies" (PhD diss., University of Maryland, 2007); Peter C. Taylor, Elisabeth Taylor, and Bal C. Luittel, "Multi-paradigmatic Transformative Research as/for Teacher Education: An Integral Perspective," in *Second International Handbook of Science Education*, ed. Barry Fraser, Kenneth Tobin, and Campbell McRobbie (New York: Springer, 2012), 373–87.
22 This table is modified from the model laid out in Michael Q. Patton, *Developmental Evaluation: Applying Complexity Concepts to Enhance Innovation and Use* (New York: Guilford Press, 2011).

23 Myles Horton and Paulo Freire, *We Make the Road by Walking: Conversations on Education and Social Change* (Philadelphia, PA: Temple University Press, 1990); Randee Lipson Lawrence, "Transformative Learning through Artistic Expression: Getting Out of Our Heads," in *The Handbook of Transformative Learning: Theory, Research, and Practice*, ed. Edward W. Taylor and Patricia Cranton (San Francisco, CA: Jossey-Bass, 2012), 471–85; Jo A. Tyler and Ann L. Swartz, "Storytelling and Transformative Learning," in *The Handbook of Transformative Learning: Theory, Research, and Practice*, ed. Edward W. Taylor and Patricia Cranton (San Francisco, CA: Jossey-Bass, 2012), 455–70.

24 Steven A. Schapiro, Ilene L. Wasserman, and Placida V. Gallegos, "Group Work and Dialogue," in *The Handbook of Transformative Learning: Theory, Research, and Practice*, ed. Edward W. Taylor and Patricia Cranton (San Francisco, CA: Jossey-Bass, 2012), 355–72; Peter Willis, "An Existential Approach to Transformation," in *The Handbook of Transformative Learning: Theory, Research, and Practice*, ed. Edward W. Taylor and Patricia Cranton (San Francisco, CA: Jossey-Bass, 2012), 212–28.

25 Kegan, "What 'Form' Transforms?"; Robert Kegan and Lisa L. Lahey, "The Real Reason People Won't Change," *Harvard Business Review*, November 2001, hbr.org/2001/11/the-real-reason-people-wont-change; Nona Lyons, ed., *Handbook of Reflection and Reflective Inquiry: Mapping a Way of Knowing for Professional Reflective Inquiry* (New York: Springer, 2010); Arjen E. J. Wals, ed., *Social Learning towards a Sustainable World: Principles, Perspectives, and Praxis* (Wageningen: Wageningen Academic, 2007).

26 World Bank Group, *World Development Report 2015: Mind, Society, and Behavior* (Washington, DC: World Bank, 2015).

27 For example, see Frederica Kolwey, "Making Migration Safer, One Strand at a Time: Wildlife Foundation's Fence Pull Program Nears 200-Mile Mark," *Jackson Hole News & Guide*, July 4, 2018, https://www.jhnewsandguide.com/news/environmental/article_7085ee51-f9a2-57ee-9a50-3d8f6b7dab75.html.

28 Robin Barrow, *An Introduction to Moral Philosophy and Moral Education* (New York: Routledge, 2007), 28.

29 Michael Bonnett, "Environmental Concern, Moral Education and Our Place in Nature," *Journal of Moral Education*, 3 (2012): 285–300.

30 Molly Absolon, "Can Conservation and Recreation Get Along?" *Jackson Hole News & Guide*, September 27, 2017, https://www.jhnewsandguide.com/special/conservation/can-conservation-and-recreation-get-along/article_fa7202cf-f422-574b-99f5-72fe2f60b4b6.html; Michael Dax, "Weakening Wilderness Act Is Antithetical to Principle of Landmark Law," *Mountain Journal*, April 10, 2018, http://mountainjournal.org/weakening-wilderness-protection-undermines-its-purpose; Martin Murie, "What Would the Muries Say about Recreation, Conservation and Wildness?" *Mountain Journal*, April 10, 2018, http://mountainjournal.org/the-muries-would-not-turn-wild-country-into-gyms; Todd Wilkinson, "Shifting Values: Are Funhog Towns 'Better' than the Ones They're Replacing?" *Mountain Journal*, April 4, 2018, http://mountainjournal.org/the-rise-of-funhog-towns.

31 Between 2009 and 2015, US Geological Survey completed a nutrient loading study for the Fish Creek Watershed in Teton County, Wyoming. Since the completion of this study, there have been changes in the stakeholder process, including changes in how stakeholders guide and contribute to research. To learn more about the stakeholder changes, visit www.tetonconservation.org. To learn more about the findings of the USGS Nutrient Loading Study, see US Department of the Interior, US Geological Survey, *Estimated Nitrogen and Phosphorus Inputs to the Fish Creek Watershed, Teton County, Wyoming, 2009–2015*, by Cheryl A. Eddy-Miller et al. (Reston, VA, 2016), https://pubs.usgs.gov/sir/2016/5160/sir20165160.pdf.

32 Wallace and Clark, *Role of Education in Conservation of the Greater Yellowstone Ecosystem*.

33 David Bohm, "Hidden Variables and the Implicate Order," *Zygon*, 2 (1985): 111–24.

34 John Dewey, *The School and Society* (Chicago: University of Chicago Press, 1915), 132.

35 By and large, college and universities are falling short of what is needed. See Chris Allen and Rob Imrie, "The Knowledge Business: A Critical Introduction," in *The Knowledge Business: The Commodification of Urban and Housing Research*, ed. Chris Allen and Rob Imrie (New York: Routledge, 2010), 1–19; Noel Castree, "Contract Research, Universities and the 'Knowledge Society': Back to the Future," in *The Knowledge Business: The Commodification of Urban and Housing Research*, ed. Chris Allen and Rob Imrie (New York: Routledge, 2010).

36 John Dewey, *Moral Principles in Education* (Boston, MA: Houghton Mifflin, 1909), 78.

37 Masashi Soga and Kevin J. Gaston, "Extinction of Experience: The Loss of Human-Nature Interactions," *Frontiers in Ecology and Evolution*, 2 (2016): 94–101; Kenneth Worthy, *Invisible Nature: Healing the Destructive Divide between People and the Environment* (Amherst, NY: Prometheus Books, 2013).

38 Robert Kegan, *In over Our Heads: The Mental Demands of Modern Life* (Cambridge, MA: Harvard University Press, 1994); Kegan, "What 'Form' Transforms?" 35–52.

39 My Yale course "Foundations of Natural Resources Policy and Management" is a research seminar for students in any subfield of environmental studies or other disciplines. The seminar's purpose is to help students develop skills for thinking more effectively and acting more responsibly in complex management and policy cases. The seminar explores comprehensive, interdisciplinary concepts and methods for thinking about problems in natural resources policy and management as well as solutions to those problems. Once students gain familiarity with the core concepts and methods of standpoint clarification and problem orientation, they apply them to particular issues in natural resources policy and management.

40 Susan G. Clark, Robert J. Begg, and Kim W. Lowe, "Interdisciplinary Problem-Solving Workshops for Natural Resource Professionals," in *The Policy Process: A Practical Guide to Natural Resource Professionals*, ed. Susan G. Clark (New Haven, CT: Yale University Press, 2002); Susan G. Clark et al., "Large Scale Conservation: Integrating Science, Management, and Policy in the Common Interest," *Yale School of Forestry & Environmental Studies, Report*, no. 24 (2010): 173–208.

41 Louis G. Lombardi, *Moral Analysis: Foundations, Guides, and Applications* (New York: SUNY Press, 1988).

42 Robert Wright, "Our Animals, Our Selves," *New York Times*, July 29, 1990.

43 Emily Mieure, "Neighbors Allege Animal Cruelty," *Jackson Hole News & Guide*, August 11, 2017, https://www.jhnewsandguide.com/jackson_hole_daily/local/article_e5910eb4-382e-5da8-840c-f5d7f68282a8.html.

44 Center for American Progress, "The Big Picture," The Disappearing West, https://disappearingwest.org/land.html#land_big_picture.

45 Susan G. Clark, *The Policy Process: A Practical Guide for Natural Resource Professionals* (New Haven, CT: Yale University Press, 2002); Harold D. Lasswell, *The World Revolution of Our Time: A Framework for Basic Policy Research* (Redwood City, CA: Stanford University Press, 1951); Harold D. Lasswell, *A Pre-View of Policy Sciences* (New York: American Elsevier, 1971), 67–69.

46 Wendell Bell, *Foundations of Future Studies* (Piscataway, NJ: Transaction, 1997), xxiii.

47 Bell, *Foundations of Future Studies*, 332.

48 Mike Koshmrl, "Y'stone's Traffic Mess Is Predicted to Worsen," *Jackson Hole News & Guide*, August 30, 2017, https://www.jhnewsandguide.com/news/environmental/article_7b62d792-5f36-5eb3-8227-91c2441bae9d.html.

49 Garry D. Brewer, "Methods for Synthesis: Policy Exercises," in *Sustainable Development of the Biosphere*, ed. William C. Clark and R. E. Munn (Cambridge: Cambridge University Press, 1986), 455–75; Garry D. Brewer, "Policy Sciences, the Environment and Public Health," *Health Promotion*, 3 (1988): 227–37.

50 Important research in the Greater Yellowstone Ecosystem comes from Dr. Andy Hanson (Montana State University), Dr. Monica Turner (University of Wisconsin), and Dr. Ann Rodman (Yellowstone National Park).

51 Valerie A. Brown, John A. Harris, and Jacqueline Y. Russell, *Tackling Wicked Problems: Through Transdisciplinary Imagination* (New York: EarthScan, 2010); Valerie A. Brown and John A. Harris, *The Human Capacity for Transformational Change: Harnessing the Collective Mind* (New York: Routledge, 2014); Brian Calvert, "Down the Dark Mountain," *High Country News*, July 24, 2017, 16–24.

52 Karen O'Brien, "Global Environmental Change II: From Adaptation to Deliberate Transformation," *Progress in Human Geography*, no. 5 (2012): 667–76.

53 Raymond De Young, "Some Behavioral Aspects of Energy Descent: How a Biophysical Psychology Might Help People Transition through the Lean Times Ahead," *Frontiers in Psychology*, 5 (2014): 1–16.

54 Pope Francis, *Laudato Si: On Care for Our Common Home*, https://w2.vatican.va/content/dam/francesco/pdf/encyclicals/documents/papa-francesco_20150524_enciclica-laudato-si_en.pdf.

55 Harold D. Lasswell and Myres S. McDougal, *Jurisprudence for a Free Society: Studies in Law, Science, and Policy* (Leiden: Martinus Nijohoff, 1992).

56 Michael W. Reisman, *Law in Brief Encounters* (New Haven, CT: Yale University Press, 1999).

57 Ronald D. Brunner, "Myth and American politics," *Policy Sciences*, 27 (1994): 1–18.

Chapter 10 The Work Ahead

1 Donella Meadows, The Donella Meadows Archives (1990), donellameadows.org.

2 Todd Wilkinson and Tom Mangelsen, *Grizzlies of Pilgrim Creek: An Intimate Portrait of 399, the Most Famous Bear of Greater Yellowstone* (New York: Rizzoli, 2015).

3 Todd Wilkinson, "399 Is the Poster Bear for All the Bruin Clan," *Jackson Hole News & Guide*, September 9, 2015, https://www.jhnewsandguide.com/opinion/columnists/the_new_west_todd_wilkinson/article_d6e70e2a-9893-5f20-a81a-cda5cee76f03.html.

4 Todd Wilkinson, "Y'stone Golden Age May Soon Be History," *Jackson Hole News & Guide*, May 11, 2016, https://www.jhnewsandguide.com/opinion/columnists/the_new_west_todd_wilkinson/article_bec34f23-a73f-5191-b6e1-24c2f04a5a9d.html.

5 US Department of the Interior, National Parks Service, *Yellowstone Resources and Issues Handbook 2015* (Mammoth, WY, 2015), 27.

6 Cory Hatch, "Grizzly Bear Attacks Visitor in Grand Teton," *Jackson Hole News & Guide*, June 14, 2007, https://www.jhnewsandguide.com/news/top_stories/grizzly-bear-attacks-visitor-in-grand-teton/article_dec1a300-6fcd-5305-9df0-7f2780cb5106.html.

7 Wilkinson, "Y'stone Golden Age May Soon Be History."

8 Wilkinson, "Y'stone Golden Age May Soon Be History."

9 Wesley J. Smith, "Personalizing Animals," *National Review*, October 26, 2012, https://www.nationalreview.com/corner/personalizing-animals-wesley-j-smith/.

10 *Organic Act*, US Code 16 (1916), §§ 1–4.

11 US Department of the Interior, National Parks Service, *Yellowstone Resources and Issues Handbook 2015*.

12 Theodore Roosevelt, "Address to the Citizens of Dickinson" (speech, Dickinson, Dakota Territory, July 4, 1886), http://www.theodore-roosevelt.com/images/research/txtspeeches/dickinson4july1886speech.pdf.

13 Paul Schullery, *Searching for Yellowstone: Ecology and Wonder in the Last Wilderness* (New York: Houghton Mifflin, 1997), 1–2.

14 US Department of the Interior, *Yellowstone Resources and Issues Handbook 2015*, 27.

15 André Gide, *Le Traité du Narcisse suivi de La Tentative Amoureuse* (Lausanne: Mermod, 1946).

16 V. Clyde Arnspiger, *Personality in Social Process: Values and Strategies of Individuals in a Free Society* (Chicago: Follett, 1961), 1.

17 Richard D. Brunner and Tim W. Clark, "A Practice-Based Approach to Ecosystem Management," *Conservation Biology*, 1 (2002): 48–58.

18 Ronald D. Brunner, "A Paradigm for Practice," *Policy Sciences*, 2 (2006): 135–67.

19 Ronald D. Brunner et al., *Finding Common Ground: Governance and Natural Resources in the American West* (New Haven, CT: Yale University Press, 2002); Ronald D. Brunner et al., *Adaptive Governance: Integrating Science, Policy, and Decision Making* (New York: Columbia University Press, 2005); Ronald D. Brunner, "Adaptive Governance as a Reform Strategy," *Policy Sciences*, 4 (2010): 301–41; Ronald D. Brunner and Amanda H. Lynch, *Adaptive Governance and Climate Change* (Boston, MA: American Meteorological Society, 2010); Toddi Steelman, "Adaptive Governance," in *Handbook on Theories of Governance*, ed. Christopher Ansell and Jacob Torfing (Northampton, MA: Edward Elgar, 2016), 538–50.

20 Modified from Pimbert and Pretty, *Parks, People, and Professionals*.

21 Susan G. Clark, *The Policy Process: A Practical Guide for Natural Resource Professionals* (New Haven, CT: Yale University Press, 2002), 85–110.

22 Tim W. Clark and Murray B. Rutherford, "The Institutional System of Wildlife Management: Making It More Effective," in *Coexisting with Large Carnivores: Lessons from Greater Yellowstone*, ed. Tim W. Clark, Murray B. Rutherford, and Denise Casey (Washington, DC: Island Press, 2005), 211–53; Murray B. Rutherford and Tim W. Clark, "Coexisting with Large Carnivores: Lessons from Greater Yellowstone," in *Coexisting with Large Carnivores: Lessons from Greater Yellowstone*, ed. Tim W. Clark, Murray B. Rutherford, and Denise Casey (Washington, DC: Island Press, 2005), 254–70; Dylan Taylor and Tim W. Clark, "Management Context: People, Animals, and Institutions," in *Coexisting with Large Carnivores: Lessons from Greater Yellowstone*, ed. Tim W. Clark, Murray B. Rutherford, and Denise Casey (Washington, DC: Island Press, 2005), 28–68.

23 Michael P. Pimbert and Jules N. Pretty, *Parks, People, and Professionals: Putting 'Participation' into Protected Area Management* (Geneva: United Nations Research Institute for Social Development, 1995).

24 Tim W. Clark, Ron Crete, and John Cada, "Designing and Managing Successful Endangered Species Recovery Programs," *Environmental Management*, 2 (1989): 159–70; Tim W. Clark and Ron Westrum, "High-Performance Teams in Wildlife Conservation: A Species Reintroduction and Recovery Example," *Environmental Management*, 6 (1989): 663–70.

25 John Dryzek, *Discursive Democracy: Politics, Policy, and Political Science* (Cambridge: Cambridge University Press, 1990), 92.

26 William Ascher and Robert Healey, *Natural Resource Policymaking in Developing Countries: Environment, Economic Growth, and Income Distribution* (Durham, NC: Duke University Press, 1990).

27 Tim W. Clark and John R. Cragun, "Organization and Management of Endangered Species Programs," *Endangered Species Update*, 4 (2002): 114–18; Tim W. Clark, "Learning as a Strategy for Improving Endangered Species Conservation," *Endangered Species Update*, 4 (2002): 119–24.

28 Susan G. Clark et al., "Large Scale Conservation: Integrating Science, Management, and Policy in the Common Interest," *Yale School of Forestry & Environmental Studies, Report*, no. 24 (2010), 173–208.

29 Clark, *Policy Process*, 85–110; Richard L. Wallace and Tim W. Clark, "Solving Problems in Endangered Species Conservation: An Introduction to Problem Orientation," *Endangered Species Update*, 4 (2002): 81–86.

30 Chris Argyris and Donald Schon, *Organizational Learning: A Theory of Action Perspective* (Reading, MA: Addison-Wesley, 1978); Lloyd S. Etheredge, *Can Governments Learn? American Foreign Policy and Central American Revolutions* (New York: Pergamon Press, 1983); Peter M. Senge, *The Fifth Discipline: The Art and Practice of the Learning Organization* (New York: DoubleDay Books, 1990).

31 Chris Argyris, *On Organizational Learning* (Malden, MA: Blackwell, 1992); Clark, "Learning as a Strategy for Improving Endangered Species Conservation," 119–24.

32 Richard L. Daft, *Organization Theory & Design* (Mason, OH: Cengage Learning, 2013); Daniel Katz and Robert L. Kahn, *The Social Psychology of Organizations* (New York: John Wiley & Sons,

1966); David A. Nadler and Michael L. Tushman, "A Model for Diagnosing Organizational Behavior," *Organizational Dynamics*, 2 (1980): 35–51.

33 L. G. Hrebiniak, *Complex Organizations* (New York: West, 1978); Ronald Westrum and Khalil Samaha, *Complex Organizations: Growth, Struggle, and Change* (Englewood Cliffs, NJ: Prentice-Hall, 1984).

34 Tim W. Clark, Richard P. Reading, and Gary N. Backhouse, "Prototyping for Successful Conservation: The Eastern Barred Bandicoot Program," *Endangered Species Update*, 4 (2002): 125–29.

35 Clark, *Policy Process*, 85–110.

36 William Ascher, *Bringing in the Future: Strategies for Farsightedness and Sustainability in Developing Countries* (Chicago: University of Chicago Press, 2009).

37 Nicholas A. Christakis and James Fowler, *Connected: The Surprising Power of Our Social Networks and How They Shape Our Lives* (New York: Little, Brown, 2009); David Easley and Jon Kleinberg, *Networks, Crowds, and Markets* (Cambridge: Cambridge University Press, 2010); Justin Farrell, "Network Structure and Influence of the Climate Change Counter-Movement," *Nature Climate Change*, 6 (2016): 370–74; Justin Farrell, "Corporate Funding and Ideological Polarization about Climate Change," *Proceedings of the National Academy of Sciences*, 1 (2016): 92–97; John Scott and Peter J. Carrington, eds., *The SAGE Handbook of Social Network Analysis* (Thousand Oaks, CA: SAGE, 2011).

38 Susan G. Clark and Richard L. Wallace, "Integration and Interdisciplinarity: Concepts, Frameworks and Education," *Policy Sciences*, 2 (2015): 233–55; Timothy Terway, "Sustained in Significance with(out) Context and Ourselves: Expert Environmental Knowledge and 'Social-Ecological-Systems'" (PhD diss., Yale University, 2018).

39 Brunner et al., *Adaptive Governance*; Brunner, "Adaptive Governance as a Reform Strategy," 301–41.

40 Brunner, ed., *Finding Common Ground*; Brunner and Lynch, *Adaptive Governance and Climate Change*; Steelman, "Adaptive Governance," 538–50.

41 Modified from Brunner, "Adaptive Governance as a Reform Strategy," 301–41.

42 Brunner, "Adaptive Governance as a Reform Strategy," 301–41.

43 Brunner et al., *Adaptive Governance*; Brunner, "Adaptive Governance as a Reform Strategy," 301–41.

44 Lynn Scarlett and Matthew McKinney, "Connecting People and Places: The Emerging Role of Network Governance in Large Landscape Conservation," *Frontiers in Ecology and the Environment*, 3 (2016): 115–25.

45 A description of this framework, its key elements, and its benefits can be found in Clark, *Policy Process*, 9–13.

46 Susan G. Clark and Richard L. Wallace, "The Integrity Problem in Higher Education: Description, Consequences, and Recommendations," *Issues in Interdisciplinary Studies*, 35 (2017): 221–47; Susan G. Clark et al., "Interdisciplinary Problem Framing for Sustainability: Challenges, a Framework, Case Studies," *Journal of Sustainable Forestry*, 5 (2017): 516–34; Richard L. Wallace and Susan G. Clark, "Barriers to Interdisciplinarity in Environmental Studies: A Case of Alarming Trends in Faculty and Programmatic Wellbeing," *Issues in Interdisciplinary Studies*, 35 (2017): 221–47.

47 To learn more on rapid learning networks, see Lloyd Etheredge, "Government Learning: An Overview," in *Handbook of Political Behavior*, ed. Samuel L. Long (New York: Plenum Press, 1981); Lloyd S. Etheredge, *Can Governments Learn? American Foreign Policy and Central American Revolutions* (New York: Pergamon Press, 1983); Lloyd Etheredge, *Humane Politics and Methods of Inquiry: Selected Papers of Ithiel de Sola Pool* (Piscataway, NJ: Transaction, 2000).

48 David Cherney, "Environmental Saviors: The Effectiveness of Nonprofit Organizations in Greater Yellowstone" (PhD diss., University of Colorado, 2011); Terway, "Sustained in Significance with(out) Ourselves and Context."

49 To learn more about the Greater Yellowstone Coalition (GYC), visit their website at http:// greateryellowstone.org/. To learn about the Yellowstone Center for Resources (YCR), see https://www.nps.gov/yell/learn/management/ycr.htm. For information about Yellowstone to Yukon (Y2Y), explore https://y2y.net/.

50 Tim W. Clark and David L. Gaillard, "Organizing an Effective Partnership for the Yellowstone to Yukon Conservation Initiative," in *Species and Ecosystem Conservation: An Interdisciplinary Approach*, ed. Tim W. Clark et al. (New Haven, CT: Yale F&ES Bulletin 105, 2001), 223–40.

51 For information on these organizations, visit their websites. Jackson Hole Conservation Alliance: https://jhalliance.org/the-alliance/; Teton Conservation District: https://www. tetonconservation.org/; Future West: www.future-west.org; Charture Institute: https:// charture.org/.

52 Cherney, "Environmental Saviors"; Terway, "Sustained in Significance with(out) Ourselves and Context."

53 William Ascher, *Bringing in the Future: Strategies for Farsightedness and Sustainability in Developing Countries* (Chicago: University of Chicago Press, 2009).

54 Michael Bonnett, "Environmental Concern, Moral Education and Our Place in Nature," *Journal of Moral Education*, 3 (2012): 285–300; Michael Bonnett, "Normalizing Catastrophe: Sustainability and Scientism," *Environmental Education Research*, 2 (2013): 187–97; Michael Bonnett, "The Powers That Be: Environmental Education and the Transcendent," *Policy Futures in Education*, 1 (2015): 42–56.

55 Ascher, *Bringing in the Future*.

56 Todd Wilkinson and Tom Mangelsen, *Grizzlies of Pilgrim Creek: An Intimate Portrait of 399, the Most Famous Bear of Greater Yellowstone* (New York: Rizzoli, 2015).

Chapter 11 Creating a New Story, The Long View

1 Gregory Bateson, *Steps to an Ecology of Mind* (New York: Ballantine Books, 1972).

2 David Bohm, "Meaning and Information," in *The Search for Meaning: The New Spirit in Science and Philosophy*, ed. Paavo Pylkkanen (Napa Valley, CA: Crucible Press 1989), 1–24.

3 John Passmore, *Man's Responsibility for Nature: Ecological Problems and Western Traditions* (New York: Charles Scribner's Sons, 1974).

4 Wendell Berry, *A Continuous Harmony: Essays Cultural and Agricultural* (Washington, DC: Shoemaker & Hoard, 1972), 83.

5 Berry, *Continuous Harmony*.

6 David Bohm, *Wholeness and Implicate Order* (London: Routledge, 1980); David Bohm and B. J. Hiley, *The Undivided Universe: An Ontological Interpretations of Quantum Theory* (London: Routledge, 1995); Lee Nichol, ed., *The Essential David Bohm* (London: Routledge, 2002); David Selby, "Reaching into the Holomovement: A Bohmian Perspective on Social Learning for Sustainability," in *Social Learning toward a Sustainable World*, ed. Arjen Wals (Wagnering: Wagnering Academic, 2007), 165–80.

7 Noel Castree, *Making Sense of Nature* (New York: Routledge, 2014), 318.

8 Robert Kegan, *In over Our Heads: The Mental Demands of Modern Life* (Cambridge, MA: Harvard University Press, 1994); Robert Kegan and Lisa Lahey, *Immunity to Change: How to Overcome It and Unlock Potential in Yourself and Your Organization* (Boston, MA: Harvard Business Press, 2009).

9 Kegan, *In over Our Heads*, 352.

10 Timothy Tate, "Even in Paradise, Everyone Needs to Heal Something, Especially the Seemingly Invincible," *Mountain Journal*, August 23, 2017; Julie Turkewitz, "National Parks Struggle with a Mounting Crisis: Too Many Visitors," *New York Times*, September 27, 2017; Todd Wilkinson, "The Big Picture: Thinking about Greater Yellowstone's Elephants in the Room," *Mountain Journal*, August 14, 2017; Todd Wilkinson, "Science and Advocacy in an Age of Uncertainty," *Mountain Journal*, September 27, 2017.

11 Jonathan Schechter, "Introduction," *Jackson Hole Compass*, July 3, 2017; Todd Wilkinson, "Unnatural Disaster: Will America's Most Iconic Wild Ecosystem Be Lost to a Tidal Wave of People?" *Mountain Journal*, September 10, 2017.

12 Bill Barmore, personal communication, October 19, 1978.

13 Franz Camenzind, "Wilderness: America's Second-Best Idea Is under Attack: Unfortunately by Some Recreationists," *Mountain Journal*, September 6, 2017; Todd Wilkinson, "Franz Camenzind Pens 'Wild Ideas,'" *Mountain Journal*, August 14, 2017.

14 Jim Lyons, "The Rush to Develop Oil and Gas We Don't Need," *New York Times*, August 28, 2017; Philip J. Nyhus, "Human-Wildlife Conflict and Coexistence," *Annual Review of Environmental Resources*, 41 (2016): 143–71; Hall Sawyer et al., "Mule Deer and Energy Development: Long-Term Trends of Habituation and Abundance," *Global Change Biology*, 23, no. 11 (2017): 4521–29; Angus M. Thuermer Jr., "CWD May Be Transmittable through Eating Game Meat," *WyoFile*, September 5, 2017; Todd Wilkinson, "Long Journeys Home," *Jackson Hole Magazine*, December 23, 2016.

15 Ronald D. Brunner, "A Paradigm of Practice," *Policy Sciences*, 39 (2006): 135–67; Ronald D. Brunner, "Adaptive Governance as a Reform Strategy," *Policy Sciences*, 43, no. 4 (2010): 301–41; Ronald D. Brunner et al., eds., *Adaptive Governance: Integrating Science, Policy, and Decision Making* (New York: Columbia University Press, 2005).

16 Susan G. Clark, *Ensuring Greater Yellowstone's Future: Choices for Leaders and Citizens* (New Haven, CT: Yale University Press, 2008), 209–22; John R. Coon, "Can We Get There from Here? Ecosystem Based Governance in the Bay of Fundy/Gulf of Maine Region" (PhD diss., University of New Hampshire, 2012); Gabriel Grant, "A Self-Determined Pursuit of Personal, Organizational, and Planetary Flourishing," (PhD diss., Yale University, 2017); Bradley Johnson, "Assessing Socio-Ecological Resilience and Adaptive Capacity in the Face of Climate Change: An Examination of Three Communities in the Crown of the Continent Ecosystem" (PhD diss., University of New Hampshire, 2011); Murray B. Rutherford, "Conceptual Frames, Values of Nature, and Key Symbols of Ecosystem Management: Landscape Scale Assessment in the Bridger-Teton National Forest" (PhD diss., Yale University, 2003); Selby, "Reaching into the Holomovement," 165–80.

17 Kegan, *In over Our Heads*.

18 Ronald A. Heifetz and Riley M. Sinder, "Political Leadership: Managing the Public's Problem-Solving," in *The Power of Public Ideas*, ed., Robert Reich (Cambridge, MA: Harvard University Press, 1990); Ronald A. Heifetz, *Leadership without Easy Answers* (Cambridge, MA: Harvard University Press, 1998).

19 Clark, *Ensuring Greater Yellowstone's Future*.

20 Susan G. Clark, *The Policy Process: A Practical Guide for Natural Resource Professionals* (New Haven, CT: Yale University Press, 2002); Clark, *Ensuring Greater Yellowstone's Future*.

21 David Cherney, "Environmental Saviors: The Effectiveness of Nonprofit Organizations in Greater Yellowstone" (PhD diss., University of Colorado, 2009).

22 David Bohm, "Fragmentation and Wholeness in Religion and in Science," *Zygon: Journal of Religion and Science*, 20, no. 2 (1985): 125–33.

23 Mike Koshmrl, "Stemming Grizzly Bloodshed," *Jackson Hole News & Guide Daily*, July 13, 2016; Mike Koshmrl, "B-T Is Near Approval of Huge Grazing Plan," *Jackson Hole News & Guide*, November 15, 2017; Kim M. Wilkinson, Susan G. Clark, and William R. Burch, "Other Voices Other, Better Practices: Bridging Local and Professional Environmental Knowledge," *Yale University School of Forestry & Environmental Studies Report*, 14 (2007): 1–58.

24 Joseph A. Tainter and Tadeusz W. Patzek, *Drilling Down: The Gulf Oil Debacle and Our Energy Dilemma* (New York: Copernicus, 2011), 97–134.

25 Bohm, "Fragmentation and Wholeness in Religion and in Science," 125–33.

26 Bohm, "Fragmentation and Wholeness in Religion and in Science," 125–33.

27 Bohm, "Fragmentation and Wholeness in Religion and in Science," 125–33.

28 Roger Pielke Jr., *Honest Broker: Making Sense of Science in Policy and Politics* (Cambridge: Cambridge University Press, 2007).

29 Joseph A. Tainter, *The Collapse of Complex Societies* (Cambridge: Cambridge University Press, 1988).

30 Tainter, *Collapse of Complex Societies*.

31 Susan G. Clark et al., "Interdisciplinary Problem Framing for Sustainability: Challenges, a Framework, Case Studies," *Journal of Sustainable Forestry*, 36, no. 5 (2017): 516–34; Richard L. Wallace and Susan G. Clark, "Barriers to Interdisciplinarity in Environmental Studies: A Case of Alarming Trends in Faculty and Programmatic Wellbeing," *Issues in Interdisciplinary Studies*, 35 (2017): 221–47.

32 Roy Scranton, *Learning How to Die in the Anthropocene: Reflections on the End of Civilization* (San Francisco, CA: City Lights, 2015); Daniel Christian Wahl, *Designing Regenerative Cultures* (Axminster: Triarchy Press, 2016).

33 David C. Geary, *The Origins of Mind: Evolution of Brain, Cognition, and General Intelligence* (Washington, DC: American Psychological Association, 2005), 3; Yuval Noah Harari, *Sapiens: Brief History of Mankind* (New York: HarperCollins, 2011).

34 Ronald D. Brunner, "The World Revolution of Our Time: A Review and Update," *Policy Sciences*, 40, no. 3 (2007): 191–219; Harold Lasswell and Abraham Kaplan, *Power and Society: A Framework for Political Inquiry* (New Haven, CT: Yale University Press, 1950); Harold D. Lasswell, *The World Revolution of Our Time: A Framework for Basic Policy Research* (Stanford, CA: Hoover Institute and Library, 1951), 1–67; Myres S. McDougal and W. Michael Reisman, *International Law Essays: A Supplement to International Law in Contemporary Perspective* (Mineola, NY: Foundation Press., 1981).

35 Judith L. Anderson, "Stone-Age Minds at Work on 21st Century Science," *Conservation Biology*, 2, no. 3 (2001): 18–25; Timothy Terway, "Sustained in Significance with(out) Context and Ourselves: Expert Environmental Knowledge and 'Social-Ecological-Systems'" (PhD diss., Yale University, 2018).

36 Amanda Lynch and Siri Veland, *Urgency in the Anthropocene* (Cambridge, MA: MIT Press, 2018).

37 David Bohm and F. David Peat, *Science, Order, and Creativity* (London: Routledge, 1987); Bohm, "Meaning and Information," 1–25; David Bohm, *On Dialogue* (London: Routledge, 1996).

38 Selby, "Reaching into the Holomovement," 165–80.

39 Bohm, "Fragmentation and Wholeness in Religion and in Science," 125–33.

40 Susan G. Clark and Richard. L. Wallace, "Integration and Interdisciplinarity: Concepts, Frameworks, and Education," *Policy Sciences*, 48 (2015): 233–55; Charles Percy Snow, *The Two Cultures* (London: Cambridge University Press, 1959).

41 Tainter, *Collapse of Complex Societies*, 123–26.

42 Susan G. Clark et al., "College and University Environmental Programs as a Policy Problem (Part 2): Strategies for Improvement," *Environmental Management*, 47, no. 5 (2011): 716–26.

43 David Brooks, "This American Land," *Jackson Hole Daily*, August 26, 2017; Tate, "Even in Paradise."

44 Clark, *Ensuring Greater Yellowstone's Future*, 139–71.

45 Jarno Rajahalme, "David Bohm's 'Thought as a System' and Systems Intelligence," in *Systems Intelligence: A New Lens on Human Engagement and Action*, ed. Raimo P. Hamaiainen and Esa Saarinen (Espoo: Helsinki University of Technology Systems Analysis Laboratory, 2008), 29–38.

46 Harold D. Lasswell, "Future Systems of Identity in the World Community," in *The Structure of the International Environment*, ed. Cyril E. Black and Richard A. Falk (Princeton, NJ: Princeton University Press, 1972).

47 Bohm, "Fragmentation and Wholeness in Religion and in Science," 125–33.

48 John Sterman, "Stumbling towards Sustainability: Why Organizational Learning and Radical Innovation Are Necessary to Build a More Sustainable World—but Not Sufficient," in *Leading*

Sustainable Change: An organizational perspective, ed. Rebecca Henderson, Ranjay Gulati, and Michael Tushman (Oxford: Oxford University Press, 2014).

49 Sterman, "Stumbling towards Sustainability."

50 Lynn White, "The Historical Roots of Our Ecological Crises," *Science*, 155, no. 3767 (1967): 1203–7.

51 David Bohm, *Thought as a System* (New York: Routledge, 1992).

52 Bohm, *Thought as a System*; Rajahalme, "David Bohm's 'Thought as a System' and Systems Intelligence," 29–38.

53 Kegan and Lahey, *Immunity to Change*.

54 Timothy Tate, "Chasing Summits and Running toward the Sun," *Mountain Journal*, October 31, 2017; Todd Wilkinson, "Unnatural Disaster: Will America's Most Iconic Wild Ecosystem Be Lost to a Tidal Wave of People?" *Mountain Journal*, September 10, 2019.

55 Derek Watkins, "Your Children's Yellowstone Will Be Radically Different," *New York Times*, November 15, 2018, https://www.nytimes.com/interative/2018/11/15/climateyellowstone-global-warming.html.

56 Richard L. Wallace and Susan G. Clark, "Environmental Studies and Sciences in a Time of Chaos: Problems, Contexts, and Recommendations," *Journal of Environmental Studies and Sciences*, 8, no. 1 (2018): 1–4.

57 Clark and Wallace, "Integration and Interdisciplinarity," 233–55.

58 David Bohm and M. Edwards, *Changing Consciousness: Exploring the Hidden Source of the Social, Political and Environmental Crises Facing Our World* (San Francisco, CA: HarperCollins, 1991), 25.

59 Peter Corning, *A Fair Society and the Pursuit of Social Justice* (Chicago: University of Chicago Press, 2011); Alan Fogel, *Developing through Relationships Origins of Communication, Self, and Culture* (Chicago: University of Chicago Press, 1993); William Ophuls, *Plato's Revenge: Politics in the Age of Ecology* (Cambridge, MA: MIT Press, 2011); Harari, *Sapiens*.

60 Ava Orphanoudakis, personal communication, April 10, 2017.

61 Carrie Nolan and Sarah M. Stitzlein, "Meaningful Hope for Teachers in Times of High Anxiety and Low Morale," *Democracy and Education*, 19, no. 1 (2011): 1–10.

62 Charles S. Peirce, "How to Make Our Ideas Clear," *Popular Science Monthly*, 12 (1878): 286–302.

63 Andrew Knight, "Knowing but Not Doing: Selecting Priority Conservation Areas and the Research–Implementation," *Conservation Biology*, 22, no. 3 (2008): 610–17.

64 John Quay and Clifford Knapp, *John Dewy and Education Outdoors* (Rotterdam: Sense, 2013).

65 Ronald D. Brunner et al., eds., *Finding Common Ground: Governance and Natural Resources in the American West* (New Haven, CT: Yale University Press, 2002); Brunner et al., eds., *Adaptive Governance*; Brunner, "Adaptive Governance as a Reform Strategy," 301–41; Toddi Steelman, "Adaptive Governance," in *Handbook on Theories of Governance*, ed. Christopher Ansell and Jacob Torfing (Northampton, MA: Edward Elgar, 2015), 538–50.

66 Kegan, *In over Our Heads*, 352.

67 Susan G. Clark, *Pathologies of Knowing (Epistemology) and Practice (Pragmatics): How to Recognize and Avoid Them. Journal of Mutlidisciplinary Research* 11 (2019): 5–30.

68 Keely Maxwell, "Getting There from Here," *Nature Climate Change*, 4 (2014): 936–37.

69 Clark and Wallace, "Integration and Interdisciplinarity," 233–55.

70 Susan G. Clark et al., "Large Scale Conservation: Integrating Science, Management, and Policy in the Common Interest," *Yale School of Forestry & Environmental Studies, Report*, no. 24 (2010): 173–208.

71 Theodore Roosevelt, "The Strenuous Life" (speech, Chicago, IL, April 10, 1899).

72 William J. Ripple et al., "World Scientists' Warning to Humanity: A Second Notice," *BioScience*, 67, no. 12 (2017): 1026–28.

73 Ripple et al., "World Scientists' Warning to Humanity," 1026–28.

74 Václav Havel, *The Power of the Powerless: Citizens against the State in Central Eastern Europe* (New York: Routledge, 1985).

75 Robert J. Brulle, "Institutionalizing Delay: Foundation Funding and the Creation of U.S. Climate Change Counter-Movement Organizations," *Climate Change*, 122 (2014): 681–94.

76 Michael Bonnett, "Normalizing Catastrophe: Sustainability and Scientism," *Environmental Education Research*, 19, no. 2 (2013): 187–97.

77 Robert Schiller, "Faith in an Unregulated Free Market? Don't Fall for It," *New York Times*, October 9, 2015.

78 Lloyd S. Etheredge, "Wisdom in Public Policy," in *A Handbook of Wisdom: Psychological Perspectives*, ed. Robert J. Sternberg and Jennifer Jordan (New Haven, CT: Yale University Press, 2005), 297–328.

79 Tainter and Patzek, *Drilling Down*, 185–214.

80 Theodore Seuss Geisel, *The Lorax* (New York: Random House, 1971).

81 Richard W. Wallace and Susan G. Clark, "The Role of Education in Conservation of the Greater Yellowstone Ecosystem: An Assessment and Recommendations for the Future" (in preparation).

82 Robert Pyle, *The Thunder Tree: Lessons from an Urban Wildland* (Corvallis: Oregon State University, 1993), 134.

INDEX